COMMON CORE BASICS

Building Essential Test Readiness Skills

SCIENCE

Mc
Graw
Hill
Education

Bothell, WA • Chicago, IL • Columbus, OH • New York, NY

Cover: Evgeny Terentev/E+/Getty Images

mheonline.com

Send all inquiries to:
McGraw-Hill Education
8787 Orion Place
Columbus, OH 43240

ISBN: 978-0-07-657552-7
MHID: 0-07-657552-7

Printed in the United States of America.

2 3 4 5 6 7 8 9 RHR 18 17 16 15 14 13

Contents

To the Student

Common Core Basics: Building Essential Test Readiness Skills, Science will help you learn or strengthen the skills and concepts you need when you take any Common Core State Standards-aligned science test. The book's instructional content is based on the National Science Education Standards of the National Academy of Sciences.

In this book, you will focus on important concepts in three broad areas of science—Life Science, Physical Science, and Earth and Space Science.

Before beginning the lessons in this book, take the Pretest. This test will help you identify which skill areas you need to concentrate on most. Use the chart at the end of the **Pretest** to identify the types of questions you have answered incorrectly and to determine which concepts or skills you need to study further. You may decide to concentrate on specific areas of study or to work through the entire book. The latter decision is recommended, as it will help you to build a strong foundation in the core areas in which you will be tested.

Common Core Basics: Building Essential Test Readiness Skills, Science is divided into three units.

Unit 1: Life Science covers the fundamentals of biology:

- **Chapter 1: Human Body and Health** teaches you about human body systems and how those systems interact. It also teaches about health and disease.

- **Chapter 2: Life Functions and Energy Intake** describes flowering plants and the energy-producing processes of cellular respiration and fermentation.

- **Chapter 3: Ecosystems** explains ecosystem structure, limitations within ecosystems, relationships among living organisms, and the impact of environmental disturbances.

- **Chapter 4: Foundations of Life** teaches you about plant and animal cells, simple organisms, invertebrates, and vertebrates.

- **Chapter 5: Heredity** explains the genetic code and teaches you about inherited traits.

- **Chapter 6: Evolution** explains the process of natural selection and how life changes over time.

Unit 2: Physical Science covers the fundamentals of physics and chemistry:

- **Chapter 7: Energy** describes energy, waves, electricity, sources of energy, and endothermic and exothermic reactions.

- **Chapter 8: Work, Motion, and Forces** explains Newton's laws of motion, forces, and simple machines.

- **Chapter 9: Chemical Properties** teaches you about matter, atoms, compounds and molecules, chemical reactions, solutions, organic chemistry, and chemical equations.

Unit 3: Earth and Space Science covers the fundamentals of Earth and space sciences:

- **Chapter 10: Earth and Living Things** describes the cycles of matter and the formation of fossil fuels.

- **Chapter 11: Earth** provides an overview of geology, oceanography, and meteorology.

- **Chapter 12: The Cosmos** teaches you about the origins of the universe, the Milky Way, the solar system, and Earth and the Moon.

In addition, *Common Core Basics: Building Essential Test Readiness Skills, Science* also includes a number of special features to guide your progress:

- The **Chapter Opener** provides an overview of the chapter content and a goal-setting activity.

- **Lesson Objectives** state what you will be able to accomplish after completing a lesson.

- **Skills** list the Core Skills and Reading Skills that are applied to lesson content. The Core Skills align to the Common Core State Standards.

- **Vocabulary** terms essential for understanding are listed on the opening page of a lesson. Bold words in the text are listed in the Glossary.

- The **Key Concept** summarizes the lesson's content focus.

- Instruction and practice are provided for **Core Skills** and **Reading Skills** in the context of lessons. Special features, including **21st Century Skills,** help you apply higher-order thinking skills to real-world examples.

- **Think About Science** questions check your understanding of content throughout a lesson.

- **Write to Learn** activities give you a purpose for writing.

- An end-of-lesson **Vocabulary Review** checks your understanding of important lesson vocabulary, whereas **Skill Review** and **Skill Practice** help you assess your learning of content and fundamental skills.

- The **Chapter Review** tests your understanding of the chapter content.

- **Check Your Understanding** charts link items to corresponding review pages.

- The **Answer Key** explains the answers for questions in the book.

- The **Glossary** and **Index** contain key terms found throughout the book and make it easy to review important skills and concepts.

After you have completed the book, take the **Posttest** to see how well you have learned the concepts and skills the book presents.

Keep in mind that learning the fundamentals of science will help you understand practical aspects of daily life and provide the framework for more advanced studies in science. Good luck with your studies!

Science

The Pretest is a guide to using this book. It will allow you to preview the skills and concepts you will be working on in the lessons. The purpose of the pretest is to help you determine which skills you need to develop to improve your understanding of science concepts. The test consists of twenty-seven multiple-choice questions that correspond to the twelve chapters of this book.

Directions: Choose the best answer to each question. The questions are based on reading passages, charts, graphs, diagrams, and illustrations. Answer each question as carefully as possible. If a question seems to be too difficult, do not spend too much time on it. Work ahead and come back to it later when you can think it through carefully.

When you have completed the test, check your work with the answers and explanations on pages 10–11. Use the Evaluation Chart on page 12 to determine which areas you need to study the most.

1. Scientists study fossilized bones by comparing them to bones of animals alive today. This helps scientists determine how the species developed over time.

   The fossil record for horses dates as far back as 55 million years. The bones of the first horse, *Eohippus*, or "dawn horse," were quite different from those of modern horses. Scientists estimate that *Eohippus* was about the size of a sheep.

   Based on the information in the passage, which statement is true about modern horses as compared to ancient horses?

   A. Modern horses also have bones that are millions of years old.
   B. Modern horses have larger bones.
   C. Modern horses are friendlier to sheep and other grazing animals.
   D. Modern horses eat grass more effectively.

2. The illustration shows nutrition facts from a typical chocolate candy.

Nutrition Facts	Amount/serving	%DV*	Amount/serving	%DV*
Serving Size 1 Bag	**Total Fat** 10g	**15%**	**Total Carb.** 34g	**11%**
Calories 230	Sat. Fat 6g	**30%**	Fiber 1g	**4%**
Fat Calories 90	**Cholest.** 10mg	**3%**	Sugars 31g	
*Percent Daily Values (DV) are based on a 2,000 calorie diet.	**Sodium** 35mg	**1%**	**Protein** 2g	
	Vitamin A**	Vitamin C**	Calcium 4%	Iron 2%
	**Contains less than 2 percent of the Daily Value of these nutrients.			

   What is the approximate number of calories in a gram of fat?

   A. 90 calories
   B. 10 calories
   C. 9 calories
   D. 1 calorie

Science

3. The diagram shows the outside edge of a cliff formed by alternating layers of sandstone and shale.

CLIFF

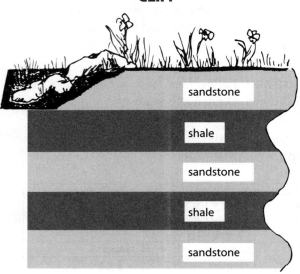

Based on the diagram, which statement is the most accurate?

- **A.** The layers of shale are older than the layers of sandstone.
- **B.** Shale erodes more easily than sandstone.
- **C.** The layers were formed by volcanic activity.
- **D.** Glaciers helped shape the cliff.

4. A star forms within a nebula, or cloud of gas and dust. Which relationship explains how gas and dust eventually form a self-igniting star?

- **A.** A cloud of greater mass results in greater molecular activity.
- **B.** The combined mass of the particles in the cloud is greater than the mass of the gases.
- **C.** As gravity pulls gas and particles closer, into a denser space, temperatures rise, causing atoms to fuse.
- **D.** The light elements blow away, leaving heavier elements among the dust and gases to ignite.

5. A fire, whether caused naturally or by human activity, can disrupt the equilibrium of an ecosystem. Soil normally absorbs water, allowing it to percolate, or flow downward through the soil. However, when soil becomes intensely hot, such as during a fire, it repels water instead of absorbing it. If rain falls after a fire, what environmental consequence can occur?

- **A.** Animals take advantage of temporary pools of water.
- **B.** Rain mixes with the soil to create heavy muds that slow the return of wildlife to the area.
- **C.** Erosion removes soil from the burned area.
- **D.** Wildlife flee to drier areas.

Science

6. American biologist Rachel Carson published *Silent Spring*, the book that helped launch the environmental movement in the 1960s. *Silent Spring* documented the use of pesticides and other potentially harmful chemicals. One particular pesticide, DDT, was killing more than just pests. It was killing birds and other small animals.

   Because of *Silent Spring*, people began to reexamine the use of DDT. Within a few years, the US government banned its use.

   Based on the information in the passage, what might the title *Silent Spring* refer to?

   A. Rachel Carson's love of nature
   B. the effect of harmful chemicals on the environment
   C. the season the book was written
   D. the struggle between the chemical companies and the government

7. Trade winds and westerlies are global wind patterns. As the diagram shows, what determines whether trade winds or westerlies form in a region?

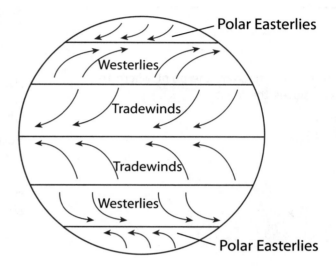

   A. the direction of Earth's rotation
   B. season of the year
   C. distance to the equator
   D. daily air temperature

8. The diagram represents one molecule of propane gas.

PROPANE

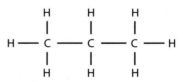

   What kind of energy is stored in the bonds of a chemical compound like propane?

   A. kinetic energy
   B. gravitational potential energy
   C. elastic potential energy
   D. chemical potential energy

Science

9. The diagram shows a simple lever.

Where is the fulcrum, the point where the lever moves the least?

A. 1
B. 2
C. 3
D. 4

 10. Some plants bend their stems or leaves in response to specific factors in their environment. This behavior is called tropism. Tropism takes many forms.

Response to light is called phototropism. A plant shows positive phototropism if its leaves and stems tend to grow toward the light.

Hydrotropism is a plant's response to water. A plant shows positive hydrotropism when its roots grow toward water.

Response to Earth's gravity is called geotropism. A plant's roots grow downward, which is an example of positive geotropism. A stem, however, grows upward, away from gravity. When a potted plant is placed on its side, the stem will turn upward within three hours. This is an example of negative geotropism.

Based on the information in the passage, what does the following illustration show?

A. positive geotropism
B. negative geotropism
C. positive phototropism
D. positive hydrotropism

Science

11. The length of the cycle of day and night can affect plant growth. The effect of this cycle is called photoperiodism.

   Some plants, such as lettuce, bloom only after a long photoperiod, meaning after many hours of daylight. These plants are called long-day plants. Other plants, such as poinsettias and chrysanthemums, bloom only after a short photoperiod, meaning after only a few hours of daylight. They are short-day plants. Still other plants, such as marigolds, are day-neutral. They bloom during either long days or short days.

   Based on the information in the passage, why don't lettuce and poinsettias bloom at the same time?

   A. A photoperiod stays the same throughout the year.
   B. A photoperiod can be either long or short, but not both.
   C. A photoperiod stays the same from location to location.
   D. Both plants require the same photoperiod.

12. Fungi and bacteria play an important role in every environment. These organisms are decomposers. They break down organic matter, or the remains of once-living things. This helps return nutrients to the soil. For this reason, fungi and bacteria are an essential part of almost every food chain.

   Which process is being described in the passage?

   A. recycling
   B. photosynthesis
   C. replication
   D. respiration

Directions: Questions 13 and 14 refer to the following passage.

> In the 1800s, Gregor Mendel worked in a monastery garden, where he made observations and conducted experiments that led to modern understandings of heredity.
>
> Mendel bred pea plants that were pure and hybrid for specific traits, such as height, pod shape, pod color, seed shape, seed color, flower position, and flower color. In one experiment, he bred a plant bearing yellow seeds with a plant producing green seeds.

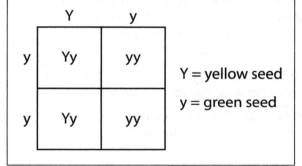

Y = yellow seed
y = green seed

13. What percentage of the offspring in this generation are likely to have green seeds?

   A. 25 percent
   B. 50 percent
   C. 75 percent
   D. 100 percent

14. If a Yy plant was bred with a Yy plant, what percentage of their offspring would have green seeds?

   A. 25 percent
   B. 50 percent
   C. 75 percent
   D. 100 percent

Science

Directions: Questions 15 and 16 refer to the following passage.

> One way that smoking damages the lungs is by destroying the cilia in the airways. Cilia are tiny hairs that trap harmful substances and help protect you from infections, such as the bacteria that can cause pneumonia. Smoking also damages alveoli, the tiny air sacs at the tips of the lungs. These sacs stretch to allow the exchange of oxygen and carbon dioxide in the capillaries that surround them. Chemicals in smoke make the alveoli less elastic. This prevents them from expanding to take in sufficient oxygen and get rid of carbon dioxide. A smoker with damaged alveoli often feels tired and winded, or unable to catch a breath.

15. Based on the information in the passage, how does smoking damage the lungs?

    A. It destroys cilia, allowing harmful substances to reach the lungs.
    B. It reduces the blood flow in the capillaries.
    C. It hardens capillary walls, preventing the exchange of gases.
    D. It enlarges the alveoli.

16. What could be one reason smokers feel winded and tired?

    A. Their lungs shrink, leaving less room for oxygen intake.
    B. Their blood pressure drops, causing the heart to pump less oxygen through the body.
    C. Cilia clog passageways, trapping harmful substances and limiting gas exchange.
    D. Alveoli lose their ability to expand, thus shrinking the area over which gas exchange can occur.

17. One of the objects in the diagram is suspended from the ceiling by two ropes.

OBJECT SUSPENDED BY ROPES

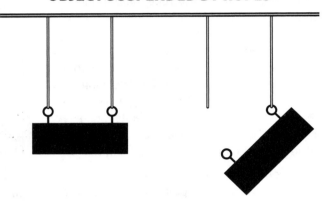

If one of the ropes is removed so that the object is suspended by only one rope, what will most likely happen?

A. The remaining rope will bear the same load as it did before the other rope was removed.
B. The load on the remaining rope will be reduced by half.
C. The remaining rope will bear twice the load than it did before the other rope was removed.
D. The load will be divided equally between the two ropes.

Science

Directions: Questions 18 and 19 refer to the passage below.

> One theory that explains how the Moon formed begins with the collision of a huge space body, about the size of Mars, and Earth. The collision resulted in a blast of matter from Earth's surface into space. In space, gravity pulled bits of matter together, forming larger pieces of matter. The matter continued to fuse, eventually creating the Moon.
>
> The massive collision between the space body and Earth pushed the angle of Earth's tilt on its axis from zero to 23.5 degrees. This tilt changed future weather patterns, as the tilt contributed to changes in seasons.

18. Scientists have found the same kind of material found in Earth's crust and mantle in samples of Moon dust. What does this evidence suggest?

    A. Space missions have accidentally carried Earth matter to the Moon.
    B. If Earth were to be slammed by another space body today, another moon would form.
    C. Earth and the Moon formed from the same dense cloud of dust and gas.
    D. The material that made the Moon began as Earth material.

19. Examine the diagram showing Earth's tilt on its axis. Use the diagram to determine the season in the Southern Hemisphere when Earth tilts away from the Sun.

    A. spring
    B. summer
    C. autumn
    D. winter

TILT OF EARTH'S AXIS

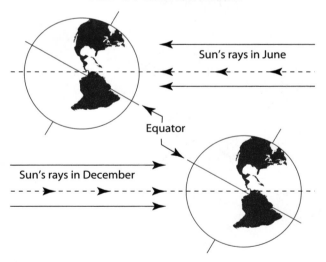

Science

20. According to the Traffic Safety Administration, wearing a seat belt while driving can reduce the risk of injuries associated with car crashes by 50 percent. What purpose do seat belts serve?

    A. They work against inertia.
    B. They work against friction.
    C. They work against gravity.
    D. They work against mass.

21. The diagram shows the typical structure of a plant and its parts.

    PARTS OF A FLOWERING PLANT

    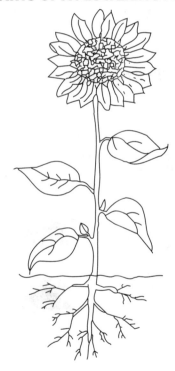

    What is the function of the stem?

    A. to protect the plant from predators
    B. to transport oxygen throughout the plant
    C. to transport water and nutrients from the roots to the leaves
    D. to attract pollinators to help the plant reproduce

Directions: Questions 22 and 23 refer to the following passage.

The water cycle is the constant movement of water between Earth's surface and the atmosphere. As water moves through the cycle, it changes from liquid form to a gas or solid state and then back to a liquid.

The Sun is the energy source that powers the water cycle. As the Sun heats oceans and other bodies of water, the liquid water on the surface escapes into the air as an invisible gas called water vapor. This process is called evaporation.

Higher in the atmosphere, water vapor may cool to form tiny water droplets. This process is called condensation. At very cold temperatures, ice crystals may form. Water in the liquid or solid state eventually falls from clouds as rain, snow, or other forms of precipitation.

22. Which event shows the process of evaporation?

    A. water flowing from a river into the ocean
    B. rain or snow falling from a cloud
    C. a storm cloud becoming larger and darker
    D. a puddle on the sidewalk drying up in sunlight

23. Based on the information in the passage, clouds are formed from which form of water?

    A. water vapor
    B. tiny water droplets or ice crystals
    C. large, heavy water droplets
    D. large, heavy snowflakes

Science

Directions: Questions 24 and 25 refer to the following passage.

A mechanical camera, unlike a digital one, needs film to capture images. Photographic film is made of plastic coated with crystals of silver bromide (AgBr), a salt. Silver bromide works on film because it is especially sensitive to light. When a camera's shutter opens, light enters, causing changes to the structure of the silver bromide crystals.

In the development process, black-and-white film is placed in a chemical bath. The bath darkens the silver particles that were struck by light. The more light that strikes the particles, the darker they become. The bath does not affect particles that were not struck by light.

The film produces a detailed "chemical picture." This picture, called a negative image, can be processed further with different chemicals to obtain a positive, or developed, copy of the photograph.

24. Some films use silver iodide (AgI) instead of silver bromide (AgBr). What property do the two chemicals most likely have in common?

    A. cost
    B. ability to dissolve
    C. color
    D. sensitivity to light

25. Based on information in the passage, what might cause a negative image to be completely black?

    A. Separated silver and bromine turned back into silver bromide.
    B. A digital camera was used.
    C. The shutter allowed too much light into the camera.
    D. The shutter allowed too little light into the camera.

26. Natural selection is a process in which the organisms that are best adapted to their environment tend to live to adulthood, reproduce, and pass their traits to offspring. Which of the following is a foundation of natural selection?

    A. Genetic variations are always present in a species.
    B. Organisms that are larger have a reproductive advantage.
    C. The rate of reproduction is faster among some species than others.
    D. Traits acquired in one generation are passed to the generations that follow.

27. The figure in the diagram is using a lever to lift a tree stump, or load. What purpose does the lever serve?

    A. It reduces the mass of the load.
    B. It reduces the force required to lift the load.
    C. It reduces the load at the same time it's reducing the force required to lift it.
    D. It eliminates the need for a compound machine with moving parts.

Answer Key

1. **B.** The passage states that scientists estimate that ancient horses were only the size of modern-day sheep, so the bones of modern horses are much larger than the bones of ancient horses.

2. **C.** The total calories from fat (90) divided by the total grams of fat (10) equals 9 calories per gram.

3. **B.** From the diagram, you can conclude that the portions of the cliff made of shale have eroded more than the portions made of sandstone. The remaining choices may be correct but are not supported by the diagram.

4. **C.** Gravity pulls dust and gases into less and less space. As density increases, meaning the matter is closer together, temperatures increase. Eventually, temperatures increase enough to cause nuclear fusion, when atoms combine, releasing energy in the form of light and heat.

5. **C.** An intense fire can make soil repel water rather than absorb it. So, when rain falls after a fire, it cannot soak into the ground. Instead, it washes away, carrying soil with it. As the process of erosion carries soil away, it also carries away valuable nutrients. Without nutrients, new plants will not grow to replace plants burned in the fire.

6. **B.** Although the other choices might be logical topics for *Silent Spring*, harmful chemicals are specifically mentioned in the passage.

7. **C.** As the diagram shows, both trade winds and westerlies blow in horizontal bands that are parallel to the equator. The distance to the equator determines whether westerlies, trade winds, or other wind patterns form.

8. **D.** Because the atoms in a chemical compound are attracted to each other, they form chemical bonds. Chemical potential energy is stored in those bonds. That energy can be converted when the compound is broken.

9. **C.** The fulcrum is the point on which the bar is resting, shown in the diagram as a triangle.

10. **D.** Positive hydrotropism is when plant roots grow toward a source of water.

11. **B.** Lettuce requires a long photoperiod, while poinsettias require a short photoperiod. A photoperiod cannot be both long and short, so the two plants cannot bloom at the same time.

12. **A.** The passage describes how fungi and bacteria break down the nutrients in organic matter and return them to the soil. This process could best be described as recycling.

13. **B.** The diagram indicates that 50 percent of the offspring would be Yy, meaning they would have yellow seeds, and 50% would by yy, meaning they would have green seeds.

14. **A.** In a cross between two plants that are hybrid for seed color, or Yy and Yy, 25 percent of the offspring would have green seeds.

15. **A.** Chemicals in smoke destroy cilia, the millions of tiny hairs that line the breathing passages. Ordinarily, cilia trap harmful substances that enter the lungs. You cough up this matter, removing it from your body. Without the cilia's protection, you become more vulnerable to harmful substances, such as bacteria that cause pneumonia.

16. **D.** Chemicals in smoke damage alveoli, causing them to lose their ability to expand. Unable to stretch, the alveoli can hold less oxygen. Smaller alveoli also limit the amount of carbon dioxide, a toxin in the bloodstream, that can leave the body. Difficulty getting oxygen and getting rid of carbon dioxide can leave a smoker feeling winded and tired.

17. **C.** With two ropes attached, each rope bears half the load. If one of the ropes is removed, the other rope must support the entire load. This means the one rope must support twice the load.

18. **D.** Finding matter in Moon dust that is the same as matter in Earth's crust and mantle suggests that Earth formed first and had cooled sufficiently to begin forming a solid surface. Something large struck the planet, sending matter from the crust and mantle into space, where it joined to eventually form the Moon.

Answer Key

19. B. When Earth tilts away from the Sun, the Sun's rays strike the Southern Hemisphere more directly, resulting in summer. It is the opposite season, or winter, in the Northern Hemisphere, where the Sun's rays strike Earth less directly.

20. A. Newton's First Law of Motion states that an object at rest tends to stay at rest, and an object in motion tends to stay in motion unless acted upon by a force. This law is also called the Law of Inertia. A seat belt works against inertia, preventing a car passenger from continuing to move forward in a straight line after a car has slowed or stopped.

21. C. The main function of the stem is the transport of materials. Stems move water from the roots to the leaves, and they also function in moving food from the leaves to the rest of the plant. Plants take in oxygen directly from the air. Nutrients enter the stem from the roots.

22. D. Evaporation involves water changing into water vapor, an invisible gas. The process can occur in any body of water, including a sidewalk puddle.

23. B. Large, heavy droplets or snowflakes would fall from the sky, and clouds often appear without precipitation occurring. The passage states that water vapor cools to become tiny water droplets or ice crystals, which are the components of clouds.

24. D. Sensitivity to light is the key property that allows silver bromide to capture an image on film. If silver iodide can also be used in film, it should have this property.

25. C. The passage states that when film is developed, light-stricken particles of silver darken into a negative image. If the negative is completely black, then all of the silver bromide must have been exposed to light, indicating that there was too much light. This result is called overexposure. In the developed image, the photograph will look very light.

26. A. No two organisms are alike in every way. There are even differences among identical twins. Genetic variations exist in every species at any given time. As changes occur in the environment, the variations in some organisms make them better able to adapt and thus to survive.

27. B. Using a simple machine like a lever does not affect the mass of an object being moved a distance by the machine, but it does reduce the force required to move it.

Evaluation Chart

Check Your Understanding

On the following chart, circle the number of any question you answered incorrectly. Next to each group of question numbers, you will see the pages you can study to learn how to answer the questions correctly. Pay particular attention to areas in which you missed half or more of the questions.

Unit	Item Number	Study Pages
Life Science	1, 2, 5, 6, 10, 11, 12, 13, 14, 15, 16, 21, 26	13–234
Physical Science	8, 9, 17, 20, 24, 25, 27	235–354
Earth and Space Science	3, 4, 7, 18, 19, 22, 23	355–435

UNIT 1

Life Science

Human Body and Health

"Eat the right foods." "Exercise regularly." "Get enough sleep." "Wash your hands often."

We certainly hear a lot of advice about how to keep our bodies working well!

The human body is a complex and unique set of structures and systems. Although humans are the most advanced of organisms, the study of human biology sheds light on the similarities among all life forms.

In this chapter you will learn about:

Lesson 1.1: Skeletal and Muscular Systems
How do your bones and muscles work together to help support your body and allow you to move? Find out about the bones, joints, and muscles in the human body.

Lesson 1.2: Digestive, Respiratory, Excretory, and Circulatory Systems
The human body is a dynamic system continuously processing nutrients, wastes, and gases. This lesson identifies the organs and processes involved in digestion, respiration, excretion, and circulation.

Lesson 1.3: Nervous, Endocrine, and Reproductive Systems
Your body reacts to many different stimuli on a daily basis. You will learn about the body systems responsible for communication and reproduction in this lesson.

Lesson 1.4: Health and Disease
Good health is a top priority for everyone. Learn about common diseases, nutrients needed by the body, and how certain drugs can help maintain good health.

Goal Setting

Why is it important to understand human biology? The more we learn about how our bodies work, the more successful we can be at maintaining good health.

To set goals for learning about human biology, make a chart with a row for each of the major body systems covered in this chapter: Skeletal, Muscular, Digestive, Respiratory, Excretory, Circulatory, Nervous, Endocrine, and Reproductive. For each system listed, write a question you want answered to learn more about that system. As you read, look for an answer to each question and write it in the second column for each topic.

Body System	Question	Answer
Skeletal		
Muscular		
Digestive		
Respiratory		
Excretory		
Circulatory		
Nervous		
Endocrine		
Reproductive		

Skeletal and Muscular Systems

KEY CONCEPT: The human skeletal and muscular systems work together for support, protection, and movement.

When it is supported by poles, a tent provides easy, portable shelter for campers. However, have you ever forgotten to bring tent poles? Without a framework, the tent collapses to the ground.

You also need a supporting framework. It's called a skeletal system. Together, your skeletal and muscular systems support and protect you. They also do what a tent can't. They let you move.

The Skeletal System

Human bones are not solid. They are more like hollow tubes that have been filled with jelly. This jellylike substance is a combination of nerves and blood vessels called **marrow**. New blood cells are produced in the marrow.

How many bones make up the human skeleton? A typical adult has 206 bones, although some people have an extra rib or extra bones in their fingers. The largest bones are the thigh bones, and the smallest bones are located in the ear. Thigh bones contribute about one fourth of a skeleton's entire weight.

Bones are made of living, growing tissue that has a unique set of needs. This is why a proper diet is important to maintain healthy bones. Bone-building nutrients include calcium and vitamin D.

Certain cells in the bone constantly break down old bone tissue, while other cells create new bone tissue. When a bone breaks, cells immediately begin to break down damaged bone tissue. Other cells begin to patch the break. Meanwhile, bone-building cells create new bone material near the break. With time, the bone can be as good as new.

THINK ABOUT SCIENCE

Directions: Fill in the blanks with the appropriate words.

1. Blood cells are produced in the _____.

2. The skeletal system is made of _____.

HUMAN SKELETON

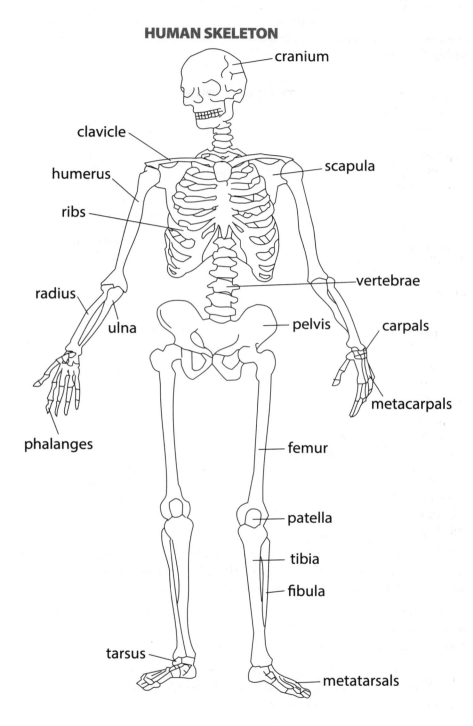

- cranium
- clavicle
- humerus
- ribs
- scapula
- radius
- ulna
- vertebrae
- pelvis
- carpals
- metacarpals
- phalanges
- femur
- patella
- tibia
- fibula
- tarsus
- metatarsals

As the body grows, the bones grow. The entire skeletal structure grows without ever interfering with other working body systems. With age, however, the bones may harden and become brittle.

The **organization**, or arrangement, of ideas in a passage is important. Recognizing and identifying the way information is organized, helps you get the most meaning from what you read.

As you read the information about joints on this page, ask yourself: *How is this information organized?* It might help you to list what you see in the diagram.

Joints

Most bones are connected to one another with tough strands of tissue known as **ligaments**. The points of connection are called joints. Most joints move. Some swing back and forth like a door. These hinged joints are at the elbows, knees, jaw, fingers, and toes. They cannot be twisted without injury.

TYPES OF JOINTS

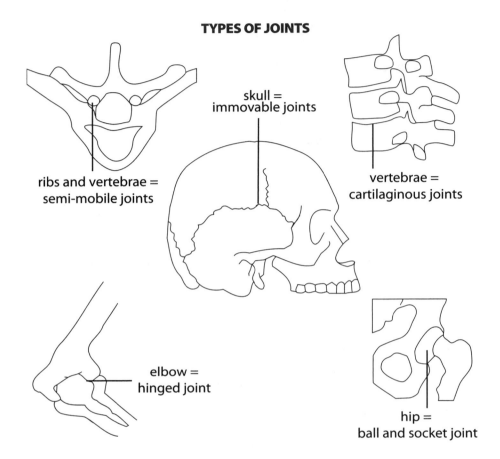

skull = immovable joints

ribs and vertebrae = semi-mobile joints

vertebrae = cartilaginous joints

elbow = hinged joint

hip = ball and socket joint

Ball and socket joints can be twisted. These joints, found in shoulders and hips, allow flexible movement in many directions. The gliding joints in the ankles, wrists, and the spinal column also allow movement in many directions, but the amount of movement is limited.

Other joints allow only a bit of movement. The ribs, for example are attached to the breastbone. If the ribs did not move, you could not breathe. Skull bones of adults do not move at all.

If bones were allowed to rub together at joints, the ends would be ground away. Instead, most bones are padded with **cartilage**. The cartilage acts as a shock absorber for the skeleton.

UNDERSTAND TEXT ORGANIZATION

When you read, you will notice that each paragraph or section is organized in a certain order, or pattern. The organization of ideas in a passage is very important. A writer decides which ideas fit best in a passage and also determines where each idea should be placed. Understanding the organization of ideas can help you make meaning as you read.

Some of the most common patterns of organization are: order of importance, time order, and cause-and-effect order. To organize by importance, a writer puts the most important ideas first. A writer using time order places ideas in the order in which events happened. If the writer organizes by cause and effect, he or she may first explain the cause and then explain the effect.

To understand organization, ask yourself: *What are the important ideas? Do they form a pattern? What pattern do they follow?*

Read the following text and identify the pattern of organization.

> Scientists have been studying the harmful physical effects of long-term space travel. When astronauts spend many weeks in space, their muscles often become weak and their bones lose calcium, an important mineral. Fatigue from space flight disturbs sleep. Pulse rates have been found to increase because the heart can shrink by about 10 percent.

The text is organized by cause and effect. The text discusses how long-term space travel (the cause) can lead to certain health problems (the effects).

Scientists use the words *voluntary* and *involuntary* to describe two kinds of muscles in the muscular system. You are able to control voluntary muscles, such as those you use to run or write. You are unable to control involuntary muscles, such as those constantly moving blood through your body.

The terms *voluntary* and *involuntary* are also used to describe nervous system responses, that is, how your brain responds to messages. When you have a physical examination, a doctor may strike your knee with a small hammer. Your leg moves forward without any direction from you. What word would you use to describe this kind of response?

The Muscular System

The muscular system allows you to walk, breathe, talk, and swallow. Some actions, such as running, are controlled by **voluntary** muscles, which means they are under your control. Other actions, such as pumping blood through your body or churning the food in your stomach, are not under your control. These actions are controlled by involuntary muscles.

There are three kinds of muscles in the body:

- **Cardiac muscles**, which are involuntary, control the heartbeat.
- **Smooth muscles**, which are involuntary, are found in the lungs, intestines, and bladder and are controlled without conscious thought.
- **Skeletal muscles**, which are voluntary, allow you to act to change the position of your body.

How Muscles Work

All muscles work in the same way; they contract. Each muscle can only pull—not push—on the bones of the skeleton. Bending a joint, such as the arm, involves the contraction of one muscle. To straighten the arm again, the first muscle relaxes, and a second muscle contracts, pulling the arm down. Most joints in the body are controlled by pairs of muscles.

MUSCLES IN OPPOSING PAIRS

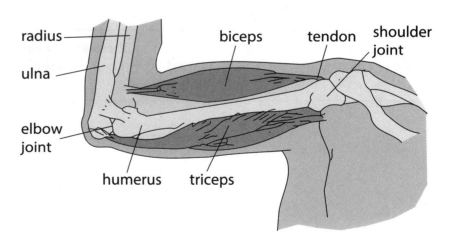

radius — ulna — elbow joint — humerus — triceps — biceps — tendon — shoulder joint

Importance of Exercise

When people hear the word *exercise*, they sometimes think its purpose is to make bigger, stronger muscles. However, exercise benefits all of the major systems in the human body, including the skeletal system. Exercise can increase the width and strength of bones, while a lack of exercise causes a decrease in bone density, size, and strength. Exercise and proper diet are especially important for older adults, whose bones can become weak and brittle.

Different kinds of exercises help the body in different ways. Some exercises, such as yoga or tai chi, increase flexibility and range of motion. This helps strengthen ligaments and reduces the strain on the joints. Other exercises, such as weight lifting, help build stronger muscle. Aerobic exercises, such as bicycling or swimming, help large muscle groups all over the body. This type of exercise also strengthens the heart and lungs.

To stay healthy and fit, exercise in each of these different ways. Just thirty minutes of exercise a day—even if it is merely a brisk walk—can lead to a longer, healthier life.

WRITE TO LEARN

Before you begin to write, think about how you can organize your ideas. Good organizational structure will help your readers understand what you are writing quickly and clearly. In a notebook, describe a physical activity or exercise you enjoy. Tell your reader the benefits of doing this activity. Choose one of these common patterns of organization: order of importance, time order, cause-and-effect order.

THINK ABOUT SCIENCE

Directions: Answer the questions below.

1. What is the purpose of exercise?

   because it keeps you healthy, and bones strong

2. Should people exercise throughout their lives? Explain.

   Because it keeps them healthy in a physical but also internal way.

3. What exercises help increase flexibility and range of motion?

   yoga and tia chi

Vocabulary Review

Directions: Match each term on the left with the correct description on the right.

1. ___E___ cardiac muscle
2. ___B___ smooth muscle
3. ___A___ marrow
4. ___F___ voluntary
5. ___C___ skeletal muscle
6. ___D___ ligaments

A. where new blood cells are produced

B. muscle that controls heartbeat

C. muscles allowing movement

D. tough tissue connecting bones together

E. involuntary muscle found in intestines and bladder

F. under one's control

Skill Review

Directions: Answer the questions below.

1. The human body is made of a huge number of tiny units called cells. Cells work together in units called tissues. Tissues work together in organs. Organs work together in organ systems.

   Which diagram below best describes the organization of the muscular system?

   A. Muscle cell ⇒ Muscle ⇒ Muscle tissue ⇒ Muscular System
   B. Bone cell ⇒ Bone tissue ⇒ Bone and muscle ⇒ Muscular System
   C. Muscle cell ⇒ Muscle tissue ⇒ Muscle ⇒ Muscular System
   D. Muscle tissue ⇒ Muscle cell ⇒ Muscular System ⇒ Muscle

2. The following diagram shows the bones and muscles of the arm. Analyze the diagram, then describe how the skeletal and muscular systems work together to make the lower arm move. Be sure to include how pairs of muscles work together.

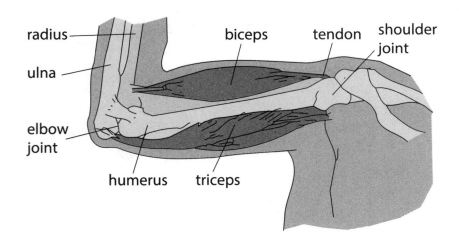

_____

_____

_____

Skill Practice

Directions: Choose the best answer to each question.

1. Which of the following is a function of the skeletal system?

    A. to provide nourishment for the body
    B. to produce new nerve cells
    C. to contract to allow movement
    D. to provide support and protection for the body

2. What type of joint is found in the shoulders and hips?

    A. ball and socket
    B. hinge
    C. gliding
    D. rotational

3. Why is it important to strengthen ligaments through exercise?

    A. They protect the ends of the bones.
    B. They attach muscles to bones.
    C. They connect bones to one another.
    D. They build muscle mass.

4. Which statement best explains how muscles work to move joints?

    A. Muscles work in groups of three to bend or straighten joints.
    B. Muscles work in pairs to bend or straighten joints.
    C. Muscles push against each other at joints.
    D. Muscles work inside joints only.

5. Muscles in the walls of arteries help push blood through the body. What type of muscle are these?

    A. involuntary muscle
    B. cardiac muscle
    C. voluntary muscle
    D. both involuntary and voluntary muscle

6. In a disease called arthritis, joints hurt when they move. Which of these events would most likely cause arthritis?

    A. a fall that breaks a bone
    B. damage to a nerve in the arm or leg
    C. damage to involuntary muscles
    D. wearing away of cartilage

Digestive, Respiratory, Excretory, and Circulatory Systems

KEY CONCEPT: To carry out life activities, cells require food and oxygen. They also produce wastes. Each system plays a role in delivering the materials that cells need and carrying away wastes they make.

For a moment, think of your body as a factory. It has a number of systems operating simultaneously, all to keep you functioning properly. Systems take in raw products, change them into forms your body can use, and get rid of wastes. Working together, your digestive, respiratory, excretory, and circulatory systems keep you functioning like a well-oiled machine.

The Digestive System

Molecules of food are too large to enter the cells of the body. To make use of the nutrients in food, the body must break down the food molecules. This process occurs in the digestive system. When food is chewed, the teeth grind and tear the food into small bits. Saliva not only keeps the food moist, it also begins to **digest**, or break down, the largest molecules.

The Stomach

When you swallow, the food goes down the esophagus and into the stomach. There, smooth muscles repeatedly contract to cause a churning motion that stirs the food. Strong acids begin to break the molecules apart. The food is like a thin soup by the time it leaves the stomach.

The Intestines

Food passes from the stomach into the small intestine. Tiny food particles move across the membrane of the small intestine into blood vessels. The blood carries the particles to cells throughout the body.

Undigested food passes into the large intestine, where much of the water in the waste is absorbed into the body. The remaining solids, called feces, leave the body through the rectum and anus.

THINK ABOUT SCIENCE

Directions: Write a short response.

Describe in your own words the pathway that a bite of food travels through the digestive system, from when it enters the mouth to when waste products leave the body.

INTEGRATE TEXT AND VISUALS

Visuals include diagrams, charts, graphs, illustrations, and photographs. In fact, they include anything that you can use your eyes to "read." Writers often include visuals to help readers interpret or understand texts. Take, for example, the diagram of the human digestive system below. It shows all of the organs that make up the digestive system and how they are connected. This could well take several pages to describe, but the diagram shows it concisely. By analyzing the diagram, you can trace the path of food from the beginning to the end of the digestive system.

There are several features that can help you analyze a diagram or visual. First, look for a title. It tells you what the visual is about in only a few words. Look for labels, too. Writers use them to point out important things they want to communicate. Some visuals have captions. Captions are also good sources of important information.

After you read the text about the digestive system, look again at the diagram. Explain how the text and visual work together.

_____

_____

HUMAN DIGESTIVE SYSTEM

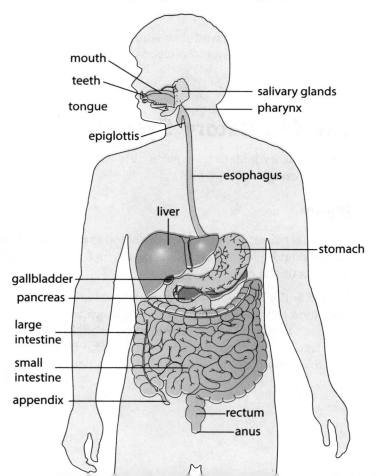

The Excretory System

Human body cells produce wastes. Most are removed as they pass through the kidneys. Inside the kidneys are millions of tiny filters that remove waste and other materials from the blood. Materials that are not waste are usually returned to the blood. Waste is carried in a liquid called urine, which passes into the bladder. The bladder pushes urine out of the body.

HUMAN EXCRETORY SYSTEM

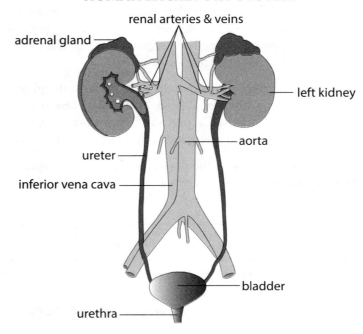

The liver also filters harmful substances, such as extra sugar, dead cells, chemicals, or drugs from the bloodstream.

The Circulatory System

The circulatory system transports nutrients, oxygen, antibodies, and wastes throughout the body.

Blood

Blood is a tissue. About half of the blood is **plasma**, a light-colored, watery liquid. Plasma carries nutrients, vitamins, minerals, and oxygen throughout the body.

The red and white blood cells and platelets are suspended in the plasma. Red blood cells carry oxygen from the lungs to the cells. White blood cells are much larger but they are able to squeeze through body tissue to reach an infection and destroy the invading cells. **Platelets** are tiny particles that form blood clots. Without platelets, a person could bleed to death.

The Heart

The heart is a fist-sized muscle divided into four chambers. The two upper chambers, called atria, collect the blood returning to the heart through the veins. The two lower chambers, called ventricles, pump blood away from the heart through the arteries.

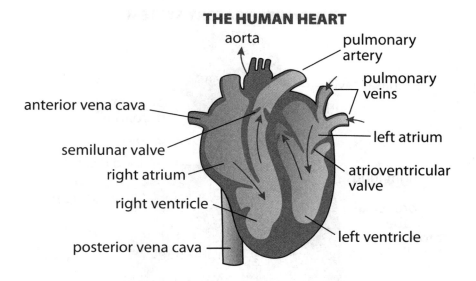

THE HUMAN HEART

aorta
pulmonary artery
pulmonary veins
anterior vena cava
left atrium
semilunar valve
atrioventricular valve
right atrium
right ventricle
left ventricle
posterior vena cava

Core Skill
Integrate Text and Visuals

Science authors use words to explain complex concepts, such as how blood moves through the human heart. Visuals can make the text easier to understand, especially if you pause during reading to compare the words you read to the visuals you see.

Look closely at the diagram of the heart. Explain the purpose of the arrows in the illustration and how they support information in the text.

The two sides of the heart work like separate pumps. The right atrium receives blood from the body. This blood has little oxygen in it. The right atrium pumps the blood down into the right ventricle. From there the blood is pumped through an artery to the lungs, where the blood releases carbon dioxide, a waste product, and picks up oxygen.

Oxygen-rich blood leaves the lungs through veins and enters the left atrium of the heart. The left atrium pumps the blood down into the left ventricle. From there the oxygen-rich blood is pumped through arteries to all parts of the body, except the lungs. The arteries branch off into smaller and smaller arteries called arterioles and capillaries. In this way oxygen reaches every cell in the body. The cells release carbon dioxide into the blood and absorb oxygen.

WRITE TO LEARN

To analyze a visual, read the title first. Then look for labels and determine what the labels point to. Choose a diagram in this chapter and write a paragraph explaining what you learned from the diagram. Then explain how the diagram connected to the text you read.

THINK ABOUT SCIENCE

Directions: Fill in the blanks below with the appropriate words.

1. Most human waste is removed by the ~~Excretory System~~ *Kidneys*

2. The *Liver* converts and stores extra sugar from the blood.

3. *Platelets* are tiny particles in the blood that form blood clots.

4. *Red* blood cells carry oxygen from the lungs to the cells.

5. The two lower chambers of the heart, called *Ventricle*, pump blood away from the heart.

6. Before circulating throughout the body, blood enters the *Heart*, where it absorbs oxygen and releases carbon dioxide.
   Lungs

As you conduct your own science research, it is important to keep in mind where you look for information. Today, billions of people use the internet, and many of these people contribute content that you are free to use. Contributors post articles, answer questions, and maintain blogs. But not all of these contributors are experts in the fields they are writing about.

Check the credentials of the authors whose work you read. Be sure they are experts whose work is reliable. Also look for additional reputable sources that support what the contributors are saying. Otherwise, the information you find may be useless to you.

The Respiratory System

The body can survive only a short time without oxygen. Cells use oxygen to burn food and release the energy in the nutrients. The waste product of this activity is carbon dioxide. The function of the **respiratory system** is to supply oxygen to the cells and remove carbon dioxide.

HUMAN RESPIRATORY SYSTEM

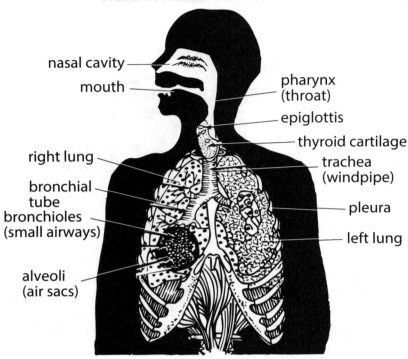

nasal cavity
mouth
pharynx (throat)
epiglottis
thyroid cartilage
right lung
trachea (windpipe)
bronchial tube
bronchioles (small airways)
pleura
left lung
alveoli (air sacs)

The hairs and the mucous lining of the nose warm, moisten, and clean the air of dust and dirt. The air enters at the rear of the mouth and then goes down the windpipe, called the **trachea**, through the bronchial tubes, and into the lungs. At the top of the windpipe is the larynx, or voice box, which allows us to make the sounds that include our speech.

In the lungs are millions of tiny air sacs. Each is surrounded by a network of capillaries. Here, oxygen passes into and carbon dioxide passes out of the blood vessels. The oxygen-rich blood heads straight for the heart, where it is pumped throughout the body. The carbon dioxide leaves the body when you exhale.

Vocabulary Review

Directions: Match each term on the left with the correct description on the right.

1. _C_ excretory system
2. _B_ platelets
3. _D_ plasma
4. _A_ respiratory system

A. supplies oxygen and removes carbon dioxide

B. help form blood clots

C. removes waste products from the body

D. liquid part of the blood

Directions: Answer the questions.

1. Explain how the circulatory system and excretory system work together to remove waste products from the body.

Food is consumed and saliva starts the breakdown of larger pieces. Passing through the throat to the stomach were it is broken down to a liquid form and pass into the smaller intestines where the body pulls the vitamines, nutrients, and minerals before it passes to the larger intestines where the fluid/water is pulled out by the body and the remaines go threw your bowels and urine.

2. Describe how the digestive system and circulatory system work together to deliver food to cells throughout the body.

the stomache breaks it all down to a liquid form before it enters the smaller intestines where the vitamines, minerals, nutrients are pulled into the blood stream and sent to cells around the body.

Check Answers

Skill Practice

Directions: Choose the best answer to each question.

1. What is the primary function of the respiratory system?

    A. removing liquid wastes from the body
    B. delivering oxygen to the blood and removing carbon dioxide
    C. breaking down food for energy
    D. filling the lungs with air

2. Which part of the blood is responsible for transporting oxygen to the cells?

    A. white blood cells
    B. plasma
    C. red blood cells
    D. platelets

3. How do cells get the energy they need for life activities?

    A. Oxygen is delivered by the respiratory system.
    B. Plasma is delivered by the circulatory system.
    C. Food is broken down by the digestive system.
    D. Excess sugar is stored by the liver.

Nervous, Endocrine, and Reproductive Systems

Lesson Objectives

You will be able to

- Recognize the organs and processes of the nervous and endocrine systems
- Differentiate between male and female reproductive organs
- Sequence the events in the development of a fetus from a fertilized egg
- Identify conclusions and supporting details

Skills

- **Core Skill:** Determine Central Ideas
- **Core Skill:** Cite Textual Evidence

Vocabulary

fetus
hormones
labor
menstrual cycle
sequence

KEY CONCEPT: The nervous system and endocrine system are responsible for communications within the body. They control many processes in the body, including those of the reproductive system.

In today's world, communication is essential and incredibly fast. Information travels through air waves and optic fibers at tremendous speeds. Such communication is essential to your survival, too. When your senses receive messages suggesting that you are in danger, your brain must receive and process those messages in time to prevent serious injury. Thankfully, you have a high-speed communication system of your own—your nervous system.

The Nervous System

The nervous system is the body's communication system. It consists of two parts. The central nervous system includes the brain and spinal cord. The **peripheral nervous system** includes all other nerves.

Nearly all messages of the nervous system involve the brain. Messages may begin with one of the major sense organs. Nerve cells in the sense organs react to changes in the environment. Information is converted to electrical and chemical signals that travel in sensory nerves. These nerves lead to the spinal cord and brain or to the brain directly. Messages carried on motor nerves act to move distant body parts, such as the arms and legs.

Nerve Cells

The shape of the nerve cell helps it function. Dendrites spread out from the cell body. The axon provides a quick path to the next nerve cell. The synapse is a small gap between cells where the electrical message changes into a chemical message.

NERVE CELL

Reflexes

Nerve cells can react to changes in the environment quickly. This reaction is called a **reflex**. Suppose you touch a flame with your hand. Nerve endings in your skin pick up the "hot" signal and immediately send it to the spinal cord. The spinal cord sends a message to your brain and hand. You jerk your hand away before you know you did it!

IDENTIFY CONCLUSIONS

Science is the study of how and why things happen as they do. To answer questions about the natural world, scientists perform investigations. Based on their results, they may create an explanation. A good explanation has three parts—a claim, evidence, and scientific knowledge. The **claim** is a statement, or a conclusion, that has been reached based on an investigation or research. Evidence and knowledge are used to support this claim.

When reading an explanation, it is important to identify what the claim is. Sometimes it is clearly stated. Other times you must use clue words or other statements that sum up the overall point of the passage. You may find that this idea is repeated throughout the passage and is also presented as the last sentence. Clue words such as *therefore, for that reason,* or *consequently* can help you identify the conclusion.

Look at the following example.

> Alcoholism is a serious disease that can have grave consequences. One of the most harmful consequences of alcoholism is a diseased liver. The liver is a key organ in the body. It removes alcohol and other harmful substances from the blood. When the liver is exposed to large amounts of alcohol on a regular basis, its tissues become damaged. A disease called *cirrhosis* is a common result.

This passage provides information about the liver and how alcohol affects it. What is the claim, or conclusion, that is stated in the passage? What are the supporting details, or the evidence, that supports this conclusion?

Remember, information can come from a variety of sources. When researching information your main goal is usually to end up with a conclusion. Sometimes just getting to the conclusion is not enough, though. In many instances you will be required to back up your conclusions with supporting evidence and knowledge gained from experience. Being able to easily spot your supporting evidence is a tool that can be applied to almost every aspect of your life.

Look at the passage about the brain on this page. Notice how the information is presented. Ask yourself, what key words and statements in this section help me understand the central idea? Using a notebook, make a graphic organizer like the one below, with the central idea in the center and the important supporting details that help verify that idea in the surrounding circles. In doing this you should be able to get a better understanding of the information presented to you.

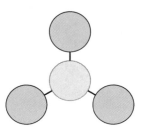

The Brain

The brain is the most complex part of the nervous system. It contains 90 percent of the body's nerve cells. All actions, except for reflex actions, are controlled by the brain. The brain is protected by the skull.

The upper portion of the brain is called the **cerebrum**. It receives, stores, and recalls all the information picked up through the senses. It is also the processing center for memory, decision making, thinking, speech, smell, taste, touch, vision, and hearing.

HUMAN BRAIN

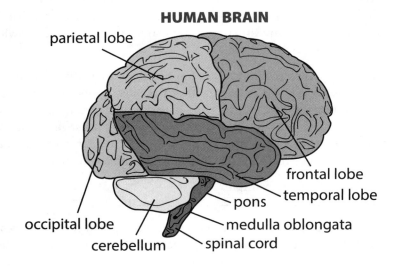

parietal lobe
frontal lobe
temporal lobe
pons
occipital lobe
medulla oblongata
cerebellum
spinal cord

The cerebrum is divided into two halves. The right cerebrum controls the left side of the body while the left cerebrum controls the right side.

Below the cerebrum is the **cerebellum**, which controls muscle coordination and balance. The **medulla** controls involuntary actions, such as the heartbeat, breathing, and digestion.

The Endocrine System

The endocrine system is a group of glands that release chemical messengers into the blood. The messengers are called **hormones**.

Hormones control the body's growth, its use of energy, and its ability to reproduce. Adrenaline is a hormone released into the body in times of danger. Some hormones control the levels of sugar or calcium in the blood. Others control the sexual maturation process of men and women.

The Reproductive System

The male reproductive system is controlled by hormones. The testes produce sperm, the male sex cells. Sperm are stored in the scrotum. The sperm leave the body through the penis.

MALE REPRODUCTIVE SYSTEM

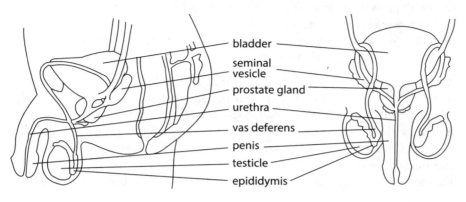

- bladder
- seminal vesicle
- prostate gland
- urethra
- vas deferens
- penis
- testicle
- epididymis

FEMALE REPRODUCTIVE SYSTEM

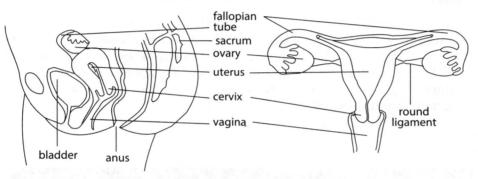

- fallopian tube
- sacrum
- ovary
- uterus
- cervix
- vagina
- round ligament
- bladder
- anus

The female reproductive system is also controlled by hormones. The ovaries produce female sex cells, or eggs. There an egg matures and is released into the uterus. If fertilized, it attaches to the wall of the uterus. If not, it passes out of the body through the vagina.

Menstrual Cycle

The growth and release of a mature egg is called the **menstrual cycle**. The cycle, roughly 28 days in length, starts when an egg begins to mature. The egg is released into the uterus. If the egg is not fertilized, uterine lining breaks down and is shed through the vagina as menstrual blood. The cycle begins again. If a mature egg and sperm unite and attach to the uterus, a pregnancy results.

THINK ABOUT **SCIENCE**

Directions: Fill in the blanks below.

1. The _____ are male reproductive organs that produce sperm.
2. The _____ are female reproductive organs that produce eggs.
3. A _____ egg attaches to the wall of the uterus.

The first two paragraphs on this page explain the events that occur during the growth of the fetus. The head ("Growth of the Fetus") at the top of the page identifies the topic. As you read each paragraph, look for clue words and phrases that signal events. Words and phrases such as *as soon as, begin, begins, during,* and *at the end of* are cues that identify actions occurring in sequence. Then look for statements that help to signal the overall point of the passage. During what period is the embryo very sensitive? During what period is good nutrition especially important?

Growth of the Fetus

As soon as an egg is fertilized, a **sequence** of changes begins—series of events occur, one following the other. The egg connects itself to the uterus through a membrane called the **placenta**. There the egg begins to divide and grow. The rapidly developing organism is called an **embryo**. During the first three months of pregnancy, the head and brain grow quickly, but the body growth is slow. The heart forms and begins to beat. Although all body systems are present, most of them cannot function yet. During this period, the embryo is very sensitive. A mother that uses nicotine, alcohol, or drugs can seriously damage it.

At the end of two months, the embryo begins to look somewhat human. At that stage it is called a **fetus**. During the next three months, the body begins to catch up with the head. Skin develops, and the mother may feel the fetus kick. During the last three months, the fetus gains a great deal of weight. Soon it can no longer move freely in the uterus. Bones, blood, and nerves develop rapidly. At this time, good nutrition—especially the addition of calcium, iron, and protein—is essential to the mother's diet.

Birth

At the end of about nine months, the mother's body produces hormones that control the baby's birth. The smooth muscles of the uterus begin to contract and relax. This movement, called **labor**, pushes the baby down the birth canal and out of the mother's body. The umbilical cord that joined the baby to the mother is cut. After a few more contractions, the placenta is forced out of the mother's body.

Vocabulary Review

Directions: Complete the sentences below using one of the following words.

fetus   hormones   labor   menstrual cycle

1. The growth and release of a mature egg every 28 days is the

   _____.

2. Chemicals released into the blood that regulate body activities are

   _____.

3. The process of _____ ends with the birth of a baby.

4. After two months, an embryo develops into a _____.

Directions: Read the passage below and choose the correct answer to the questions that follow.

In many sports, players are required or encouraged to wear helmets. In many states, motorcyclists are required by law to wear a helmet. A helmet helps protect the brain from serious injury. Brain injuries can lead to long-term disabilities and even death.

1. The job of a helmet is most like the job of which body part?

   A. brain
   B. skull
   C. eye
   D. spinal cord

2. If a patient showed loss of coordination and balance following an injury to the head, what part of the brain was most likely injured?

   A. medulla
   B. cerebellum
   C. right cerebrum
   D. left cerebrum

Directions: Look at the diagram of the brain and spinal cord on page 32.

3. Use the diagram to describe the pathway of nerves used to pick up a pencil from the floor.

Skill Practice

Directions: Choose the best answer to each question.

1. Which action involves the shortest path through the nervous system?

   A. jerking a foot away from a sharp tack
   B. kicking a soccer ball
   C. using one foot to scratch the other
   D. using toes to pick up a dime

2. How are the nervous and endocrine systems alike?

   A. Both act at the same speed.
   B. Both act to send messages.
   C. Both use the same chemicals.
   D. Both use the same body parts.

3. When does the thickened uterine lining shed as menstrual blood?

   A. after a fertilized egg connects to the uterus
   B. after an unfertilized egg is released into the uterus
   C. when the uterus is diseased
   D. when smooth muscles of the uterus contract

4. Which of the following describes the role of the placenta?

   A. It fertilizes a new egg.
   B. It releases an egg into the uterus.
   C. It connects an embryo to the uterus.
   D. It breaks down the uterine lining.

Health and Disease

Lesson Objectives

You will be able to

- Identify common diseases and their causes

- Discuss the types of nutrients used by the body

- Relate different types of drugs to their effects on the body

Skills

- **Core Skill:** Evaluate Conclusions

- **Core Skill:** Compare and Contrast Multimedia Sources

Vocabulary

acquire
antibiotic
calorie
drug
immunity
over-the-counter
prescription
symptom
well-balanced diet

KEY CONCEPT: To promote wellness and avoid common diseases, it is important to maintain a well-balanced diet and avoid any substances that change the normal functioning of the body.

Given regular oil changes, the right gasoline, and proper maintenance, an automobile can keep running for well over 100,000 miles. However, if a harmful substance were put into the car, it might stop running altogether and end up in a junkyard!

Cars and your body have some things in common. They are both machines, and with proper care and maintenance, and by avoiding harmful substances, your body can remain healthy and functional for many years.

Health and Disease

Good health habits include a balanced diet, adequate amounts of sleep and exercise, cleanliness, and a positive mental attitude. Maintaining these habits can help people avoid many health problems. However, very few people go through life without getting some type of disease. Many of these diseases, such as measles, mumps, polio, and the flu, are easily avoided by having proper vaccinations.

Disease

Diseases have many causes. Germs such as bacteria and viruses can enter the body and multiply, causing **infectious diseases.** Some infectious diseases are **acquired** (delivered or obtained) or transferred from one person to another. These are called **communicable diseases**. Other diseases, such as arthritis, are caused by body parts wearing out with age. Many diseases are triggered by lifestyle choices. Eating a diet of fatty foods can lead to heart disease. Smoking or chewing tobacco products can cause lung and heart disease.

Some infectious diseases can be acquired only once. The reason is that the body develops an **immunity,** an ability to resist the disease, that protects it from further infection. An adult who had measles as a child will not acquire it again. Hygiene habits such as regularly washing hands help people avoid many germs.

STDs

Sexually transmitted diseases (STDs) are caused by germs that travel through sexual contact. The chart on the next page shows symptoms, diagnosis, and treatment of the most common STDs. All STDs can be prevented by avoiding exposure to them.

SEXUALLY TRANSMITTED DISEASES

Disease	Symptoms	Treatment
Chlamydia	Pain; itching; discharge; inflammation; scarring; infertility	Antibiotics
Gonorrhea	Discharge; itching; pain; swelling; abnormal bleeding	Antibiotics
Syphilis	Organ damage; mental problems; blindness; deafness; heart failure; death	Antibiotics
Herpes	Sores that ooze and scab; fever; swollen glands; body aches	No cure; antiviral medications to treat symptoms
AIDS	Flu symptoms; swollen glands; skin rash; makes people susceptible to infectious diseases and cancers	No cure; anti-retroviral drugs slow virus growth and organ damage

COMPARE MULTIMEDIA SOURCES

Printed books have long been the primary source of information for learners. Textbooks provide explanations on various subjects such as science or history. Visuals, including photographs and diagrams, are often included to help clarify topics. Many textbooks are authored by experts in their fields.

However, technology has changed the way content is delivered. **Multimedia** refers to multiple forms of communication. It includes textbooks, plus the internet, software, CDs, DVDs, podcasts, television, and more. Words that were once only available through print can be heard. Visuals that were once static, or still, can move by animation. Information is regularly communicated through video and other sources. Some multimedia sources are also much easier to update and keep current than printed materials.

With so many multimedia sources available, sometimes it is best to compare them to determine which kind best suits your purpose. Read each example below. Then explain which multimedia source would be best to use and why.

1. You want to learn about the process of cell division. Would you use a picture or an animation?

2. You want to learn about current treatments for heart disease. Would you use a textbook or the internet?

_____

_____

_____

Symptoms are indicators of something out of the ordinary, or a disorder. However, they are subjective indicators, meaning they are not by themselves final indicators. Doctors may use symptoms as clues that suggest a specific disease, but only after collecting testable evidence, do they draw conclusions.

In science, conclusions are always based on evidence. For example, scientists have concluded that a number of factors are associated with cardiovascular disease. Explain how you can be confident that their conclusions are valid.

Cardiovascular Disease

Cardiovascular diseases affect the heart and the blood vessels. A common cause of this disease is the buildup of fatty deposits inside blood vessels. These deposits can make the arteries stiffer and more resistant to blood flow. This forces the heart to pump harder, raising the blood pressure.

High blood pressure can lead to a sudden failure of the heart, called a **heart attack**. A sudden failure of the blood supply in the brain causes an event called a **stroke**. Smoking, obesity, stress, lack of exercise, and a high-fat diet can increase the risk of cardiovascular disease. A high-salt diet can make high blood pressure worse.

Cancer

Cancer is the uncontrolled growth and spread of abnormal cells in the body. Cancer may appear in the breasts, lungs, brain, bones, skin, or other organs. The seven warning signs of cancer, shown in the chart below, have saved many lives. Some doctors add rapid weight loss and extreme fatigue to those warning signs.

Early Warning Signs of Cancer
• Unusual bleeding or discharge
• A lump or thickening in the breast or elsewhere
• A sore that does not heal
• A change in bowel or bladder habits
• Hoarseness or cough that continues
• Indigestion or difficulty in swallowing
• A change in size or color of wart or mole

Researchers are learning more and more about cancer. Many cancers can now be treated successfully if they are found and treated in the early stages. Surgery to remove tumors is now routine. Radiation and a variety of chemicals are used to destroy cancerous cells that cannot be removed surgically.

Diabetes

Diabetes is a disease in which the body does not produce enough of a hormone called insulin. **Insulin** helps the body use and store the sugar glucose. Symptoms of diabetes include frequent urination, extreme thirst, and fatigue. Uncontrolled diabetes can lead to blindness, diseased legs and arms, and ultimately, death.

Many people inherit a tendency to develop diabetes. Being overweight and not getting enough exercise greatly increase the risk of diabetes. The disease can be controlled through diet and exercise, as well as insulin treatments.

Cirrhosis

In cirrhosis, the cells of the liver are replaced by scar tissue. The scar tissue prevents the liver from functioning properly. People with the disease may bruise easily, have nosebleeds, and vomit blood.

Alcoholism and poor nutrition are the major risk factors for cirrhosis. Correcting the diet and avoiding alcohol can stop the disease, but the liver cannot be repaired. Without treatment, cirrhosis leads to death.

Arthritis

As people age, many get arthritis. They have some pain or difficulty with their joints. When the joints become swollen and stiff, the pain can be quite severe. Bony deposits may develop over the joints, locking them in place. Knees, hips, fingers, and toes are common sites for arthritis.

Heredity and obesity contribute to the disease. Although there is no cure for arthritis, the pain can be reduced with medicine, special exercises, and the application of heat.

THINK ABOUT SCIENCE

Directions: Match each disease or condition on the left with the symptoms on the right.

Disease/Condition

1. arthritis C
2. syphilis E
3. high blood pressure F
4. cancer D
5. diabetes A
6. cirrhosis B

Symptoms

A. frequent urination, extreme thirst, and fatigue
B. bruises, nose bleeds, vomiting blood
C. stiff and swollen joints
D. sore that does not heal
E. organ damage, heart failure, death
F. fatty deposits build up inside blood vessels

WRITE TO LEARN

To write a summary, you need to identify the main idea and supporting details. This helps you better understand what you read. In a notebook, write a summary of a news story you find in a newspaper or online. Be sure to include only necessary information.

WORKPLACE CONNECTION

Information technology (IT) is a rapidly expanding field. IT specialists are responsible for designing and developing computer software and network systems, including the internet. They may apply their skills to basic computer applications, computer-generated animations, or even support systems for space programs.

There are a wide variety of job opportunities for people with IT training. Educational requirements range from high school graduates to PhDs.

How could you apply training as an IT professional to developing multimedia resources for science and education? Write a job description that explains one possible career opportunity.

Nutrition and Diet

The body needs nutrients to carry out the functions of maintenance, growth, repair, and reproduction. As a result, a **well-balanced diet** is essential. **Nutritionists** study foods that the body needs to stay healthy. Six types of nutrients are vital to life. They are carbohydrates, fats, proteins, vitamins, minerals, and water.

Good nutrition means eating the right foods in the proper proportions. A meal should include a range of different foods that provide a balance of carbohydrates, proteins, vitamins, and minerals. Every packaged food sold in the United States is required by the Food and Drug Administration to bear a nutrition facts label that lists the percentage of nutrients contained in it. Some states require that restaurant chains provide nutritional information on all menu items to consumers.

Vital Nutrients

Carbohydrates are the starches and sugars that are the main energy source for the body. They are supplied in large amounts in grains, fruits, and vegetables.

Like carbohydrates, **fats** provide a concentrated dose of energy. Large amounts of fats are present in oils, dairy products, nuts, and meats. The body needs only very small amounts of fat.

In the body, **proteins** are broken apart into amino acids. The body can manufacture most of the amino acids it needs to make new cells and to repair old cells. Nine of these acids, however, cannot be made and must be supplied in our food every day. To work, these essential amino acids must be present in the body at the same time.

Proteins from animal sources, such as fish, eggs, cheese, and meat contain all of the essential amino acids. Combinations of certain vegetables, such as corn and beans, can combine to provide the essential amino acids. Vegetarians must learn to plan their meals carefully to make sure that all the essential amino acids are available to the body at the same time.

Vitamins and minerals in small amounts are needed to regulate the body's activities. Each has a specific function. Some break down fats into proteins; some build bones. Others allow nerves to carry messages or muscles to contract.

Water is the body's most important nutrient. It is used for almost every bodily function. Without water, people can survive for only a few days.

Calories

Calories are a measure of energy value. Everyone needs calories for energy. Not everyone, however, needs the same number of calories. People who are very active need more calories than those who get little exercise. Younger people generally need more calories than older people do.

Nutritionists know that some foods are more valuable than others. Milk, for example, provides just a bit of almost everything the body needs. A candy bar, on the other hand, provides many calories with few nutrients. Nutritionists call these "empty" calories. They may provide quick energy, but they are of no lasting benefit to the body.

A Balanced Diet

No single food provides all the nutrients that the body needs to stay healthy. This is why it is important to eat a balanced diet, which is made of a variety of healthful foods.

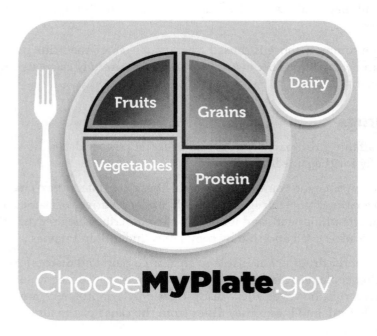

The new food guide shows what a healthy and balanced diet might look like. It includes the five groups of food that include essential nutrients that promote good health. These are: fruits and vegetables, grains, proteins, dairy, and oils. Oils are not a food group, but used sparingly, they are important nutrients. As a general rule, you should make half your plate fruits and vegetables. You should select whole grains, such as wheat flour, oatmeal, and brown rice, at least half the time. Following these guidelines provides an excellent foundation for a diet that promotes wellness.

THINK ABOUT SCIENCE

Directions: Answer each question below.

1. Why is it important to eat a balanced diet? *To stay healthy*

2. What are "empty" calories? *Calories that provide few nutrients*

3. What do the parts of the food guide represent?
   Fruit, veg, grains, protein, Dairy, and oils

To be a successful student or employee, it is important to remain flexible and adaptable. This is particularly true as advances in technolology continue so rapidly. Less than 30 years ago, people who worked with print books typed on typewriters. As computers rapidly became common in workplaces, people had to be flexible and adapt to the new technologies by learning new skills.

As a student, how have you adapted to changes in technology? Describe some of the changes that have occurred since you began school.

Drugs

Drugs are any kinds of chemicals that affect the body, mind, or behavior. Some drugs are quite mild and short lasting. Other drugs are stronger, but are safe when used properly. Their distribution is carefully regulated and available only under a doctor's order. A few drugs are both strong and addictive and may easily cause physical or mental harm. These drugs are illegal or are closely regulated.

Any drug can be abused. One common abuse is caused when people share medicine prescribed by a doctor. This practice can be dangerous because many drugs are carefully calculated for a specific person's age and weight.

Over-the-Counter Drugs

A few mild drugs, such as caffeine, are a part of foods and beverages. Most drugs, however, are packaged and sold specifically for their medical effects.

Some mild drugs are sold to anyone who wants to buy them. These are called **over-the-counter** drugs. The same over-the-counter drug is often sold under a variety of names. For example, the names aspirin, acetaminophen, and ibuprofen each describe certain drugs that relieve pain. Each drug is also advertised and sold by companies under brand names.

A wide variety of over-the-counter drugs are available to treat mild pain and discomfort. Different kinds of drugs help people recover from a headache, sore throat, upset stomach, skin rash, and other problems. As a rule, however, these drugs treat only symptoms. Symptoms are the effects that someone feels from an illness. They differ from the cause of the illness.

Prescription Drugs

The symptoms of some illnesses do not improve on their own, even with the help of over-the-counter drugs. When suffering continues or seems very serious, people should consult a physician, nurse, or other health care provider. The provider may recommend a **prescription** drug, a type of drug that is very carefully regulated.

Only pharmacists are allowed to sell prescription drugs, and they do so only under a doctor's order. The order, called a prescription, tells the amount and strength of the drug, as well as precisely how the patient should use it.

One reason that prescription drugs are tightly controlled is because they can be easily abused. Taking a few doses of a strong painkiller can help a patient manage a serious illness. Taking the same drug for many weeks can lead to a dangerous addiction.

Delivery of Drugs

A drug can be delivered to the body in many ways. Drugs to treat rashes are usually spread on the skin. Many drugs for the lungs are breathed in. Other drugs are taken orally, meaning they are swallowed. These drugs are broken down in the digestive tract, much like food. Then the drug is absorbed into the blood and carried throughout the body.

Some drugs, however, are too delicate for the digestive tract. They may need to be delivered intravenously, or through a needle directly into the bloodstream. People who suffer from diabetes take insulin in this manner.

In hospitals, many drugs are delivered through an intravenous tube. The term **intravenous** means "within the veins." An intravenous tube acts like a highway into the bloodstream. It allows doctors and nurses to provide drugs very quickly and efficiently.

Inventing New Drugs

There are more drugs available today than at any time in human history. Drugs are available to treat almost every disease in the dictionary. One reason is that scientists continue to research new drugs and improve existing ones.

Sometimes drugs are discovered by accident. In 1928, English scientist Alexander Fleming noticed something that he did not expect. Mold was killing the bacteria he was trying to grow. Rather than toss away his experiment, he investigated what was happening. This led him to discover penicillin, the first antibiotic. An **antibiotic** is a drug that helps the body kill bacteria.

Since Fleming's time, penicillin has been used to treat bacterial infections all over the world. So have many other antibiotics that scientists discovered. Unfortunately, bacteria become resistant to antibiotics over time. This is why new antibiotics must continue to be invented and why doctors keep the newest, strongest antibiotics in reserve.

Unfortunately, no drug exists to cure one of the worst diseases that people now suffer from. Drugs can help people live with AIDS, but there is no drug that will cure it. AIDS is caused by a virus, not bacteria, so antibiotics are useless against it.

THINK ABOUT SCIENCE

Directions: Answer the questions below.

1. What is the difference between over-the-counter drugs and prescription drugs? *prescription drugs consist of a stronger dose*

2. When should people consult a health care provider about an illness? *If the over the counter meds don't help the symptoms*

3. Why do scientists continue to develop new antibiotics? *Because Bacteria becomes resistant to antibiotics*

Vocabulary Review

Directions: Circle the word in the parentheses that best completes the sentence.

1. (Over-the-counter, (Prescription)) drugs are available only under a doctor's order.

2. A good wellness plan would include a (calorie, (well-balanced)) diet.

3. Anyone can buy a(n) ((over-the-counter), prescription) drug to treat a common cold or headache.

4. The effects someone feels of an illness are ((symptoms), antibiotics).

5. An ability to resist infection is a(n) (drug, (immunity)).

6. A measure of the energy value of a food is a ((calorie), symptom).

Delivery of Drugs

A drug can be delivered to the body in many ways. Drugs to treat rashes are usually spread on the skin. Many drugs for the lungs are breathed in. Other drugs are taken **orally**, meaning they are swallowed. These drugs are broken down in the digestive tract, much like food. Then the drug is absorbed into the blood and carried throughout the body.

Some drugs, however, are too delicate for the digestive tract. They may need to be delivered through a needle directly into the bloodstream. People who suffer from diabetes take insulin in this manner.

In hospitals, many drugs are delivered through an **intravenous** tube. The term intravenous means "within the veins." An intravenous tube acts like a highway into the bloodstream. It allows doctors and nurses to provide drugs very quickly and efficiently.

Directions: Research two other sources for information on drug delivery systems. Each source should be in a different media format such as an animation, video, or website.

1. Compare the two resources you found to the reading passage above. Which delivers *more* content? Which makes the content easier to understand? Explain why.

_____

_____

_____

2. Suppose you were a nursing student. Explain which multimedia resource would be best to learn about how to deliver drugs using different methods.

_____

_____

_____

3. Why would a nurse need to be flexible and adaptable in his or her career? Consider what you have learned about new drug inventions and methods for delivery.

_____

_____

Skill Review (continued)

Directions: Read the passage. Then answer the questions that follow.

A common cause of skin cancer is excess exposure to sunlight. The ultraviolet (UV) radiation in sunlight can lead to mutations, or changes to genes, in skin cells. Certain mutations cause cells to become cancerous, meaning they grow out of control. Skin cancer can be prevented by applying sunscreen. Sunscreen blocks harmful UV radiation.

4. What conclusion can you draw from this passage?

    A. Ultraviolet (UV) radiation is a part of sunlight.
    B. Most mutations cause cancer.
    C. Sunscreen is a cure for cancer.
    D. Skin cancer is preventable.

5. Who is the most likely author of this passage?

    A. a doctor who studies skin cancer
    B. a scientist who studies oceans and beaches
    C. a physicist who studies UV radiation
    D. the manager of a beach resort

Directions: Questions 6 and 7 refer to the Nutrition Facts chart.

6. According to the nutrition facts found on the food label, how many calories are in one serving of this food?

    A. 90
    B. 80
    C. 100
    D. 180

7. How much total fat is in one serving of this food?

    A. 3 grams
    B. 1 gram
    C. 0 gram
    D. 5 grams

Nutrition Facts

Serving Size 1 cup (50 g)

Amount Per Serving	
Calories 180	Calories from Fat 9
	% Daily Value*
Total Fat 1 g	2%
Saturated Fat 0.15 g	1%
Trans Fat 0 g	
Total Carbohydrate 40.54 g	14%
Protein 5 g	

Skill Practice

Directions: Choose the best answer to each question.

1. What do all infectious diseases have in common?

    A. All affect adults only.
    B. All affect children only.
    C. All are sexually transmitted.
    D. All are caused by germs.

2. Grains, fruits, and vegetables are the best sources for which nutrient?

    A. carbohydrates
    B. proteins
    C. fats
    D. amino acids

3. Choose one human body system. Write a conclusion about how to maintain the health of that system. Use evidence to support your conclusion.

Review

Directions: Choose the best answer to each question.

1. Which statement explains why the same accident may cause a broken bone in an older adult, but not in a child?

   A. The adult skeletal system is made of bone only, not cartilage.
   B. Bones become thicker and stronger with age.
   C. Bones lose marrow with age.
   D. Bones weaken and become brittle in old age.

2. Which phrase best describes the role of the skeleton in the human body?

   A. a growth source
   B. an energy source
   C. a protective shell
   D. a supportive structure

3. Why are fixed joints best for connecting the bones of the skull?

   A. They allow the brain to grow to any size.
   B. They help the skull protect the brain.
   C. They allow the skull to bend under certain circumstances.
   D. They help the skull repair brain injuries.

4. What change is caused by tearing a ligament?

   A. painful digestion of fatty foods
   B. immature blood cells entering the bloodstream
   C. a muscle disconnecting from a bone
   D. bones at a joint moving slightly out of place

5. The lining of the small intestine performs a unique job in the digestive system. What is the best model for this job?

   A. a bath towel that soaks up water
   B. a hose that carries water from place to place
   C. a hammer that breaks things into small pieces
   D. glue that holds things together

6. Smooth muscle, not skeletal muscle, lines the walls of the stomach. Why are smooth muscle best for helping the stomach work?

   Smooth muscles help the stomach because

   A. the stomach moves on its own, without you deciding how or when.
   B. the stomach never moves.
   C. the stomach moves in a rhythym like a heart beating.
   D. the stomach is C-shaped.

Review

7. Which symptom is most likely a sign of kidney failure?

    A. high levels of wastes in the blood
    B. low levels of wastes in the blood
    C. watery, light-colored urine
    D. thick, dark-colored urine

8. Which human body system is affected by strokes?

    A. the endocrine system
    B. the cardiovascular system
    C. the excretory system
    D. the reproductive system

9. What role does the upper respiratory system (nose, trachea, lungs) play in the process of respiration?

    A. It delivers oxygen to the cells where it is needed for respiration.
    B. It starts the process of glycolysis.
    C. It delivers oxygen to the cells where it is needed for fermentation.
    D. There is no relationship.

10. A sore that does not heal, a continuous cough, a lump or thickening under the skin, and a change in bowel habits are all early warning signs of what disease?

    A. cardiovascular disease
    B. diabetes
    C. cirrhosis
    D. cancer

11. What is the best way to prevent AIDS and other STDs?

    A. practice good hygiene
    B. develop immunities
    C. avoid direct contact
    D. use antibiotics

12. A food label states that one serving of food contains 124 calories. What does this value of calories indicate about the food?

    A. nutritional content
    B. carbohydrate content
    C. energy content
    D. protein content

13. Which illness would best be treated by an antibiotic?

    A. arthritis
    B. skin cancer
    C. a viral infection
    D. a bacterial infection

14. What is the function of the cerebrum?
    A. muscle coordination and balance
    B. involuntary actions, such as breathing
    C. decision making
    D. sensory perception

Directions: Questions 15–17 are based on the following passage.

A pregnant woman's consumption of alcohol can cause great harm to her fetus. Fetal Alcohol Syndrome (FAS) is the term doctors use to describe children born to mothers who drink. FAS may only cause slight behavioral problems. More often, unfortunately, mental disabilities or physical deformities may also occur as a result of FAS.

When a pregnant woman drinks, so does her unborn child. The alcohol flows into the fetus's bloodstream long before its body systems are able to handle the alcohol. As a result, the alcohol acts as a poison. Even a small amount of alcohol—especially in the first few months of pregnancy—can be enough to damage the baby's developing organs.

The damage done to a FAS fetus is not evident until it is born. The FAS baby may show any combination of the following symptoms:

- low birth weight
- weak heart or kidney problems
- learning difficulties
- slow development

Even when a newborn does not have the characteristics of FAS, the pregnant mother's alcohol consumption may become apparent as the child grows older. Researchers call these less-apparent consequences fetal alcohol effects or modified FAS. These include:

- mental retardation
- abnormally high levels of activity
- inability to concentrate
- abnormal sleep patterns

It has been estimated that FAS is the third most common birth defect. The tragedy is that FAS is easily prevented.

15. What is the main idea of this passage?

   A. Pregnant women should limit their alcohol consumption.
   B. Babies with FAS have low birth weight.
   C. Alcohol can result in low birth weight.
   D. Drinking alcohol while pregnant can harm a fetus.

16. Which of these details supports the main idea?

   A. FAS is easily prevented.
   B. Characteristics of FAS can occur as a child grows older.
   C. Consuming alcohol during pregnancy harms the fetus.
   D. Pregnant women should never drink.

17. Why does alcohol work as a poison in fetuses?

   A. The fetus absorbs more alcohol per pound of body weight.
   B. Body systems equipped to deal with alcohol are not fully formed.
   C. The alcohol remains in the body until birth, when it triggers a series of harmful effects.
   D. Alcohol replaces water in the fetus's cells.

Directions: Questions 18–19 are based on the following passage.

Organ donation is an opportunity to improve, or even save, the life of a fellow human being. Most organ donations involve the removal of organs and tissues from a person who has recently died. Sometimes living donors can donate because it is possible to live a normal life with only one of a pair of organs, or with a part of an organ. For example, people can donate one kidney or a part of a liver.

The first successful organ transplant occurred in 1954, when a man dying of kidney disease received a kidney from his sister. Doctors can now transplant kidneys, hearts, livers, lungs, intestinal organs, eyes, skin, and bone. Doctors can sometimes help up to fifty patients from one donor.

It is easy to become an organ donor. Anyone at any age can become an organ donor and give the gift of life.

---

18 Which condition makes it possible for a living person to donate a kidney?

  A. The donor is willing to use life-saving machines.
  B. The body can survive with only one kidney.
  C. Some people can survive without a kidney.
  D. Some people are born with an extra, nonfunctioning kidney.

19. What can be inferred about an organ donor who helped fifty patients?

  A. The donor died at an old age due to serious illness.
  B. The donor died suddenly, and had been in excellent health.
  C. The donor gave a variety of organs throughout his or her life.
  D. The donor gave a variety of organs just after death.

Check Your Understanding

On the following chart, circle the number of any item you answered incorrectly. Next to each group of item numbers, you will see the pages you can review to learn how to answer the items correctly. Pay particular attention to reviewing those lessons in which you missed half or more of the questions.

Chapter 1 Review

Lesson	Item Number	Review Pages
Skeletal and Muscular Systems	1, 2, 3, 4	16–23
Digestive, Respiratory, Excretory, and Circulatory Systems	5, 6, 7, 9, 18, 19	24–29
Nervous, Endocrine, and Reproductive Systems	14, 15, 16, 17	30–35
Health and Disease	8, 10, 11, 12, 13	36–45

Application of Science Practices

CHAPTER 1: HUMAN BODY AND HEALTH

Question

How do various nutrients play a role in the functioning of the different systems in the human body?

Background Concepts

The human body consists of interconnected systems that work together to maintain overall health. These systems include the skeletal, muscular, digestive, respiratory, excretory, circulatory, nervous, endocrine, and reproductive systems.

Each of these systems requires various nutrients, such as vitamins, minerals, proteins, carbohydrates, and fats, which are either created by our bodies or provided by the foods we eat. Each system has unique nutrient requirements for proper functioning. Without critical nutrients, a system cannot function effectively.

Investigation

1. Draw a chart that lists each of the systems of the human body in the first column. Use the second column to indicate the critical nutrients required for each system. In the third column, list possible sources of those nutrients. Follow this example.

System	Critical Nutrients	Sources of Nutrients
Skeletal	Calcium Vitamin D Vitamin C Iron	Calcium: found in dairy products, iron, kale, okra, collards, soy beans, some fish Vitamin D: fortified dairy products, orange juice, soy milk, and cereals; cheese; egg yolks Vitamin C: guava, red and green sweet peppers; kiwi; oranges; grapefruit; Brussels sprouts; cantaloupe Iron: mollusks; squash and pumpkin seeds; beans; dark leafy greens; dark chocolate; cocoa powder; tofu; whole grains; liver; some lean meats
Muscular		
Digestive		
Respiratory		
Excretory		
Circulatory		
Nervous		
Endocrine		
Reproductive		

Application of Science Practices

2. Research materials in a library or on the internet or contact a health care professional such as a doctor, nurse, or dietician to help you identify the critical nutrients for each system and the sources of these nutrients.

3. In your research, note diseases that can result if body systems don't get all of the critical nutrients they need. Look for examples of diseases that affect more than one body system.

Interpretation

Which nutrients are critical to more than one of the systems of the human body?

_____

_____

Which diseases affect more than one system of the human body? What does this tell you about how systems function?

_____

_____

Answer

How do various nutrients play a role in the functioning of the different systems in the human body? Which critical nutrients can come only from the food we eat? Which, if any, nutrients can come only from the human body itself?

_____

_____

Evidence

What specific evidence did you find in your research that supports the connection between nutrients and the health of various systems?

_____

_____

Life Functions and Energy Intake

All living things require energy to do work. Plants get the energy they need by producing their own food. Other organisms get the food they need by eating other organisms, including plants.

Plants make their own food in a process called photosynthesis. The prefix *photo-* means "light," and the base word *syn* means "together." During photosynthesis, carbon dioxide and water react in the presence of light energy from the Sun. The reaction produces food in the form of a sugar. It also releases oxygen as a waste product.

Organisms that eat other organisms to get the energy they need must digest their food. Digestion is the process of breaking down food into its simplest parts, or molecules. The body breaks down molecules in the presence of oxygen to release the energy locked in the molecules' chemical bonds. This process, called respiration, is the opposite of photosynthesis.

Some organisms live in environments where oxygen is unavailable, making respiration impossible. Instead, these organisms depend on the process of fermentation to get the energy they need for work.

In this chapter you will learn about:

Lesson 2.1: Flowering Plants
Flowering plants vary greatly in appearance. This lesson discusses their basic parts and describes plant reproduction and photosynthesis.

Lesson 2.2: Respiration
Respiration, which occurs in the presence of oxygen, breaks apart sugar molecules to capture the energy trapped in the molecule's chemical bonds. The process occurs in the presence of oxygen and yields high-energy molecules called ATP.

Lesson 2.3: Fermentation
Fermentation has the same purpose as respiration, but it is conducted in the absence of oxygen. It yields fewer molecules of ATP, making it less efficient than respiration.

Goal Setting

You may already know a great deal about the processes of photosynthesis, respiration, and fermentation. There may also be more you want to know.

Think about some questions you already have about these processes. Record them in the chart below. Then, as you read, return to the chart. Write the answers to your questions.

As you read and learn, you may find that you have additional questions. Return to this chart as often as you like. Record your questions and then return to write the answers when you know them.

Questions	Answers

Flowering Plants

KEY CONCEPT: Flowers contain male and female reproductive structures and attract pollinators that transfer reproductive materials.

Have you ever opened a magazine and smelled an advertisement? Some companies make colorful, attractive, and sometimes scented advertisements to get your attention.

Some plants use the same advertising strategies. Think about flowers' bright colors, patterns, and fragrances. These "advertisements" attract pollinators, such as birds and bees, that can spread pollen from one flower to another.

Flowering Plants

Earth is home to hundreds of thousands of species of flowering plants. In recent studies, some scientists claim there may be as many as 400,000 different species.

Flowering plants share common parts. Observe a flowering plant, and you're likely to see one or more flowers, a stem that supports the flowers, and leaves along the stem. If you were able to look beneath the ground, you would also likely see a root or root system. Tiny root hairs often branch off the roots.

As you continue reading, return to this page to refer to the diagram. Use its labels to help you remember the plant parts common to flowering plants.

PARTS OF A FLOWERING PLANT

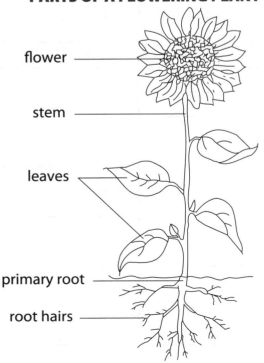

flower

stem

leaves

primary root

root hairs

INTEGRATE TEXT AND VISUALS

You can think of **visuals** as pictures that support, interpret, or extend an explanation. Graphs, charts, diagrams, illustrations, and photographs are all examples of visuals that you may use to understand written information. Read the following paragraph. Then examine the visual.

Leaves possess such a variety of shapes, patterns, edges, and means of attachment that scientists have developed a special vocabulary to refer to the individual features. Consider leaf shape, for example. Leaves are either simple or compound. Simple leaves have a single blade, while compound leaves have smaller leaflets that make up the whole.

NORTHERN US TREE LEAVES

Simple Leaf	Simple Leaf	Compound Leaf
Lobed edges	Double-toothed edges	Smooth or slightly toothed edges

How does the diagram support what you read in the text?

Roots

After a seed is planted in the ground, the **roots** are the first part of the plant to develop. Roots anchor the plant in the ground. The plant will stay in that spot for its entire life.

The roots also absorb water and food from the soil and transport them to the stem. In some plants, food is stored in the roots. Most roots are covered with threadlike structures called **root hairs**. These hairs increase the amount of food and water the plant can absorb from the soil.

Not all roots are the same. Some plants, such as grass, have a mass of roots that spread out wide and grow near the surface of the ground. Other plants, such as the dandelion and carrot, have one major root that grows deep into the ground. This deep root is called a **taproot**.

Stems

Stems have two basic functions. They support the plant and they transport food and water throughout the plant. In some parts of the stem, water and nutrients travel up from the roots to the leaves. In other parts of the stem, food manufactured in the leaves moves down to the roots.

Tree trunks are stems that can grow quite tall and strong. Other stems are small and grow close to the ground. Some stems even grow underground. Onions and tulip bulbs are underground stems that also store food.

Leaves

Leaves are green because so many of their cells contain **chlorophyll**, a green **pigment**, or colored substance, that enables light to be absorbed by the plant's cells. It is in these cells, the chloroplasts, that food is made. Leaves also contain tiny openings, called **stomates**, that allow gases from the air to enter and exit the leaf. Both the chloroplasts and the stomates are essential to life on Earth.

Cacti, which grow in dry deserts, have leaves that are thin needles. This reduces loss of water through the leaves. The broad, flat leaves of some water-dwelling plants help the plants float.

Flowers

Flowers are actually a specialized part of the stem. They are made of four different types of leaves called sepals, petals, stamens, and carpels.

Sepals enclose the flower bud. **Petals** have different colors and markings that attract specific insects, birds, and mammals. For example, some flowers have markings that guide insects to the center of a flower. Some flowers also produce fragrances to attract pollinators that have a better sense of smell than sight. Drawing organisms to flowers in this way is essential to plant reproduction. **Reproduction** is the process by which organisms generate new individuals of the same kind.

Stamens and carpels are reproductive structures. **Stamens** are the male parts of a plant and contain pollen. **Carpels** are the female parts of a plant and contain ovules, which can develop into seeds.

Flowers vary tremendously from plant to plant. Each type of flower may have a unique shape, color, markings, and fragrance. Flowers also may be simple or composite. Simple flowers, such as buttercups, are single flowers with many petals. Composite flowers, such as red clover and sunflowers, are made of clusters of many individual flowers, which may appear to be a single flower. Such flowers are sometimes called **blooms**. Because of their clustered arrangement, composite flowers produce many fruits. Each fruit carries seeds inside it.

THINK ABOUT SCIENCE

Directions: Write letters on the lines below to match each plant structure to its function.

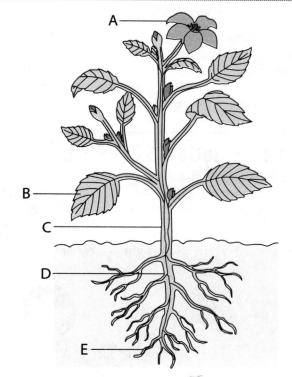

1. ___C___ holds up the plant
2. ___B___ makes food
3. ___D___ anchors plant
4. ___E___ soaks up water
5. ___A___ makes seeds

Photosynthesis

Photosynthesis is the process that plants use to make food. The equation below will help you to understand it. It involves both sunlight and chlorophyll, the pigment that gives leaves their color.

PHOTOSYNTHESIS FORMULA

carbon dioxide + water + sunlight $\Rightarrow$ sugar + oxygen

$6CO_2 + 6H_2O$ in the presence of sunlight $\Rightarrow C_6H_{12}O_6 + 6O_2$

As the Sun shines on a leaf, chlorophyll absorbs the Sun's light energy. At the same time, water moves into the leaf from the stem and roots. Carbon dioxide, a gas from the air, enters the leaf through tiny holes.

Cell parts called chloroplasts perform the reactions of photosynthesis. Using the energy of sunlight, they combine carbon dioxide and water into glucose, a simple sugar. Oxygen is left over and is released into the air. Both plants and animals use that oxygen to break down food and release its energy, a process called **respiration**.

THINK ABOUT SCIENCE

Directions: Label the diagram to identify the steps of photosynthesis. Then write a paragraph to explain each step. Use the visual to support your words.

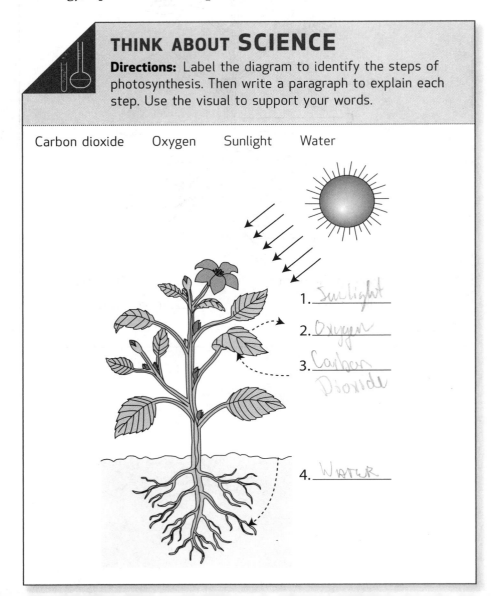

Carbon dioxide   Oxygen   Sunlight   Water

1. Sunlight
2. Oxygen
3. Carbon Dioxide
4. WATER

Reproduction in Flowering Plants

Most people notice and admire the petals of a flower. The petals, however, serve mainly to protect the plant's reproductive structures. In the diagram of the flower, find the stamens that are just inside the petals. Stamens are the flower's male reproductive structures. Tiny grains of pollen are produced on the tip of the stamens. **Pollen** is the male reproductive material of flowering plants. These pollen grains hold one half of the plant's genetic information.

One or more carpels form the pistil. The **pistil** is the female reproductive structure of the flower. The pistil is covered with a sticky fluid. When a pollen grain lands on the top of the pistil, a tube grows down to the swollen base of the pistil. The base contains an egg cell, which holds the other half of the plant's genetic information. The pollen grain travels down the tube and fertilizes the egg. Fertilization combines the genetic information from the male and the female cells.

If **pollination**, the process by which plants transfer reproductive materials from one plant to another, succeeds, the egg cell at the base of the pistil develops into a seed. The seed may develop into a fruit. The fruit falls to the ground or is eaten by a bird or an animal. Some seeds, such as dandelion seeds, simply blow away. If the seed lands in a place where it can root, it will grow into a young plant.

Bees are the pollinators for many types of plants. Bees are attracted to the bright colors and sweet smells of flowers. Bees also gather the flower's nectar to make honey. As the bees explore the flower, they brush up against the stamens. The pollen sticks to the bees' legs. When the bees enter the next flower, the pollen falls onto the pistil.

Other insects, birds, animals, the wind, and even rain can pollinate flowers. Some types of plants, such as cotton, pea, and tomato plants, can even pollinate themselves.

ANGIOSPERM FLOWER

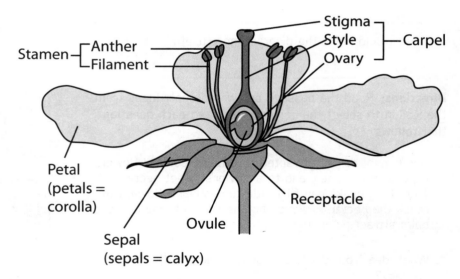

Stamen — Anther / Filament

Stigma / Style / Ovary — Carpel

Petal (petals = corolla)

Sepal (sepals = calyx)

Ovule

Receptacle

THINK ABOUT SCIENCE

Directions: Complete each sentence.

1. According to the diagram of the flower, the flower's female reproductive structure is the Carpel .

2. Fertilization combines the genetic information from the male and female.

Vocabulary Review

Directions: Match each word on the left with its correct meaning on the right. Write the correct letter on the line.

1. ___A___ pollination
2. ___F___ stamen
3. ___D___ reproduction
4. ___E___ chlorophyll
5. ___C___ pistil
6. ___B___ photosynthesis

A. transfer of reproductive materials from plant to plant

B. food-making process in plants

C. female reproductive structure in flowering plants

D. creating more of the same species

E. a substance that allows the plant's cells to absorb light

F. male reproductive structure in flowering plants

Skill Review

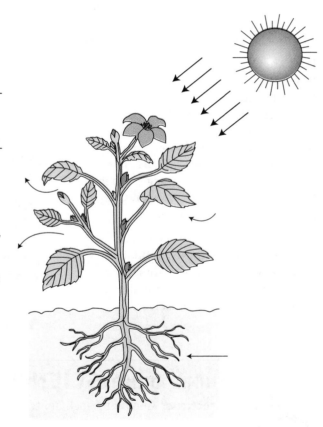

Directions: Use the diagram to answer the questions.

1. What do the arrows show in the diagram?

   Sun raise / sunlight

2. What would make the diagram more useful?

   _____

Directions: Read the passage below and apply ideas from the lesson to select the correct answer to each question that follows.

Hummingbirds are attracted to brightly colored flowers, especially red ones, and to a flower's sweet nectar. Their long beaks can fit inside tube-shaped flowers to reach the nectar. Flowers such as trumpet vine and bee balm attract hummingbirds.

3. What idea from the lesson best applies to this passage?

   A. A flower's petals often attract pollinators.
   B. Petals serve mainly to protect a plant's reproductive structures.
   C. Pollination occurs when nectar is transferred from one plant to another.
   D. Flowers attract as many pollinators as possible.

4. What statement would best describe a trumpet vine and bee-balm plant?

   A. They pollinate themselves.
   B. Bees are not attracted to them.
   C. They attract many different pollinators.
   D. They have tube-shaped flowers and sweet nectar.

Skill Practice

Directions: Choose the best answer to each question.

1. Plants release oxygen during photosynthesis. This activity permits which of the following human processes?
   A. respiration
   B. digestion
   C. circulation
   D. elimination

Directions: Questions 2 and 3 refer to the following passage.

> Like all living things, plants respond to the world around them. Unlike animals, their responses are often too slow for us to watch.
>
> Instead of moving quickly from one place to another, the plant may respond by growing toward or away from something. Over time, we may notice the change. This response is called tropism. By observing a plant over time, we can observe the effects of tropism.
>
> For example, no matter how a seed lands on the ground, the root grows down. The stem grows up. By experimenting, scientists have learned that this happens because the plant responds to Earth's gravity.
>
> Another example you might have noticed is that a plant grows toward light. If the direction of the light changes, the plant's growth pattern will also change. In a few days, the leaves will turn to the light.
>
> In addition, many plants respond to touch. The Venus flytrap uses touch to capture an insect. The pea plant responds to touch when it climbs up a fence.

2. What is the main idea of this passage?
   A. Roots always grow down.
   B. Plants respond to light.
   C. Plants grow toward or away from things.
   D. Plants respond to things in the world around them.

3. Which of the following is a detail that supports the main idea?
   A. Plants responses are too slow to observe.
   B. A plant grows toward light.
   C. A pea plant responds to touch when it climbs a fence.
   D. both B and C

Respiration

KEY CONCEPT: When you breathe, you respire, or bring oxygen into your body. This process is called respiration. There's another kind of respiration, too. Within your cells, microscopic structures use the oxygen that you respire to release the energy locked in food molecules.

You know that you breathe in and out all day long. When you breathe in, oxygen goes into your lungs. When you breathe out, carbon dioxide leaves the lungs. At rest, the average adult breathes in and out about 18–20 times per minute. This rate increases with exercise or other activities. You may wonder why we breathe in oxygen and breathe out carbon dioxide. You may also wonder why we breathe faster and deeper during exercise. These are good questions, and their answers are related to your body's energy demands.

Lesson Objectives

You will be able to
- Relate respiration to energy
- Identify step-by-step scientific procedures
- Describe the role of oxygen in the process of respiration
- Explain the process of cellular respiration

Skills

- **Core Skill:** Determine Meaning of Terms
- **Core Skill:** Follow a Multistep Procedure

Vocabulary

aerobic
cellular respiration
glycolysis
initiative
mitochondria
procedure
process

The Need for Energy

When a car runs low on fuel, you put more gas in it. The car needs energy to move, and that energy is locked within the molecules of the gasoline in the car's tank. **Molecules** are atoms connected by chemical bonds.

Once you turn the ignition, a process begins. A **process** is a series of actions that leads to a result. In this process, parts of the car's engine begin to work, and oxygen is mixed with gasoline. The spark plugs ignite the gas. This causes explosions that provide the force to move the car. As you drive the car, waste products—mostly carbon dioxide and water—leave the car through the tailpipe. About 25 percent of the energy locked within the molecules of gasoline is used to move the car. The rest is lost as heat. That's why your car heats up as you drive it.

Gasoline comes from fossil fuels. What do you know about fossil fuels? They are made from the remains of once-living plants and animals. When these organisms died, all of the energy stored within them was buried. Millions of years later, we unearth these materials to capture the energy stored in their fossilized remains. We call that energy "fossil fuel."

Your cells are not completely unlike a car engine. Your fuel is the food you eat. This food, either directly or indirectly, got energy from the Sun. The Sun is the source of all energy on Earth, past and present.

Eating food is like putting gasoline in a car. Your body works to release the energy stored in food molecules, and just as in a car's engine, your body needs oxygen to do the job effectively.

Opposite Processes

All cells need energy to survive, meaning they need food and oxygen. When plants perform **photosynthesis**, they create and store food molecules for their own use. The food may be stored in the leaves, stems, or roots, and it is in the form of organic compounds called **carbohydrates**, or sugars. These sugars are the fuel—the source of energy for a plant cell. Like all substances, the sugars stored in a plant's parts are made of molecules. The chemical bonds that hold these molecules together store the energy produced during photosynthesis. To release the energy, the bonds must be broken. This process of breaking the bonds of food molecules to release energy is called **cellular respiration**. Cellular respiration is an **aerobic** reaction, meaning that oxygen must be present for it to occur.

In photosynthesis, carbon dioxide and water combine in the presence of sunlight to make sugar and oxygen. The process looks like this:

carbon dioxide + water + sunlight ⟶ sugar + oxygen

In cellular respiration, sugar and oxygen produce high-energy molecules, as well as waste heat, carbon dioxide, and water. The process looks like this:

sugar + oxygen ⟶ energy + heat + carbon dioxide + water

Therefore, photosynthesis is considered the opposite process of cellular respiration. Essentially, that's because photosynthesis removes carbon dioxide from the atmosphere, and cellular respiration puts it back in.

CELLULAR RESPIRATION

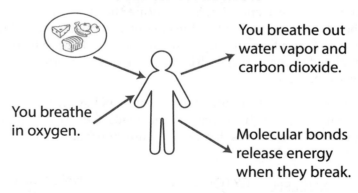

You breathe in oxygen.

You breathe out water vapor and carbon dioxide.

Molecular bonds release energy when they break.

Cellular respiration occurs in your body's cells.

Science has a special collection of vocabulary terms that other scientists recognize and use regularly. It is often possible to understand an unfamiliar science word by finding the meanings, and the **etymology**, or the origin, of a word or root word. To understand a word, you also need to understand the prefixes, suffixes that make up the word. For example, the word, *aerobe*, refers to an organism that can only live in the presence of oxygen. Adding the *-ics*, meaning activity, creates the word *aerobics*—activities or exercises that increase oxygen, and subsequently respiration.

Use the following roots to define each science word.

a- "not" + *-oxi* meaning oxygen = anoxic, without or lacking oxygen

• _____

From the Greek root *kinesis* "movement" + *-logy* "study of or branch of knowledge" = **kinesiology**, which means the study of human movement.

• _____

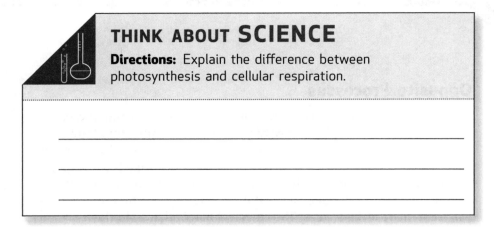
Core Skill
Follow a Multistep Procedure

Laboratory investigations and activities are an important part of science. They often require you to follow a **procedure**, or set of written instructions. The instructions follow a series of steps, which usually are numbered. The first step begins with the number 1, the next step begins with the number 2, and so forth. Sometimes, picture clues accompany a procedure. Numbered steps, drawings, photos, and diagrams can make a procedure easier to follow. Read the following procedure.

Examining Microorganisms in Pond Water

1. Wear safety glasses and an apron to protect your eyes and clothes.
2. Use an eyedropper to take a sample of pond water.
3. Place a drop of the sample on a microscope slide.
4. Place a cover slip over the drop.
5. Place the slide under a microscope.
6. Observe any organisms that are visible in the pond water.
7. Record your observations.

Why is it important to follow steps in a procedure?

Mitochondria

Plant and animal cells, as well as cells in microscopic organisms called **protists**, are **eukaryotic** (you-CARE-ee-ah-tik). A eukaryotic cell has a **nucleus**, or structure containing its hereditary information. Outside the nucleus and within the cell's membrane is the **cytoplasm**, a fluid substance that contains **organelles**, or "little organs." Organelles each have jobs to do, and the organelles essential to the process of cellular respiration are the **mitochondria**.

In illustrations and microscopic slides, mitochondria appear to have the same cylindrical shape. However, time-lapse videography shows that they change their shapes, move about in the cell, and divide in two.

A single mitochondrion has two outer membranes. The outer membrane is smooth, unlike the inner membrane, which is folded. The folds are called **cristae** (KRISS-tee). The cristae form a boundary around the cell's **matrix**, a fluid substance that contains **enzymes**, or proteins. Enzymes increase the speed of the chemical reactions that occur in the mitochondria during cellular respiration.

The purpose of cellular respiration is to capture the energy held in the chemical bonds of sugar molecules. The energy is ultimately stored in molecules of ATP, or adenosine triphosphate. Because the production of ATP is completed in the mitochondria, the structures are often called the "powerhouses" of a cell.

Some cells have thousands of mitochondria. Others have far fewer. The number depends on how much energy a cell needs. Cells that require a lot of energy, such as muscle cells, have many mitochondria.

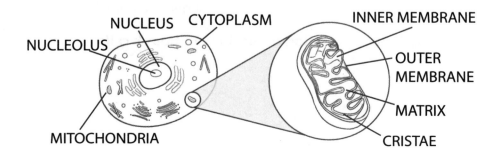

Inside the Cell

In your body's cells, the process of cellular respiration begins with a molecule of sugar called glucose. The molecule enters a cell through the cell's membrane. There, the glucose molecule undergoes **glycolysis**. The prefix *glyc-* or *glyco-* "means carbohydrate and sugar" and the root *lysis* means "to disintegrate or dissolve." The word *glycolysis* then, can mean "to split, break, or disintegrate carbohydrates or glucose." All together, there are ten steps in glycolysis, and each step involves an enzyme.

Glycolysis happens in the **cytosol**, inside the cell's cytoplasm. Energy is necessary for the process to begin. That energy comes from two molecules of ATP. It may seem odd to use energy to make energy, but as you will see, the investment provides a huge payoff.

A molecule of glucose has six carbon atoms in its structure. This molecule splits to make two three-carbon sugars. Chemical changes act upon the new molecules, ultimately changing them into two molecules of a compound called **pyruvate**. Glycolysis is complete. The entire process generates four molecules of ATP. Two of these molecules can be said to replace the molecules that were used to start the process, leaving a net gain of two molecules of ATP.

By now, half of the steps that occur during cellular respiration are finished, but more than 75 percent of the energy belonging to the glucose molecule remains locked inside the two molecules of pyruvate. Further chemical changes must unlock that energy, and if oxygen is present, those changes occur inside the mitochondria.

Inside the mitochondria, the pyruvate molecules undergo further chemical changes that include the movement of electrons. High-energy electrons shuttle through the cristae, where an enzyme uses the energy in those electrons to build molecules of ATP. By the conclusion of the cellular respiration process, the mitochondria will generate almost 38 molecules of ATP. Given that the process began with an investment of two molecules of ATP, the generation of so many energy-storing molecules is something of an energy bonanza.

The process can be described in simple terms as:

sugar + oxygen $\longrightarrow$ energy + heat + carbon dioxide + water

The waste products carbon dioxide and water leave your body when you breathe out, or exhale.

THINK ABOUT SCIENCE

Directions: Summarize the purpose of cellular respiration.

_____

_____

21st Century Skill
Initiative and Self-Direction

Initiative is taking steps toward putting your goals in action, whether it is learning something new, a self-improvement project, or starting a new job. Nothing begins without initiative. But initiative isn't enough in itself. Self-direction is also essential. Reaching a goal means defining, prioritizing, and completing tasks on time. For scientists, it means identifying a problem or question, gathering information about that problem, and developing a **hypothesis**, or testable explanation for their problems. Investigations follow, which may lead to more questions, more hypotheses, and more experimentation. Perhaps you recognize this process as the scientific method.

You can use initiative and apply self-direction to reach a personal goal. Start by choosing a goal. Then record the steps you will follow to reach it, list the resources you need for each step, and finally, mark the date you complete each step.

Goal:		
Step	Necessary Resources	Completion Date
1.		

Bioremediation

All living things, even the smallest single-celled organisms, eat food for the same purpose. They eat to provide nutrients and energy for their cells. However, not all organisms depend on the same kinds of food molecules for cellular respiration. For example, some bacteria naturally feed on chemical substances that are environmentally harmful to other living things, including humans. Within their cells, these microorganisms get the energy and nutrients they need for growth and survival from breaking down pollutants, such as fuels or industrial chemicals. In this process of cellular respiration, bacteria give off carbon dioxide and water as waste products, just as your body's cells do. However, the natural process is slow. That's when bioremediation becomes necessary.

Bioremediation is a branch of biotechnology. It is the application of biological processes to solve problems within industry, medicine, and the environment. The prefix *bio-* means "living," and the base word *remediate* means "to correct or solve a problem." So bioremediation refers to the use of microorganisms, or chemicals produced by those microorganisms, to solve a problem.

Scientists use natural and developed cells that feed on pollution. In the environment, these custom-made cells, which include microorganisms, plants, and some fungi, get the energy they need and benefit the environment at the same time.

One of the earliest examples of bioremediation occurred in 1992. Almost two decades earlier, huge quantities of jet fuel from a military base had seeped through the soil and entered the community's water supply. Scientists studying the soil and water had found bacteria that digested molecules of fuel and gave off carbon dioxide and water as waste, just as your body's cells do in the process of cellular respiration. However, the bacteria couldn't consume fast enough to solve the environmental disaster. So, scientists added nutrients to the soil, which increased the bacterial population and digestion rate. Within a year, 75 percent of the pollutant was gone from the groundwater supply. Nearer the sites where nutrients were put into the groundwater, there were no signs of the pollutant at all.

THINK ABOUT SCIENCE

Directions: Explain how cellular respiration eliminated a serious environmental hazard.

_____

_____

©Natalie Fobes/Corbis

Groundwater forms when **precipitation** such as rain and snow soaks into the ground. Water travels downward, through soil, sand, gravel, and rock, until the ground is saturated, meaning it cannot hold any more water. The top of this saturated area is called the water table. In the United States, people depend on this groundwater for freshwater.

As the jet fuel example shows, more than precipitation drains into groundwater. So do fertilizers, pesticides, liquid waste from landfills, and even human sewage. Naturally occurring bacteria living in the groundwater will reduce the pollution into less toxic compounds. This process is most effective in areas of groundwater where bacteria are plentiful and where pollution levels are low. The process is speeded up through bioremediation. This involves stimulating the natural bacteria by injecting nutrients and possibly carbon compounds needed by the bacteria into the groundwater. The bacteria consumes the harmful pollutants at a much faster rate.

Bioremediation is only one of several methods used to remove harmful materials from water. One of its advantages is cost. Bioremediation can be up to 90 percent less expensive than other water-cleaning technologies. However, there is no perfect solution to cleaning up groundwater pollution. Bringing water back to drinking water standards is very difficult. Scientists are continuing to research bioremediation to improve its performance. The protection of groundwater sources is vitally important to every living organism.

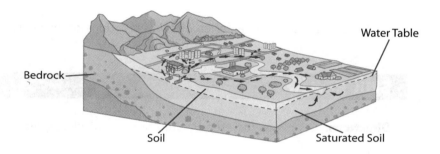

Vocabulary Review

Directions: Choose the correct word to fill in the blank.

aerobic	cellular respiration	glycolysis	initiative
mitochondria	procedure	process	

1. It takes _____ to put a plan into action.

2. The process of splitting or disintegrating glucose is called _____.

3. A _____ is a series of ordered steps.

4. A(n) _____ reaction is one in which oxygen must be present.

5. Energy is released from food in the process of _____.

6. Cellular respiration occurs in a cell's _____.

7. A _____ is a series of related actions or changes.

Skill Review

Directions: Complete the step-by-step process of glycolysis.

1. A molecule of _____ enters a cell.

2. The energy locked in two molecules of _____ is used to start the process.

3. A six-carbon sugar molecule is split into two _____-carbon sugar molecules.

4. These sugar molecules are altered to become _____.

5. If _____ is present, the pyruvate molecules move into the mitochondria.

6. _____ and _____ are given off as waste.

Skill Practice

Directions: Read the passage. Choose the best answer to questions 1 and 2 that follow.

> All of the energy a car needs to move is locked in the molecules of gasoline. Once you turn the ignition, things start to happen. Parts of the car's engine begin their work, and oxygen is mixed with gasoline. The spark plugs ignite the gas. Explosions occur, which provide the force to move the car. As you drive, waste products—mostly carbon dioxide and water—exit the tailpipe. About 25 percent of all of the energy locked in molecules of gasoline is used to move the car. The rest is lost as heat energy. That's why your car's engine feels hot to the touch after it's driven.

1. Which statement best describes the similarities between car engines and cells?

   A. They use the same source of fuel.
   B. They produce different waste products.
   C. They require a spark to start the breakdown of fuel.
   D. They require the presence of oxygen to break down and use the fuel.

2. Which statement about energy is true for both car engines and cells?

   A. It is renewable.
   B. It is extracted from fossil fuels.
   C. It comes from nonliving sources.
   D. It can be traced back to energy from the Sun.

3. Label and complete each process.

   The process of _____:

   carbon dioxide + water + sunlight ⟶ _____ + oxygen

   The process of _____:

   sugar + oxygen ⟶ energy + heat + _____ + water

4. Explain the role of glycolysis in cellular respiration.

   _____

   _____

   _____

5. Explain why cellular respiration must be completed in the mitochondria if the energy locked in food molecules is going to be put in a form useful to a cell.

   _____

   _____

   _____

Fermentation

KEY CONCEPT: Fermentation is a process that produces energy within a cell in the absence of oxygen.

Have you ever played an intense round of basketball or run full-speed for a long distance? Activities like these often leave you with burning muscles. You may also experience shortness of breath. What causes the burn, and how is it related to breathing? Both are good questions. In order to answer them, we need to examine the process of fermentation.

Lesson Objectives

You will be able to

• Relate fermentation to energy

• Relate the absence of oxygen to fermentation

• Explain the process of fermentation

Skills

• **Core Skill:** Apply Scientific Processes

• **Core Skill:** Compare and Contrast Information

Vocabulary

accountability
anaerobic
compare
contrast
fermentation
productivity
research

Fermentation

Cellular respiration is an aerobic process that depends on oxygen to break apart sugar molecules for the purpose of producing ATP. Fermentation also breaks apart sugar molecules for the purpose of producing ATP. However, it does so without the presence of oxygen, meaning it is an **anaerobic** process. Plus, it generates only two molecules of ATP for each molecule of glucose, making it a less efficient means of energy production.

There are different types of fermentation, and they can be identified by three things during the process. One is the kind of molecules that form as a result of glycolysis. Another is the different enzymes—complex proteins that speed up chemical reactions during the process. The third is the kind of waste the process creates.

Two common kinds of fermentation are lactic acid fermentation and alcohol fermentation. Both are anaerobic processes, but lactic acid fermentation occurs in animal tissue such as human muscle, while alcohol fermentation occurs in plant and yeast cells. Both are used in food production. You may already be familiar with some fermented products, including yogurt, cheese, wine, beer, soy sauce, salami, and pepperoni. Fermentation also causes breads to rise and to produce their assorted shapes and textures.

Fermentation increases the nutritional value of food.

Lactic Acid Fermentation

When you exercise, your muscles work harder than normal, and you breathe more heavily to take in the additional oxygen your muscle cells need for cellular respiration. However, after prolonged exercise, you may "run out of breath;" you may not be able to take in as much oxygen. In the absence of oxygen, the process of lactic acid fermentation replaces cellular respiration in your body's muscle cells.

As in cellular respiration, glycolysis occurs in lactic acid fermentation, too. The process generates two molecules of pyruvate, which then change to lactic acid. Lactic acid then becomes lactate, a waste product. The buildup of lactate in your muscles causes a burning sensation and muscle fatigue. When you slow down, your muscle cells begin to recover. More oxygen enters the cells, allowing cellular respiration to resume. The muscle cells release lactate into the bloodstream, where it travels to the liver. The liver changes the lactate back into glucose that is stored in the body or enters the bloodstream for cells to pick up.

Here is a simplified description of the process of lactic acid fermentation:

One molecule of sugar (glucose) $\longrightarrow$ **2 molecules of pyruvate (which are later changed to lactate) + 2 molecules of ATP**

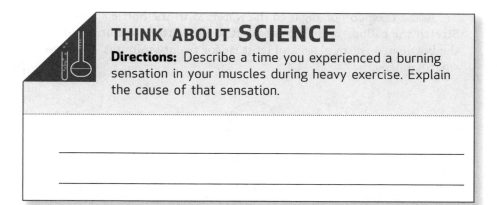

THINK ABOUT SCIENCE

Directions: Describe a time you experienced a burning sensation in your muscles during heavy exercise. Explain the cause of that sensation.

_____

_____

Core Skill
Compare and Contrast Information

The "Blow Up a Balloon" experiment is often used to demonstrate the fermentation process. Do some research to find other descriptions and demonstrations of this same experiment. How many different examples can you can find in a variety of media? Note how the experiment is presented in each. You may find print versions presented as a step-by-step process, a video demonstration, or a version that contains both pictures and print.

While studying each version of the experiment, take notes on how the experiments are presented. Then write a summary that tells how each one was similar and different. In other words, **compare** and **contrast** the level of detail, the visual style, and other factors important to the presentation of the experiment. Finally, tell which version was most useful to you for your understanding and explain why.

Alcohol Fermentation

Unlike human muscle cells that produce lactate during fermentation, yeast cells change glucose into ethyl alcohol, or **ethanol**. The process produces carbon dioxide as a waste product.

When yeast cells are added to bread dough, they spread out. During fermentation, the yeast cells release carbon dioxide as waste. Bubbles of carbon dioxide form in the dough and cause the bread to rise. When the bread is baked, the yeast cells die, and the burst bubbles leave the holes you see.

Here is a simplified description of the process of alcohol fermentation:

1 molecule of glucose $\longrightarrow$ 2 molecules of ethanol + 2 molecules of carbon dioxide + 2 molecules of ATP

Conduct Your Own Investigation

Blow Up a Balloon

You need only a few materials to conduct this investigation. Begin by getting safety glasses and an apron. Then collect a small, clear, plastic soda bottle, a balloon, water, sugar, and a packet of baker's yeast.

Fill the bottle with warm water. The water must be warm, not hot, because yeast are living organisms and hot water will kill them. Next, pour the baker's yeast into the warm water. Swirl the bottle gently.

Add a teaspoon of sugar to the water. Swirl the bottle again. Stretch the balloon and then attach it to the bottle's opening. Put the bottle in a warm place and let it sit for twenty minutes. Observe what happens.

Ethanol: An Alternative Energy Source

Fossil fuels are a nonrenewable resource. It took millions of years for them to form. So, once they are gone, they cannot be replaced. Fossil fuel production and use also harm the environment.

Scientists have been seeking alternatives to fossil fuels. One such alternative is ethanol. Ethanol can be produced through the fermentation of plant crops, such as corn. Recall that when plants perform photosynthesis, they produce food in the form of glucose. Energy, captured from the Sun, is stored in the bonds of the glucose molecules, which are stored in the plant.

When you eat a plant such as corn, enzymes in your saliva break down the carbohydrates in corn into simpler sugars. Your body then breaks down the sugars during cellular respiration.

To make fuel, the corn is cooked and enzymes are added. The result is a simple sugar to which yeast cells are added. During alcohol fermentation, the yeast cells feed on the sugar and produce ethanol. One acre of corn can produce more than 300 gallons of ethanol. The ethanol holds much of the energy that was in the sugar, making it a good source of fuel.

For every 85 parts ethanol, 15 parts gasoline are added. The result is a fuel that decreases the demand for fossil fuel.

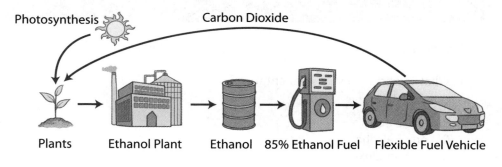

Photosynthesis   Carbon Dioxide

Plants   Ethanol Plant   Ethanol   85% Ethanol Fuel   Flexible Fuel Vehicle

THINK ABOUT SCIENCE

Directions: Use the diagram above to explain the relationship between photosynthesis and the production of ethanol.

_____

_____

_____

_____

_____

WRITE TO LEARN

Your muscle cells can use the process of lactic acid fermentation to get the energy they need to sustain heavy activity for a limited time. High rates of fermentation can last for one to three minutes. During that time, lactic acid continues to accumulate in your blood stream, causing uncomfortable burning sensations. What benefit could be gained from this discomfort? How does the production of lactic acid in muscle tissue affect athletes and others who rely on muscle endurance? Do some research on the relationship of lactic acid to muscle endurance and write a paragraph about what you learned.

Vocabulary Review

Directions: Fill in the blank with the correct word.

accountability   anaerobic   productivity
fermentation   research

1. Fermentation is a(n) _____ process, meaning there is no oxygen present.

2. A close, careful study of a topic or problem involves _____.

3. Yeast starts the _____ process that makes bread rise.

4. The responsibility people take for their work and actions is called

   _____.

5. Your _____ will improve when time is used efficiently to produce results

Skill Review

Directions: Choose the best answer to each question.

1. What is absent during fermentation?

   A. oxygen
   B. lactic acid
   C. ethyl alcohol
   D. carbon dioxide

2. What fermentation product is used to produce an alternative fuel?

   A. corn
   B. yeast
   C. ethanol
   D. lactic acid

3. What causes bread to rise?

   A. bubbles of carbon dioxide
   B. bubbles of ethyl alcohol
   C. the presence of sugar
   D. dying yeast

4. What causes a burning sensation in muscle cells during heavy exercise?

   A. a build-up of lactic acid
   B. a build-up of ethyl alcohol
   C. a build-up of carbon dioxide
   D. a build up of sugar molecules

Skill Review (continued)

Directions: Conduct the following investigation.

Observe Fermentation

Step 1. Gather the following materials:

- 4 plastic bags
- sugar
- room-temperature water
- measuring spoons
- a marker

- 4 packages (or 8 teaspoons) baker's yeast
- warm water
- salt
- measuring cup

Step 2. Label the bags:

1. sugar + warm water
2. sugar + room-temperature water
3. sugar + salt + warm water
4. warm water

Step 3. Add 2 teaspoons of sugar to bags 1, 2, and 3.

Step 4. Add 1 teaspoon of salt to bag 3.

Step 5. Add ½ cup of warm water to bags 1, 3, and 4.

Step 6. Add ½ cup of room-temperature water to bag 2.

Step 7. Seal the bags, squeezing out as much air as possible.

Step 8. Put the bags in a warm space and observe them for 15–20 minutes.

Directions: Use the investigation you completed to complete each task.

5. Describe your observations.

_____

_____

_____

_____

6. Explain your observations. Use the word _fermentation_ in your explanation.

_____

_____

Skill Practice

1. Use the following Venn diagram to compare and contrast lactic acid and alcohol fermentation. Include details from your readings and from your investigations.

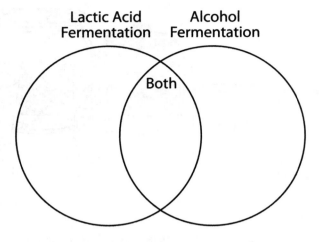

_____

_____

2. Why does fermentation occur?

_____

_____

3. What are two products of the alcohol fermentation process?

_____

_____

Skill Practice (continued)

4. Predict the effect of water temperature on the production of carbon dioxide during fermentation. Justify your answer.

_____

_____

_____

5. What would you expect to happen if you added yeast to a sealed bag containing honey mixed in warm water? Justify your answer.

_____

_____

_____

6. Why do you think scientists hold each other accountable for their work and processes?

_____

_____

_____

7. Explain the purpose of maintaining records of observations and measurements during a scientific investigation.

_____

_____

_____

Directions: Choose the best answer to each question.

1. Why do you think roots are the first parts of a plant to grow?

    A. The force of gravity pulls upon the plant, causing the roots to develop first.
    B. They anchor the plant and absorb nutrients for growth.
    C. Flowers need something on which to grow.
    D. Photosynthesis occurs primarily in the roots.

2. What is the purpose of glycolysis?

    A. to provide oxygen to body cells
    B. to release the energy trapped in chemical bonds
    C. to generate ATP
    D. to produce sugar molecules

3. Which organelle is the site of respiration within a cell?

    A. nucleus
    B. cell membrane
    C. nucleolus
    D. mitochondria

4. The pistil is covered with a sticky fluid. What purpose do you think the fluid serves?

    A. Insects feed on it.
    B. It provides structural support.
    C. Pollen grains stick to it.
    D. It contains pigments required for photosynthesis.

5. What must be present for respiration to occur?

    A. water
    B. carbon dioxide
    C. oxygen
    D. heat

6. The process of photosynthesis produces food for a plant. Which process breaks down the food to capture energy locked in chemical bonds?

    A. respiration
    B. fermentation
    C. evaporation
    D. pollination

7. What purpose does an enzyme have during chemical processes within a cell?

   A. It speeds up chemical reactions.
   B. It replaces the need for oxygen.
   C. It traps heat given off during a chemical reaction.
   D. It removes waste from a cell.

8. What is the purpose of lactic acid fermentation in your muscle cells?

   A. to provide energy in the presence of too much oxygen
   B. to provide energy in the presence of little or no oxygen
   C. to take over the liver's responsibility during heavy exercise
   D. to reduce stress on the respiratory system during heavy exercise

9. What substance do yeast form during fermentation?

   A. ethanol
   B. lactate
   C. glucose
   D. water

10. What makes fermentation different from respiration?

    A. It occurs in the presence of oxygen.
    B. It occurs in the absence of oxygen.
    C. It occurs in the presence of ethanol.
    D. It occurs in the absence of ethanol.

11. Where does alcohol fermentation occur?

    A. in plants and yeast cells
    B. in animal tissues, such as muscles
    C. in a plant's root tissues
    D. in water where algae grow

12. What is a waste product of photosynthesis?

    A. oxygen
    B. carbon dioxide
    C. ethanol
    D. glucose

13. What is within every grain of pollen?

    A. all of a plant's reproductive information
    B. half of a plant's reproductive information
    C. materials that attract pollinators
    D. immature plant seeds

14. What is the purpose of a stomate?

    A. It stores sugar molecules for the plant to use as food.
    B. It holds a plant's genetic information.
    C. It lets gases move in an out of a leaf.
    D. It is the energy powerhouse of a plant cell.

Review

Directions: Read the text. Then complete the activity.

15. Examine the diagram of a flowering plant. Identify the name and purpose of each labeled item.

PARTS OF A FLOWERING PLANT

A. _____

B. _____

C. _____

D. _____

E. _____

16. Successful pollination leads to the production of new plants. Methods of pollination vary. List five possible means of pollination and explain how each process works.

    A. _____

    _____

    _____

    B. _____

    _____

    _____

    C. _____

    _____

    _____

    D. _____

    _____

    _____

    E. _____

    _____

    _____

17. Imagine being asked to explain the processes of photosynthesis and respiration to a younger student who is unfamiliar with the concepts. Write a brief explanation of each process and the relationship between them.

    _____

    _____

    _____

    _____

    _____

    _____

    _____

    _____

    _____

18. Use the flowchart to show the steps of cellular respiration.

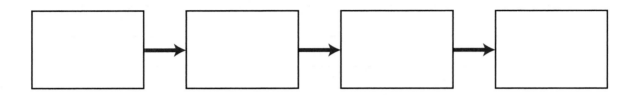

19. Write a blog post explaining the advantages of bioremediation.

_____

_____

_____

_____

_____

_____

_____

_____

_____

_____

_____

_____

_____

_____

_____

_____

_____

_____

_____

_____

20. In an experiment to prove that yeast give off carbon dioxide as a gas during fermentation, a student attaches a balloon to a clear jar that holds warm water and healthy yeast cells. The balloon remains unchanged. What reason could you give for the balloon's failure to inflate?

_____

_____

_____

_____

_____

_____

_____

_____

_____

_____

_____

_____

Check Your Understanding

On the following chart, circle the number of any item you answered incorrectly. Next to each group item numbers, you will see the pages you can review to learn how to answer the items correctly. Pay particular attention to reviewing those lessons in which you missed half or more of the questions.

Chapter 2 Review

Lesson	Item Number	Review Pages
Flowering Plants	1, 4, 12, 13, 14, 15, 16	54–61
Respiration	2, 3, 5, 6, 7, 17, 18, 19	62–69
Fermentation	8, 9, 10, 11, 20	70–77

CHAPTER 2: LIFE FUNCTIONS AND ENERGY INTAKE

Question

How can you demonstrate that plant respiration occurs and what is produced during this process?

Background Concepts

During cellular respiration, plants and animals break down molecules of glucose to release energy stored in each molecule's chemical bonds. Chemical reactions that occur during the process also produce water and carbon dioxide.

Aquatic plants, or plants that live partially or completely underwater, produce carbon dioxide and water during respiration. The carbon dioxide the plant gives off combines with water to create carbonic acid, which increases the acidity of the water.

Investigation

Materials required

aquatic plant	knife or food processor
head of purple cabbage	strainer or coffee filter
boiling water	aluminum foil
a heat-proof bowl	2 glass jars with lids

1. Ask someone who owns an aquarium for a piece of an aquatic plant, or purchase an aquatic plant at a nursery or pet store.

2. You can use purple cabbage and water to make your own acid indicator. Because you will be chopping and working with heat, wear protective clothing.

3. Chop the cabbage into small pieces and put the pieces in a bowl that can withstand high temperatures. Carefully add boiling water until the cabbage is completely covered. Let the cabbage sit for at least 10 minutes. The pigment in the cabbage will turn the water purple.

4. Pour the water through a strainer or coffee filter to strain out cabbage pieces. You will be left with purple water.

5. Put the aquatic plant in one glass jar. Put no plant in the second jar. Add the purple cabbage water to both jars. Seal the jars and cover them with aluminum foil. Let both jars sit for 3 days.

6. Remove the aluminum foil from the jars and observe the color of the liquid in both jars. If the cabbage water has been exposed to an acidic substance, it will be pink. If it has been exposed to an alkaline substance, or base, it will be green.

Application of Science Practices

Interpretation

What color changes occur?

_____

_____

What does the color change of the liquid in the jar with the aquatic plant tell you?

_____

_____

Answer

What is produced during plant respiration?

_____

_____

If water were the only product of plant respiration, what color would you expect to see?

_____

_____

What color would you expect to see if the product of plant respiration was an alkaline substance, such as ammonia?

_____

_____

Evidence

How do you know that the color of the liquid in the jar with the aquatic plant is not caused by something in the water itself?

_____

_____

Ecosystems

All life depends on healthy ecosystems that can provide the resources organisms depend on for survival, including clean water, air, and soil, as well as food. Biodiversity is also an important element in maintaining healthy ecosystems. The greater Earth's biodiversity, or abundance of different kinds of life on Earth, the more resources, opportunities for medical discoveries, and adaptive responses to natural disasters and human activities.

In this chapter you will learn about:

Lesson 3.1: Ecosystems
An ecosystem is a community of living things adapted to a specific environment. There are a variety of ecosystems on land and in water. Biomes are major ecological communities, identified by where they exist geographically. This lesson discusses the organization of ecosystems and describes the biomes of the world.

Lesson 3.2: Carrying Capacity
Imagine inviting friends to lunch. You prepare enough food and set the table for eight people. Your planned lunch will satisfy the needs of eight people in all. But then each of your friends invites another friend. You don't have enough resources to feed or seat them all. You might say that your lunch community has exceeded its carrying capacity. This lesson talks about carrying capacity, or the environmental limitations that determine how many organisms an ecosystem can support.

Lesson 3.3: Symbiosis
The prefix *sym* means "together," and the prefix *bio* means "life." As organisms on Earth, we live together with other organisms. But some organisms have special ways of living together. This lesson introduces you to different kinds of symbiotic relationships.

Lesson 3.4: Disruption
Both nature and human activity can lead to environmental disturbances and major disruptions. Fires and floods can alter ecosystems in a short time. So can the introduction of exotic or non-native species to an ecosystem, where the species competes for the same resources as native species. In some cases, humans remove ecosystems entirely to create city structures. This lesson introduces you to natural and unnatural ecological disruptions.

Lesson 3.5: Environmental Issues
Controlling pollution and keeping Earth's environment habitable are important issues for everyone. In this lesson, you will learn how human activities affect the environment and investigate the difference between renewable and nonrenewable resources.

Goal Setting

In the particular geographical location in which you live, you may be aware of clues to the original ecosystem.

Explore your environment, or use print and online resources to learn more about the ecosystem that existed before the place in which you live was established. Describe the original ecosystem. Then describe how the ecosystem has been altered.

The original ecosystem:

_____

_____

_____

_____

_____

_____

_____

_____

_____

_____

_____

Alterations to the ecosystem:

_____

_____

_____

_____

_____

_____

_____

_____

_____

_____

_____

Ecosystems

KEY CONCEPT: Within an ecosystem, organisms interact with one another and with nonliving things in their environment.

Your school is not just a building. It is a community made up of populations such as teachers, students, and custodians, all of whom interact with each other. When you interact with someone, you act in close personal relationship with that person. You have an effect on each other. A population also interacts with nonliving things. Each population at your school has a role in making the school successful.

In nature, all living things interact with other living and nonliving things in their environment.

Communities of Living Things

The part of Earth in which life exists is called the **biosphere**. The biosphere covers the entire surface of Earth and includes the atmosphere. An **ecosystem** is a smaller part of the biosphere. It is made up of all the living organisms in one area (of any size), as well as the nonliving parts of the environment, such as water and rocks.

Every living thing depends on other living things and many nonliving things. In a healthy ecosystem, all the parts are in balance, meaning the populations of plants and animals are not rising or falling in huge numbers. Even healthy ecosystems are always changing, at least in small ways. When old animals die, they make room for their young to live. When a tree falls, young seedlings begin growing in its place.

Organisms in the Environment

An **environment** is the living and the nonliving surroundings of an organism. All the organisms in a certain environment make a **community**. The community of living things where you live may include squirrels, birds, trees, and humans. A **population** is all the organisms of one type. Squirrels and pigeons, for example, belong to the same community, but to two different populations. Within a community, each population has its own place to live, or **habitat**. It also fills its own role or job, which is called a **niche**.

THINK ABOUT SCIENCE

Directions: Choose the word in column 1 that is described in column 2.

1.	biosphere / niche	where life exists on Earth
2.	ecosystem / habitat	home for living things
3.	environment / community	all the organisms
4.	niche / population	a role or job

ANALYZE AUTHOR'S PURPOSE

An author's purpose is the reason why he or she writes a certain passage of text. Purposes for writing vary. Some authors write to inform or explain. Others write to entertain, persuade, or elaborate upon ideas.

To determine an author's purpose for writing, it is helpful to read titles and subtitles and examine illustrations, photographs, and diagrams. Getting the "big picture" gives you some idea of the author's reason for writing. Predicting an author's purpose *before* reading helps you be more aware as you read. You can look for evidence that either supports your prediction or causes you to revise it.

Brief passages may not include titles, subtitles, and visual clues. In these cases, you must read and then pause to ask yourself *Did the author present facts or opinions? Did the author use words or present ideas aimed at evoking, or bringing out, emotion? Did the author attempt to convince me to do something?*

Read the following passages. Analyze the author's purpose for writing each passage.

1. Feeding relationships are a part of every ecosystem. Green plants make their own food using water, carbon dioxide, and energy from the Sun. Many animals eat these green plants. These animals, in turn, become food for other animals. As long as these feeding relationships are stable, all the animals in the ecosystem have enough food to live and grow.

2. Human actions are always harmful to ecosystems. For example, consider a meadow near a forest. Rabbits may live in the meadow, where they eat grasses and flowers. The rabbits are food for foxes that live in the forest. When humans pave over the meadow, the rabbits die, and the foxes starve. Humans must stop all development of wilderness areas in order to protect fragile ecosystems.

In the first passage, the author explains feeding relationships in an ecosystem. How would you describe the author's purpose for writing this passage?

_____

In the second passage, the author describes the negative effects of human behavior on ecosystems. The author uses the words *always* and *must*. How would you describe the author's purpose for writing this passage?

_____

Energy Cycles

Energy from the Sun is the source of almost all energy on Earth. Green plants (the producers) trap energy during photosynthesis and convert it into glucose. Herbivores, such as cows and deer, eat the green plants. Carnivores, such as bears and falcons, eat other animals. Omnivores, such as humans and raccoons, eat both plants and animals.

The **food chain** is the path that energy follows as it moves through the community. It starts with green plants, the producers, and moves on through herbivores. Large carnivores are usually at the top of the food chain. When an organism dies, decomposers break apart the body, returning the nutrients to the soil. All the many food chains within a community form a **food web**. A change in the population of one organism of the food web will affect every other organism.

THE FOOD CHAIN

VOLE
(eats grasshoppers)

HAWK
(eats vole)

DEAD ORGANISMS AND DROPPINGS
(bacteria and fungi break down droppings and dead organisms into nutrients)

SUN

GRASSHOPPER
(eats grass)

NUTRIENTS
(phosphorous, water, nitrogen, calcium . . .)

PLANT LIFE
(organisms capable of self-nourishment through photosynthesis)

As demonstrated in the food chain above, each living thing is food for the next organism. For example, plants become food for grasshoppers, grasshoppers become food for voles, and voles become food for hawks. The hawks' droppings become food for bacteria, which turn it into nutrients for grass.

THINK ABOUT SCIENCE

Directions: Answer the questions below.

1. What type of organism is the basis of any food chain?

   _____

2. What would happen if there were no decomposers?

   _____

3. The diagram shows one aspect of an ecosystem—the food chain. What would happen if the number of grasshoppers decreased?

   _____

Biomes

A large group of ecosystems with similar climates and communities is known as a **biome**. Scientists disagree about exactly where one biome ends and another begins. Most agree, however, that there are at least six distinct biomes throughout the world. These are deserts, tundras, grasslands, tropical rain forests, temperate forests, and oceans.

Deserts

Lack of rainfall is the major characteristic of a desert biome. Most of the southwestern United States is a desert biome. Deserts tend to be hot during the day. At night, however, surface heat is rapidly lost. The desert at night can become quite cool.

Desert plants and animals are adapted to harsh, dry conditions. They have adapted to survive with very little water. Most desert animals are active at night when the temperature is cooler. Many of the plants, such as the cactus, store large amounts of water in their roots or stems. Some desert plants have extra deep roots. Other plants have tiny leaves or a waxy covering to reduce water loss from evaporation.

Tundras

If you were to move a desert to an extremely cold area, the result would be a tundra. There are two kinds of tundras: arctic and alpine. The arctic tundra is in the far northern parts of the world. Alpine tundras are at the tops of high mountains.

In many places on the tundra, ice and snow cover the ground for long periods. The soil may never totally thaw. Polar bears, as well as migrating animals, such as birds, reindeer, and caribou, live in the tundra. Small plants, such as lichens and mosses, grow close to the ground where strong, icy winds will not damage them.

Grasslands

Grasslands get more rainfall than deserts but not enough to support large trees. They have long, hot summers, cold winters, and high winds. The soil is generally fertile. The dominant plants are a wide variety of grasses. Animals tend to be small, but some large grazing animals are also common.

America's prairie is made up of grasslands that once were home to huge herds of bison. Today, there are cattle ranches on these grasslands. This area is also known as "the world's breadbasket" because of the huge quantity of grain produced there.

Tropical Rain Forests

In a tropical rain forest, the weather is mild all the time. Both sunlight and rain are abundant. Surprisingly, the soil is not very fertile. Because of the abundance of sunlight, moisture, and minerals from decomposing plant matter, however, trees grow tall. Vines and other plants grow in the trees. Huge varieties of insects, snails, birds, snakes, frogs, and small mammals are at home in the tropical rain forest.

Core Skill
Analyze Author's Purpose

One way to analyze an author's purpose is to look at how the text is structured. In science, the text structure helps the author explain a concept or a process. First, read the main head and the subheads beneath it. These heads can give you clues about the author's purpose. Then read the paragraph below the main head. This paragraph introduces the topic and prepares the reader for the subheads below it. Finally, determine the relationship between the main head and the topics in the subheads below it.

As you read this page and the next page, ask yourself *What are the subheads? How are the subheads related to the text in the opening paragraph under the main head?*

Science writers usually present clear, precise language, supported by data and directly related visuals. Their purpose is usually to educate and inform.

Temperate Forests

Temperate forests have distinct seasons. Temperature and rainfall vary with the seasons. Trees that lose their leaves in the fall are most common in southern temperate forests. Farther north, evergreens are the most common trees. Birds and mammals are abundant throughout the forest, but because so much has been cut to make room for farms and homes, the number of animals has greatly decreased. Much of the United States is now—or once was—temperate forest.

Oceans

The ocean biome covers two-thirds of Earth's surface. Climates vary, depending on location. Most of the plants and animals, such as algae, plankton, fish, and whales, live near the surface of the ocean where sunlight reaches.

Some organisms exist deep below the ocean surface where sunlight cannot penetrate. Most of these animals scour the ocean bottom for organic matter that sinks. Some, however, live near hot vents in volcanic areas. Rather than converting the Sun's energy to produce food, these creatures produce food by converting the energy in chemicals that rise from beneath the ocean floor.

Freshwater Areas

A wide variety of organisms live in lakes, ponds, streams, and rivers. Many, such as algae, can be too tiny to see, whereas seaweed can grow into a thick mass. Animals that live in or near freshwater include insects, amphibians, and fish. Many birds build their nests along the shores and hunt for fish or other animals in the water. A few mammals, such as beavers, also live here.

THINK ABOUT SCIENCE

Directions: Match the name of the biome on the left with its description on the right.

_____ 1. desert

_____ 2. tundra

_____ 3. grassland

_____ 4. tropical rain forest

_____ 5. temperate forest

_____ 6. ocean

A. cold and dry

B. home to algae and plankton

C. distinct seasons, trees

D. infertile, tall trees, rain, sunlight

E. hot summers, cold winters, fertile

F. dry and hot during the day

Protecting Biomes

With the possible exception of deserts, all of the biomes you read about have been greatly changed by human activities. Humans have cut down vast tracts of both temperate forests and tropical rain forests. Wild grasslands have been changed into farms and ranches as well as cities and suburbs. Many freshwater lakes and rivers are being drained, at least partially, so that humans can use the water.

As a result of these and other activities, Earth is losing its wide variety of living things. This variety is called **biodiversity**. Biodiversity is important in every biome and ecosystem. Scientists argue that protecting biodiversity is vital to the health of life across the planet.

Ecology

Ecology is the study of how organisms interact with one another and with the world around them. The predator-prey relationship is one important interaction. **Predators** hunt and kill other organisms—their **prey**. Organisms are often both predator and prey. Consider a snake, for example, being caught by an eagle. The snake may have just eaten a frog, which in turn, had just eaten a mosquito.

One community role is more important than all the others. Plants and green algae are **producers**, meaning they make their own food. Directly or indirectly, producers provide food for all the other organisms in an ecosystem. Animals eat either producers or other animals. They are called **consumers**.

Decomposers and **scavengers** also play an important role. They feed on dead organisms. Without them, the decaying remains of dead organisms would litter the Earth.

Reading Skill
Understand Text

Science texts often use specific words, or jargon. To understand science texts, it is helpful to identify jargon before beginning to read. Skim the text and highlight unfamiliar words. Also highlight words that appear in bold or italic type. Take time before reading to define these words. If possible, write their meanings in your own words directly in the text. Then, as you read, include your definitions in your reading. They will help you understand the text more completely.

Vocabulary Review

Directions: Complete the sentences below using one of the following words:

biome   biosphere   ecosystem   environment   food chain   interact

1. A(n) _____ is defined by living organisms interacting with each other and with nonliving things in an area.

2. The _____ is the part of Earth where life exists.

3. One way that organisms _____ is a predator-prey relationship.

4. A(n) _____ shows the flow of energy through an ecosystem.

5. The _____ includes the living and nonliving surroundings of an organism.

6. A large group of ecosystems with a similar climate and communities is a(n) _____.

Directions: Read the passage below. Combine the information you find here with what you read in the lesson to choose the best answer to each question.

> The tundra is sometimes thought of as a cold desert. You might guess it would be home to very few living things. During summer, however, the tundra is covered in lichens and small plants, such as mosses and grasses. You also will find traveling herds of caribou and reindeer, as well as some wolves, foxes, and lemmings. Many birds migrate to the tundra for the summer. Bacteria and fungi live in the soil. A few arctic foxes and musk oxen live in the tundra throughout the year.

1. According to the passage, which best describes the tundra?

   A. a cold desert
   B. an ecosystem as diverse as a rain forest
   C. an ecosystem as diverse as a dry desert
   D. a diverse but limited ecosystem

2. Which group of organisms best describes the tundra community?

   A. all of the plants mentioned in the passage
   B. all of the animals mentioned in the passage
   C. all of the plants and animals mentioned in the passage
   D. all of the bacteria mentioned in the passage

Directions: Read the passage below and choose the correct answer to each question that follows.

> Healthy ecosystems stay in balance. One way nature maintains this balance is through predator-prey interactions. In 1972, the Marine Mammal Protection Act made it illegal to kill seals. Now, in areas along the eastern United States, seal populations are increasing dramatically. Seals prey on fish such as cod and halibut. In turn, great white sharks prey upon seals.

3. What do you predict would happen to the population of great white sharks if people began hunting seals again?

   A. Their population would stay the same.
   B. Their population would increase slightly.
   C. Their population would increase greatly.
   D. Their population would decrease.

4. How do you predict the seal populations would change if sharks began hunting them more aggressively?

   A. Their population would stay the same.
   B. Their population would increase slightly.
   C. Their population would increase greatly.
   D. Their population would decrease.

Skill Practice

Directions: Read the passage below and choose the correct answer to each question that follows.

Native species are those plants and animals that have inhabited an area for a very long period of time. Some native species were a part of their environment long before humans. They are well-established in their particular ecosystem. Each has a vital role, such as a tree providing shelter for small mammals and birds. Each native species is also an important part of the food chain. Through healthy predator-prey relationships, the ecosystem maintains balance.

As humans' ability to travel has advanced, so has their ability to transport different species from one location to another—sometimes across the world. When a plant or animal is transported from another place and introduced into a new area, it can threaten, or even cause the collapse of, the existing balance in an ecosystem. These non-native species often have no effective predators. Without predators, their populations can quickly soar, crowding out the native species as they compete for sunlight, water, food, and space.

In 1859, twenty-four rabbits were transported to Australia and released on private property. In a short time, the property owner could no longer contain the rabbits, and their population increased and spread across the continent. Without any natural predators, the rabbit population grew into the millions. By 1950, their numbers had reached 500 million. The rabbits overgrazed the natural vegetation, leaving nothing for species native to the continent to eat. This reduced Australia's wool and meat production. Despite efforts to control rabbit populations, they remain a serious problem today.

1. According to the passage, what has been the effect of advancements in transportation on native species?

   A. It has destroyed habitats as roads and railways are built.
   B. It has made it easier to introduce non-native species into certain areas.
   C. It has created pollution that damages the environment.
   D. It has caused only minimal changes.

2. What type of interaction would help control rabbits in Australia?

   A. native vs. non-native relationships
   B. producer-consumer relationships
   C. interactive relationships
   D. predator-prey relationships

3. Which statement best summarizes the problem that rabbits caused in Australia?

   A. The rabbits competed with native animals for resources.
   B. The rabbits spread disease.
   C. The rabbits were not adapted for their new environment.
   D. The rabbits preyed on native animals.

Carrying Capacity

KEY CONCEPT: A habitat's limited ability to support the living things within it is called its carrying capacity. Carrying capacity is shaped by limiting factors in the environment.

You live in and play a role in an ecosystem. An ecosystem is a community, or collection of populations of different plants and animals that share a physical environment. In an ecosystem, species live together and interact. Terrestrial ecosystems occur on land. Aquatic ecosystems exist in freshwater and salt water. The largest terrestrial ecosystems are called biomes, and they include different kinds of grasslands, deserts, forests, and alpine or mountain biomes. Biomes and aquatic ecosystems have different chemical and physical characteristics. They also have different species of organisms. Think about the plants and animals you see every day in your surroundings. You and they are part of the same ecosystem.

Carrying Capacity

Every organism and **population**, or group of organisms of the same kind, has a **habitat**—a place where they live naturally. A habitat, however, can support only a limited number of organisms.

Imagine inviting some relatives to dinner. You prepare enough food for six people, but two aunts and an uncle hear about your dinner party and drop by unexpectedly. There is a limited amount of food to offer to your guests, and if any more relatives appear, there will be even less food for everyone. There are similarities between your experience and a habitat's carrying capacity.

A habitat has a limited number of resources that organisms in that habitat must share, meaning it has a carrying capacity. **Carrying capacity** is the number of organisms within a population that a habitat can support without losing its resources or damaging them severely.

Food, water, shelter, and space are some of the critical resources in every habitat. In fact, each is a limiting factor within a habitat. A **limiting factor** is any factor that limits the growth, the numbers of, or the distribution of organisms within an ecosystem.

Each species, or kind of organism, has a range of **tolerance**. In other words, there are lower and upper limits of any factor that a species can tolerate, or allow, and still survive. For example, all organisms require water for survival. Too little water limits the number of organisms that can survive in a habitat, but too much water can do the same.

Relationships in a Habitat

Every living organism competes with other living organisms for limited resources, including water, food, shelter, and space. Competition exists among organisms within the same species. It also occurs between species. Take squirrels and blue jays in a forested habitat as an example.

Squirrels and blue jays eat many of the same kinds of food. Acorns are a particular favorite. Each species competes for the acorns. As the acorn supply decreases, both animals must find other food resources, such as berries or seeds. The species will compete for these resources, too. Ultimately, there may not be enough food to support both populations of animals. Some squirrels and blue jays may leave to find food in a different habitat. Others may die. In either case, populations shrink because the habitat's carrying capacity has been **exceeded**. It has gone beyond the natural limits by which all organisms can be sustained.

Other relationships besides competition among species limit a population's growth. Predator-prey relationships are an example. A **predator** hunts for, captures, and eats all or part of its **prey**. The red fox, for example, lives in a variety of habitats, including forests, grasslands, mountains, deserts, and even human neighborhoods. Among other things, red foxes prey upon rodents such as squirrels, rabbits, and mice.

When there is an **abundance** of rodents, meaning the population of rodents is high, food is not a limiting factor, and the fox population grows quickly. The increased pressure on rodents, however, causes their populations to fall. Food supply becomes a limiting factor for the fox population. Foxes must compete for fewer food resources. Some foxes may move elsewhere in search of food. Others may suffer from disease or starvation. Consequently, the fox population declines. It moves toward **equilibrium**, or a balancing point. The fox and rodent populations may increase and decrease a number of times before equilibrium is established.

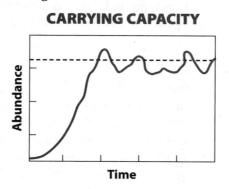

CARRYING CAPACITY

Abundance

Time

Core Skill
Cite Textual Evidence

For more than 50 years, scientists have studied the relationship between moose and wolves in Isle Royale National Park, an island in Lake Superior. The island supports three packs of wolves, with numbers totaling from 18 to 27. The moose population ranges from 700 to 1,200.

Moose arrived on the island in the early 1900s. No wolves lived on the island at the time, and the moose population grew rapidly. The only limiting factor was food, and periods of starvation caused the animals' abundance on the island to rise and fall.

Then in the 1940s, wolves reached the island by crossing an ice bridge that stretched from the Canadian mainland.

What do you think happened to the moose population after wolves appeared on the island? Cite evidence from the text "Relationships in a Habitat" on this page to support your answer. To cite the textual evidence, first reread the lesson text and look for information on the relationships among species in a habitat. Then, when you find the information you need, highlight it, put a star next to it, or circle it.

THINK ABOUT SCIENCE

Directions: If the dashed line in the graph represents equilibrium within a habitat, what does the movement of the line representing abundance over time indicate about equilibrium?

_____

_____

Reindeer on the Pribilof Islands

Like you, scientists write to share their ideas and work. Their writing often includes **jargon**, or terms that are specific to a subject area. Reading text that includes jargon can be difficult. However, you can use a special strategy to help you understand what you read.

Parsing is an effective reading strategy. To parse a sentence means to break it into parts and analyze each part to make sense of the whole. To begin, it is helpful to identify the jargon in what you're reading. Define the terms, and if possible, replace the jargon with familiar synonyms or brief descriptions. Then analyze each sentence piece. When you understand its meaning, repeat the process with the remaining pieces. Then reassemble the sentence.

Read the following example of scientific writing. The jargon has been underlined. What would you do first if you were going to use the strategy of parsing to understand the text?

Carrying capacity in oceanic ecosystems has been a topic of renewed vigor as a result of increasing anthropogenic pressure in certain coastal environments. Increases in coastal populations are correlated to sharp decreases in biodiversity, suggesting that the world's carrying capacity is limited.

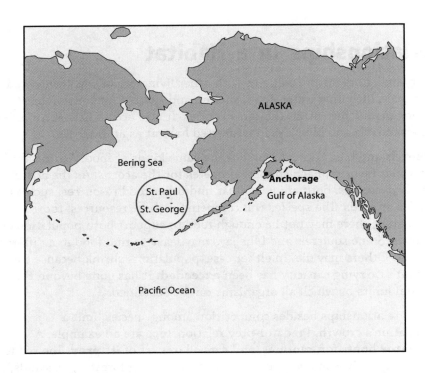

The reindeer of the Pribilof Islands offer a good example of what happens when a population exceeds a habitat's carrying capacity. There are four volcanic islands within the group called the Pribilof Islands. Two of the largest islands are St. Paul and St. George. These rocky islands off the coast of Alaska are covered with **tundra**, which is flat, treeless, Arctic land. Tundra is the coldest and driest of all biomes. Permafrost, or permanently frozen ground, exists most of the year, melting only in summer. So little liquid water and poor soil limit what grows in the tundra.

Lichens are among the few organisms that can survive in the tundra. While they are able to colonize bare rock, they grow slowly because they get their nutrients from air, rain, and melted water.

In 1911, government officials introduced 25 reindeer to St. Paul Island. Ten years later, the population had grown to 250 reindeer. The number continued to grow until there were about 2,000 reindeer on the island in 1938. Then a number of factors combined to reduce the population drastically. Illegal hunting and harsh winter weather affected the population, but lichens proved to be the most important limiting factor.

Lichens are a major source of food for reindeer. As the reindeer population grew, so did the demand for lichens. Lichens did not grow fast enough to continue feeding the reindeer. The reindeer exceeded the island's carrying capacity, and their numbers dropped dramatically. By 1950, only eight reindeer remained alive on St. Paul Island.

THINK ABOUT SCIENCE

Directions: Draw and label a graph to show the change in reindeer population on St. Paul Island from 1911 to 1950. Explain what caused the island's carrying capacity for reindeer to be exceeded.

_____
_____
_____
_____
_____
_____
_____

REINDEER POPULATION ON ST. PAUL ISLAND FROM 1911 TO 1950

Number of Reindeer

Years

Dietary Diversity

Migration and starvation are possible responses to scarce food resources, but so is dietary **diversity**, or variety. Scientists have observed that when a preferred food becomes hard to find, animals usually **diversify**, or broaden their diets to include different kinds of foods. Sea otters living in California's coastal waters are no exception.

In their research, scientists learned that red sea urchins, a favorite food among sea otters, are plentiful in the waters around San Nicolas Island. The sea otters living there feed on these spiny sea animals, as well as marine snails and crabs. Within the population, there is little diversity in the eating habits of individual sea otters.

Fewer red urchins live in the waters off the Central Coast. So it was no surprise to scientists that the diets of sea otters living along the Central Coast were more diversified than those of the San Nicolas sea otters. However, studies of *individuals* in the population show more specialized diets, and that was surprising to scientists. The finding suggests that food webs are more complex than scientists think. Wildlife managers who monitor animal populations may need to change their focus from the dietary habits of the group to the dietary habits of the individual, making their studies of food limitations even more challenging.

KELP FOREST FOOD WEB

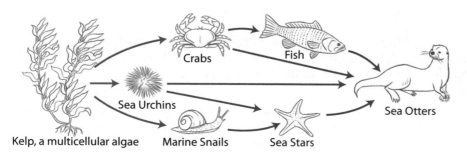

Crabs · Fish · Sea Urchins · Marine Snails · Sea Stars · Sea Otters · Kelp, a multicellular algae

WORKPLACE CONNECTION

Wildlife Management

One of the primary purposes of wildlife management is to maintain the health and size of wildlife populations. Wildlife managers study ecosystems with an eye on balance among wildlife, the environment, and human behaviors.

Because wildlife managers must understand animal populations and the physical environment well, they have extensive academic training. For example, they learn the principles and practices of **conservation biology**. This branch of biology focuses on the preservation and protection of **biodiversity**, or the number of different species of living things. Greater biodiversity is an indication of greater environmental health.

Wildlife managers assume a variety of responsibilities. Within the US National Park Service, for example, wildlife managers take actions to protect the safety of both wildlife and humans, as the two interact in national parks. They also monitor wildlife populations within parks and create plans for dealing with populations that grow too large or too small.

Wildlife biologists also work within the national parks. As part of their research, they may mark and track animal movements. Think about what scientists could learn by tracking such movements.

In nature, species tend to reproduce until a limiting factor stops their population growth. Because humans put pressure on environmental resources, they can affect an environment's carrying capacity.

Think about the communities around where you live. How can you apply the **concept**, or general idea, of carrying capacity to urban planning? What ideas related to population and limiting factors do you think should be considered by planners who may be looking at developing a neighborhood or a housing complex?

Vocabulary Review

Directions: Fill in the blank with the correct word.

carrying capacity exceed equilibrium habitat jargon limiting factor   population

1. The number of individuals of one kind found in a habitat form a _____.

2. The place where organisms live and interact is called a _____.

3. A _____ is food, water, space, or any other thing that restricts population growth.

4. Vocabulary that is specific to areas of study is called _____.

5. The maximum population of a species that an environment can support is the environment's _____.

6. The balance between a population and the resources it needs for survival is called _____.

7. A rise in the number of small rodents can cause the population of owls to _____ its limits.

Skill Review

Directions: Choose the best answer to each question.

1. Which two species represent a competitive relationship?

    A. black bears and lichens
    B. squirrels and field mice
    C. oak trees and blue jays
    D. polar bears and rattlesnakes

2. Which two species represent a predator-prey relationship?

    A. squirrel and acorns
    B. field mouse and grass seeds
    C. fox and rabbit
    D. reindeer and lichens

Directions: Answer the questions.

3. Give four examples of limiting factors for a squirrel.

    _____, _____, _____, _____

4. Imagine that a small population of beavers lives at a pond near a state highway. There are proposals for a new highway construction project. What questions would you ask for the purpose of an investigation into how the project may affect the beaver population?

    _____

    _____

Skill Review (continued)

Directions: Complete the diagram.

5. Use the Venn diagram to compare and contrast equilibrium and carrying capacity.

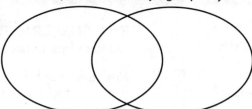

Skill Practice

Directions: Read the following passage. Answer the questions that follow.

For a period of 4.5 years, scientists studied Osceola and Okefenokee populations of black bears in north Florida and southeast Georgia. The scientists identified the kinds of natural foods the bears consumed over that period. They also measured the shape and size of the bears' seasonal ranges, or areas they inhabited in different parts of the year.

Measures showed ranges among female Osceola black bears to be similar to the ranges of other North American black bear populations throughout the year. However, the range of Okefenokee females was about twice as large as the range of Osceola females. The greatest change in range occurred each autumn.

1. Propose an explanation for the greatest seasonal difference in range for female Osceola and Okefenokee bears.

_____

_____

2. If the average yearly range for female Osceola black bears was about 30.3 square kilometers, what was the average range for female Okefenokee black bears?

_____

_____

Directions: Read the text and graph. Answer the questions that follow.

Sometimes people introduce a species to an environment. This species becomes an invasive species, competing for limited resources and putting pressure on other species. Some invasive species are so successful that competing species die out completely.

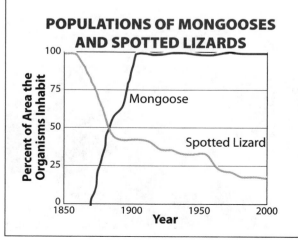

POPULATIONS OF MONGOOSES AND SPOTTED LIZARDS

3. How would you describe the relationship between mongooses and spotted lizards? Use the lines on the graph to justify your description.

_____

_____

_____

4. List two research questions that scientists could investigate to learn more about the spotted lizards remaining in the habitat.

_____

_____

_____

Symbiosis

KEY CONCEPT: The term *symbiosis* describes specific kinds of relationships between organisms in the same environment.

Have you ever had an allergic reaction to something, such as pet fur or dust? That was your body reacting to something in your environment. In every ecosystem, there are biotic and abiotic factors in constant states of interaction. Biotic factors are the living organisms that interact with each other and with the abiotic, or nonliving, parts of their environments. Abiotic factors include temperature, sunlight, water, and available nutrients. Each organism within an ecosystem affects the organisms around it. The organism may compete for resources or even prey upon others. It can alter the physical and chemical environment, too, changing the resources that other organisms also depend on.

Lesson Objectives

You will be able to
- Define symbiosis
- Describe mutualism, commensalism, and parasitism
- Give real-world examples of each type of symbiotic relationship

Skills

- **Core Skill:** Identify Hypotheses
- **Reading Skill:** Summarize Text

Vocabulary

antibodies
host
mutualism
parasite
summarize
symbiosis

Interactions Among Living Things

Because all living things within an ecosystem interact with each other and with nonliving things in their environment, ecosystems are complex and always changing. Some scientists who study ecosystems focus on specific relationships between organisms. A well-studied example is the relationship between the Canadian lynx (a species of wildcat) and the snowshoe hare. Two hundred years ago, Canadian trappers observed changes in population between the two animals. As the population of lynx rose, the population of hares fell. The lynx population then fell and hare populations rose. For a long time, scientists believed that the number of lynx determined the fluctuating numbers of hares. But today, scientists know that the density of hares changes regardless of the lynx population. The question is *why*.

Scientists study other kinds of interactions between animals, too. You may be familiar with the word **symbiosis**. The prefix *sym–* means "together," and the base word *bios* means "life." *Symbiosis*, then, means "living together." But all organisms within an environment could be said to be living together. The term *symbiosis* is used in science to describe three special relationships.

Mutual Symbiosis

You may have watched bees buzzing from flower to flower in a summer garden. The bees hover over flowers, extending their long tongues into the heart of flowers to soak up nectar to take home to feed the hive. As bees gather food, pollen sticks to their hairy

Martin Ruegner/Photodisc/Getty Images

legs. The tiny grains of pollen contain cells that are necessary for a plant to reproduce.

As bees move from flower to flower, the pollen attached to their bodies travels, too. Pollen rubs off of the bee's body and falls into flowers. Inside the flowers, changes in the pollen play a special part in the plant's reproductive process. Thus two things happen. First, flowers feed bees, and bees help flowering plants reproduce. The relationship between the two organisms is mutually beneficial—the relationship benefits both species. This is an example of mutual symbiosis, or **mutualism**. A relationship that is **mutual** is one that is directed toward and received by each other equally.

There are many examples of mutualism in the natural world. Some include organisms too small for us to see.

Termites and Bacteria

The cell walls of plants are stiff and strong. This strength comes from the fibers of **cellulose** that form the cell walls. Cellulose is a carbohydrate that most animals—even animals that eat only plants—cannot digest. That's because their bodies don't produce the enzyme to do the job. However, some plant-eating organisms, like termites, have a mutualistic relationship with microscopic organisms that can do the job for them.

Termites are the **host** for microscopic bacteria and protists. The host is the organism on which another organism lives or feeds on. Most protists are single-celled, but there are multicellular protists also. These organisms, which are like plants, animals, and fungi in some ways but not in others, are independent organisms that interact with their environments as other living things do.

At first, scientists thought that both bacteria and protists helped termites digest cellulose. Now they know that the protists themselves have a mutualistic relationship with the bacteria. Like the termite, protists can't digest cellulose without assistance. They benefit from their relationship with bacteria but seem to have no effect on termites.

21st Century Skill
Critical Thinking and Problem Solving

Investigation is the foundation of science. It involves asking questions, seeking answers, proposing explanations (hypothesizing), and designing investigations to test hypotheses.

To be successful, an investigation requires a combination of creativity, step-by-step planning, and methodical behavior. A good investigation must also be replicable, meaning that other scientists can conduct the same investigation and expect similar results.

Many universities post descriptions and results of ongoing research on their websites. Choose a university that interests you. Search for information regarding science investigations their faculty and students are conducting. What are their scientific questions? What are their hypotheses? What methods are they using to find the answers?

THINK ABOUT **SCIENCE**

Directions: What is the fundamental condition for a relationship between two organisms to be considered a form of mutualism?

_____

_____

Acacia Trees and Ants

Some plants produce chemical defenses, or substances that hungry insects and mammals find distasteful. Other plants, like the acacia tree, do not. Although the trees form nasty thorns, the thorns alone aren't enough to deter hungry pests. The trees attract animals that strip leaves from their

branches, slowing tree growth. Slower growth gives a competitive advantage to other vegetation, which can grow faster and taller, blocking sunlight for the acacias.

Acacias, however, have a mutually beneficial relationship with ants that serve as effective defenders. These ants live inside the thorns that grow at the base of a tree's leaves. The thorns provide a home for ants. The ants bite animals that attempt to eat a tree's leaves. Their bites are annoying enough that even elephants avoid acacias that serve as hosts to ants.

Oxpeckers, Rhinos, and Zebras

Imagine having a personal pest-control service. An oxpecker is an African bird that feeds on ticks and other **parasites**, or harmful organisms. It lands on rhinos and zebras, where it feasts on any parasites it finds. This pest-control service is one benefit for rhinos and zebras, but there is another. When it senses danger, an oxpecker screams a warning as it flies upward.

Humans and *E. coli*

Your body has its own mutualistic relationship with bacteria called *Escherichia coli*, or *E. coli*, which live in your digestive system. As a host, you provide these microbes with protection, nutrients, and a means of moving. In return, the bacteria help you digest food and absorb vitamins. They make the walls of your small intestines slightly acidic, which slows or prevents harmful bacteria from colonizing there. *E. coli* also stimulate your immune system to produce **antibodies**. These are proteins that kill harmful microbes or prevent them from colonizing.

Parasitic Symbiosis

The word *predator* may lead you to think of animals you've seen on television sprinting across African savannahs to capture helpless prey. But not all predators are so large or even able to run. Some are microscopic parasites that depend on other organisms for their survival in a kind symbiotic relationship called **parasitism**. These parasites meet their nutritional requirements through their hosts. However, in parasitism, what benefits the parasite harms the host.

There are two kinds of parasites. **Endoparasites** live inside an organism's body. **Ectoparasites** live on the host's skin. There is also a third kind of parasitic relationship, in which an insect lays its eggs on a living host. The eggs hatch, and the young parasites devour the host. Consider the following examples of parasitic symbiosis.

The Tick

Ticks are not insects but **arachnids**, tiny invertebrates related to spiders. They have biting mouthparts that help them attach firmly to a host's skin, sucking blood. A tick may feed for several days. In that time, a variety of **pathogens**, or disease-causing organisms, can pass from the tick into the host's bloodstream.

Ticks cannot run, fly, or jump, but they can crawl. As animals walk through grasses or near shrubs, ticks waiting on the tips of the plants crawl from the plants to the unsuspecting hosts. The American dog tick, also called a wood tick, feeds on humans and a variety of medium-to-large mammals.

The Tapeworm

There are many different kinds of tapeworms that infect both **vertebrate** and **invertebrate** species—animals with and without a spinal column. Among vertebrates, including humans and domestic animals, these endoparasites infect the liver and digestive tract.

A tapeworm has no mouth, but its head has suckers and sometimes hooks, which it uses to attach to the host. Food is absorbed through the worm's outer covering.

The worm, which is self-reproducing, can produce as many as 40,000 unhatched young, or **embryos**. In a human, these embryos leave the body through waste, or feces. If the infected waste reaches water supplies, other organisms, including humans, ingest the embryos. Once in an organism's digestive tract, the embryos transform into larvae. The larvae bore through the host's intestinal wall and enter a blood vessel. The blood carries it to muscle tissue, where the larvae form a **cyst**, or protective outer covering. If that cyst is eaten, the larvae attach to the new host's intestine and quickly develop into adults. Tapeworm cysts sometimes find their way into the digestive tracts of humans and animals by way of infected or poorly cooked meat.

Not all people infected by tapeworm cysts show symptoms of disease. Some, however, experience many of the same symptoms associated with flu, including vomiting, headache, and weakness. About 70 percent of people with the disease also develop **seizures**, which are temporary electrochemical changes in the brain.

THINK ABOUT **SCIENCE**

Directions: Read the text and answer the question that follows.

Parasites are commonly foodborne or waterborne. **Foodborne** parasites enter the body through impure food. **Waterborne** parasites enter the body through polluted water. Humans are infected when they drink, bathe in, wash clothes in, or prepare food in contaminated water.

Are waterborne and foodborne parasites classified as endoparasites or ectoparasites? Explain your answer.

_____

_____

Orchids and trees represent a commensal relationship. In a tropical forest, trees compete for sunlight. Their leaves form a canopy over the forest floor, blocking sunlight and making it difficult for competing plants to survive. Some plants, like orchids, have adapted to this lightless condition. Orchids attach themselves to trees, close to the top where sunlight penetrates. At these heights, orchids are exposed to sunlight, and they do not need heavy rains for water. Orchids are able to meet their water demands by extracting water moisture from the air. The plants have another adaptation that increases species survival. They produce microscopic seeds. Winds above the canopy carry the seeds away, where they can land and begin to grow new orchids in the canopy.

Reread the passage above and then write a summary of the passage. To **summarize**, choose the most important ideas of the passage and restate these ideas in your own words. As you read a second time, ask questions, such as *what*, *why*, and *how*. Identify key words and the most important ideas. Then state the main idea of the passage clearly in two or three summarizing statements.

Commensal Symbiosis

Commensal symbiosis, better known as **commensalism**, occurs when one species benefits from a host, but the host is unaffected. In other words, the host is neither helped nor harmed by the relationship. Here are some examples of commensalism.

The Pseudoscorpion and Beetles

The pseudoscorpion (soo-doh-SKOR-pe-uhn) is a small eight-legged invertebrate, like its distant relative, the scorpion. But unlike a scorpion, the pseudoscorpion has no tail. To move from place to place, this tiny animal attaches itself to the underside of a beetle's wings. When the beetle flies, the pseudoscorpion hitches a ride to new territory.

Cattle Egrets and Livestock

Cattle egrets are small white birds that follow grazing cattle, horses, and even farm tractors. As livestock or machinery move through grassy fields, they disturb insects. Cattle egrets, which move among the feet of livestock or perch upon their backs, catch these insects. The egret gets food, but the livestock receive no benefit.

Sharks and Remoras

A remora is also called a suckerfish. It has a flat **appendage**, or external structure, on the top of its head. The appendage works somewhat like a vacuum, allowing the remora to attach itself snugly to the underside of a swiftly moving shark. Although a remora is capable of swimming, hitching a ride with a shark or other fish larger than itself has benefits. The remora uses its tiny, sharp teeth to eat whatever food debris falls from its host's mouth without causing any harm or benefit to the host.

THINK ABOUT SCIENCE

Directions: Read the text and answer the question that follows.

For successful reproduction, plants must spread their seeds. Burdocks are weeds that grow along roadsides. The Great Burdock has large, spiny seed heads called burs. The spines are tipped with hooks that snag the fur, hair, or clothing of passing organisms, such as dogs, deer, or humans. When the burs fall off or the animal pulls them off, the seeds may have opportunities to grow.

What makes the burs of the Great Burdock and some animals a good example of commensalism?

_____

David Littschwager/National Geographic/Getty Images

Vocabulary Review

Directions: Match the word to its correct definition.

1. _____ antibodies
2. _____ host
3. _____ mutualism
4. _____ parasite
5. _____ symbiosis

A. living together

B. an organism that harms or kills the host

C. proteins that fight off or destroy pathogens

D. a relationship that benefits both organisms

E. an organism that supports another organism

WRITE TO LEARN

Directions: A female dog and newborn puppy have a specific kind of symbiotic relationship. Think about the needs of both the mother and the puppy. What kind of symbiotic relationship do they have? Justify your answer.

Skill Review

Directions: Read the description. Then complete the activity.

1. In this chapter, you learned about three kinds of symbiosis. Use the concept map to describe the different kinds of symbiosis. Give examples of each relationship.

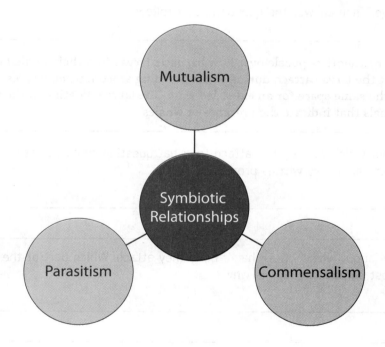

Skill Review (continued)

Directions: Read the passage. Then answer the questions that follow.

> As larvae, barnacles are free-floating organisms, but they attach themselves to a surface and soon change into adult form. As adults, they are sedentary, meaning they are unable to move independently. They remain attached to rocks, boats, the shells of other organisms, and even the bodies of whales.

2. What two benefits exist for barnacles that attach themselves to a whale?

_____

_____

3. Different kinds of barnacles live on the bodies of different kinds of whales. What conclusion can you draw from this information?

_____

_____

Skill Practice

Directions: Read the passage. Then answer the questions that follow.

> Given the size of the ocean, scientists puzzle over how barnacle larvae find their whale hosts. Some scientists think that the larvae attach during whale breeding season, when whales collect in groups and remain in the same space for an extended time. The larvae, floating in the water, may pick up chemical signals that indicate the presence of whales.

1. Barnacles are choosy about which whales they attach to. What question could you ask in an investigation into the chemical signals whales produce?

_____

_____

2. Barnacles are also choosy about where on a whale's body they attach. Which part of the body do you think attracts the most barnacles? Explain why.

_____

_____

3. Scientific research on the barnacle and whale relationship is scarce. What could make it difficult for scientists to investigate questions about the relationship?

_____

_____

Skill Practice (continued)

4. A fungus grows on part of the roots of an orchid plant, where it eats food the plant makes during photosynthesis. The fungus also absorbs nutrients from the soil that it passes to the plant. What kind of relationship do the orchid and fungus have? Use text evidence to support your answer.

_____

_____

_____

5. Leaf-cutter ants chew off bits of leaves and carry them to their dens. In the den, ants chew the leaves into a paste. They release a small amount of fecal matter, or body waste, on the paste. Afterward, a fungus grows on the paste. Note: The fungus is the only food the ants eat. What kind of relationship do the ants and fungus have? Use text evidence to support your answer.

_____

_____

_____

6. Hummingbirds suck nectar, a sugary liquid that plants such as the trumpet vine, trumpet honeysuckle, and the firecracker plant produce. What benefit comes to plants that produce nectar to attract birds?

_____

_____

_____

7. Scientists have found a nectar-sucking species of mites in the nasal cavities of hummingbirds. What kind of relationship do you think these mites and hummingbirds have? Explain your answer.

_____

_____

_____

_____

8. What additional information would you need to determine if your answer in Question 7 is correct?

_____

_____

_____

_____

Disruption

Lesson Objectives

You will be able to
• Identify laws of ecology

• Give examples of environmental disruptions

• Explain the consequences of disruptions

Skills

• **Core Skill:** Determine Meaning of Terms

• **Core Skill:** Cite Textual Evidence

Vocabulary

abiotic
biodiversity
biotic
degradation
destruction
endangered
fragmentation
invasive species
threatened

KEY CONCEPT: A disruption is a change that greatly alters an environment. Disruptions transform environments. In some cases, one ecosystem can temporarily or permanently replace another. In other cases, an ecosystem can become degraded, making it unfit for living things. Still other ecosystems are destroyed altogether.

Take a walk outside and look around. No matter where you live, no matter where you are, the organisms that surround you make up an ecosystem. You may see plants, animals, rocks, soil, and water. An ecosystem also includes microorganisms that you don't see. Within an ecosystem, organisms interact with one another and with their physical and chemical environment. Although no ecosystem remains static, or unchanging, a healthy ecosystem maintains equilibrium, meaning it keeps balance among its species. These species depend on the stable functions of each other and the water, gases, and essential chemicals that cycle through every ecosystem.

The Laws of Ecology

An ecosystem is composed of biotic and abiotic factors. **Biotic** factors include all living organisms, from single-celled microbes to huge land and water mammals. **Abiotic** factors are the nonliving components of an ecosystem, including water, gases in the air, and minerals in rocks and soil. **Ecologists** are scientists who study the interactions between the biotic and abiotic factors within an ecosystem.

Decades ago, ecologist Barry Commoner described four basic laws of ecology that apply to all ecosystems.

1. Everything is connected to everything else. What happens to one organism within an ecosystem affects all organisms in some way.

2. Everything goes somewhere. Nature doesn't create waste, and it doesn't throw things away. What exists remains in existence in one form or another.

3. Nature knows more than humankind. Things are always changing within nature. When humans cause those changes, however, the changes are likely to be harmful to the system.

4. Everything has limits. That is, there is only so much nature that humans can exploit, or take advantage of. When humans use natural resources, those resources eventually change from useful to useless forms.

Responding to Change

As Commoner's third law of ecology tells us, nature is always changing. And as his first law says, all living and nonliving elements within an ecosystem are connected. What happens to one affects all.

Within an ecosystem, organisms live in equilibrium, or balance, responding to the natural changes that are constantly occurring within any system. Some changes, however, are larger than others. All ecosystems respond to **disruptions**, which are breaks or interruptions in normal events. Nature causes some of these disruptions. Humans cause others.

Fire and Floods as Disruptions

Fire, both natural and caused by human behavior, is a common disruption. Natural fires can both harm and help a forest ecosystem. For example, after a wildfire, soils absorb nutrients from the charcoal and ash left after vegetation burns. The soil is warmer, too, encouraging microbial activity. However, intense heat can also cause soil particles to repel, or shed, water instead of allowing the water to soak in. After a fire, the water-resistant soil causes soil **erosion**, as rainwater carries soil away.

Some animals are affected more than others during a fire. Small animals, insects, and sick or old organisms may die. Larger animals are normally able to flee to safety. The fires, which destroy food resources, make it impossible for these animals to return immediately after the fire. This, however, gives other organisms opportunities for survival. **Scavengers**, or organisms that feed on dead plant and animal matter, take advantage of new food resources. Also, areas once thick with trees are laid bare, making it easier for predators to find prey.

Raymond Gehman/Corbis

New plants begin to grow, and increased quantities of seeds on the forest floor encourage birds to feed. The burning of trees and other plants make more nutrients, light, and water available to survivors. With such resources suddenly available, new and surviving plants grow more quickly. Among the benefits of fire is the destruction of harmful plant parasites, like mistletoe, that rob trees of essential nutrients. Another benefit is seed production, as some plants require the heat of wildfires to open their seeds.

Like fire, flooding can cause both harmful and beneficial consequences. Those consequences depend on where floods occur, how long they last, how deep and swift the waters are, and how sensitive the environment is to such disruption.

Floods can lead to the loss of plant and animal life, as well as loss of soils. They can also lead to new life. Some plant seeds remain **dormant**, or inactive, until heavy rains occur. Then these seeds sprout, leading to new growth. Some insects and reptiles that remain inactive during periods of hot or dry weather emerge from their resting state.

THINK ABOUT SCIENCE

Directions: Think about the effects of fire and floods on habitats. List some of the advantages and disadvantages of such disturbances.

Advantages	Disadvantages
_____	_____
_____	_____
_____	_____
_____	_____
_____	_____

Introduced Species as a Disruption

Since the 1700s, about 40 percent of the species on Earth have become **extinct**, meaning they have died out completely. Scientists think that the human introduction of foreign species into environments that are not natural to these species has probably contributed to the massive extinction over recent centuries.

An **invasive species** is an organism that humans move from its natural environment to a new, foreign environment in which it causes harm. Sometimes, the introduction of an invasive species is accidental. At other times, it is done for a purpose.

An example of an accidental introduction comes from shortly after World War II, when military troops shipped cargo from Papua New Guinea to the

Pacific island of Guam. Until then, Guam had no snakes other than a kind of blind, worm-like snake that fed on termites and ants. However, it is likely that a brown tree snake hitched a ride with the cargo. By the early 1960s, tree snakes inhabited more than half of the island. After only a few more years, they had colonized the entire island. As the tree snake population grew, native bird populations shrank. By the time the US Fish and Wildlife Service began listing species as endangered or threatened in 1984, most of the native forest bird species were extinct.

In the late 1800s, a group of Americans organized a club for the purpose of bringing birds mentioned in the plays of William Shakespeare to the United States. In 1890 and 1891, the club's members released about 100 starlings into Central Park. The species flourished. Today, their population exceeds 200 million, and the birds live across the country. Scientists continue to study their effect on native species.

In the 1930s, the US Department of Agriculture imported a Japanese plant called kudzu, and they paid southern farmers to plant it as a means of controlling soil erosion. At first, the plant's rapid growth was considered a success story. However, today some people call kudzu "the plant that ate the South." Able to grow as much as two feet per day, kudzu blankets other vegetation, blocking sunlight. The plants beneath the kudzu die, eliminating potential food and shelter for native animals. Kudzu roots also penetrate the soil, where they affect water levels throughout ecosystems. Kudzu is such a successful invasive species that it has been labeled as one of the 100 worst invasive species on Earth.

THINK ABOUT **SCIENCE**

Directions: Answer the question below.

What makes human beings an example of an invasive species?

_____

_____

_____

Habitat Loss as a Disruptive Force

Some scientists identify the disruptive effect of human activity on natural habitats as the greatest threat to Earth's **biodiversity**. *Diversity* means "variety," so biodiversity refers to various forms of life, including the variety of species and genes. Agriculture, forestry, mining, urban growth, and the pollution that comes with these human efforts have led to massive habitat loss. The International Union for Conservation reports that human interference has caused the rate of species extinction to increase 1,000 times its natural rate.

21st Century Skill
Media Literacy

Scientists began associating starlings with the decline of native bird species as early as 1921. They blamed starlings for the decline of the Eastern bluebird that builds nests in tree cavities. Research has continued, but it hasn't all come to the same conclusion.

For example, in one report, scientists state that the starling actually has had little effect on native bird species, including the Eastern bluebird. The only declining bird population, they say, that can be attributed to the presence of starlings is that of the sapsucker.

Among the skills linked to media literacy is the ability to analyze and evaluate media. With the wealth of available media growing daily, it is critical that readers accept that not all information is reliable. Some information, as in the example in the first paragraph, may be based on incomplete or outdated research. Only further research and analysis can lead to conclusions that are more likely to be accurate.

What are some media sources you could use to gather more information about the effects of European starlings on native bird species?

Habitat loss includes habitat **degradation**, or the loss of habitat due to pollution or the introduction of invasive species. In habitat **fragmentation**, remaining wildlife areas are separated and divided into sections by roads, dams, and other structures. The remaining sections are often too small or they restrict access to larger areas where species can find food and mates. Fragmentation also makes it less likely that migratory animals will have the places they need to rest and feed along their routes.

In habitat **destruction**, habitats are destroyed. People use machines to cut or knock down trees, to fill wetlands with soil in preparation for building, and to scoop sediments and soils from river bottoms for the purpose of building waterways and dams and reclaiming land. Some powerful examples of habitat destruction are found in the United States. According to the US Fish and Wildlife Service, more than 85 percent of forest habitats have been permanently destroyed or logged. More than 75 percent of forests growing along waterways such as streams have been destroyed. In Michigan, 99 percent of mature oak and beech-maple forests are gone. In Oregon, nearly all of the state's temperate rain forest has been destroyed. Across the country's prairies, 95 percent of grasslands have been planted with crops or destroyed. In the Southwest, where desert conditions exist, cattle have overgrazed more than 90 percent of sagebrush habitats.

Leading causes of habitat destruction include agriculture, the conversion of land to building sites and parking lots, and water-development projects, such as dams. They also include pollution, particularly of freshwater resources. In some places, untreated sewage and other human waste enter water resources. So do metals and acids from mines and fertilizers and pesticides from farms.

Photodisc/Getty Images

Vocabulary Review

WRITE TO LEARN

The IUCN produces Biodiversity Indicators, or mathematical measures of biodiversity. People can use these measures to understand factors that affect biological and genetic diversity. Explain how such information could be helpful to decision makers in government.

Directions: Fill in the blanks with the word that best fits the sentence.

abiotic biodiversity biotic degradation destruction endangered
fragmentation invasive species threatened

1. A species that is presently abundant in its native habitat but is likely to become extinct without protection is classified as _____.

2. Rocks, minerals, soil, and water are examples of _____ factors in an ecosystem.

3. An _____ is accidentally or deliberately introduced to a non-native ecosystem.

4. Species may be unsuccessful in finding enough space and food, as well as finding mates in a habitat that has undergone _____.

5. Rodents, mammals, and bushes are examples of _____ factors in an ecosystem.

6. Paving land to make parking lots is an example of habitat _____.

7. _____ describes the variety of all forms of life on the planet.

8. When human waste enters freshwater, it can cause habitat _____, making it unfit for living things.

9. A species with so few members that an ecological disturbance could cause it to disappear entirely is _____.

Skill Review

Directions: Read the passage. Then answer the questions that follow.

In 1859, Thomas Austin released about one dozen European rabbits from his property in Australia. Today, the descendants of these rabbits live throughout the continent, from coastal plains to deserts. The rabbits compete with native species for food and other resources, perhaps causing the extinction of several ground-dwelling mammals. They also damage vegetation, stripping bark from trees and eating seeds and seedlings, preventing regrowth. In some places, rabbits have eliminated all vegetation, leaving only bare rock behind.

1. Use the example of wild European rabbits in Australia to explain the term invasive species to someone unfamiliar with the concept.

_____

_____

_____

Skill Review (continued)

2. Describe the relationship between wild European rabbits and habitat loss.

_____

_____

_____

3. Officials have undertaken a number of steps to control the populations of wild European rabbits, including the use of poisons. Use Commoner's second law of ecology to write an argument opposing the use of poisons.

_____

_____

_____

_____

4. Write two questions you might ask officials in charge of controlling the movement of species in and out of Australia today.

_____

_____

_____

Skill Practice

Directions: Read the passage. Then answer the questions that follow.

The glassy-winged sharpshooter is a successful invasive species. This leaf-hopping insect, a native to the southeastern United States, may have accompanied a delivery of ornamental or agricultural plants to California. In its new, non-native home, the sharpshooter threatens the health of a number of plants, including grapes. That is because the insect carries disease-causing bacteria that it injects into plant fluids as it's feeding. Another sharpshooter feeds from the same plant and ingests the bacteria, which multiply in its mouthparts. When the second insect feeds on a new leaf, it spreads the bacteria, and the process continues. It takes very few insects to spread the disease, and at present, there is no cure for the disease or effective way of controlling the insect population.

1. Imagine you are speaking to a group of wine producers, who grow grapes for making wine. Explain the problem to them in terms they will understand.

_____

_____

_____

Skill Practice (continued)

2. Given that there is no cure for the disease spread by the glassy-winged sharpshooter, what solution would you propose to a wine producer who found a few insects on grape leaves?

_____

_____

_____

3. How has a global economy led to problems such as the glassy-winged sharpshooter?

_____

_____

_____

Directions: Think about what you learned about the kinds of habitat loss.

4. Use the following table to identify, define, and give one example of each kind of habitat loss.

Kind of Loss	Definition	Example

Environmental Issues

KEY CONCEPT: Increased human population makes increased demands on Earth's resources and adds to pollution in the environment.

In a system, different parts interact to make something work as a whole. You have first-hand knowledge of a system—you! The human body is a system made up of parts working together. When one part is harmed or diseased, it can make the whole person ill.

Ecosystems are also made up of parts working together. When one part of an ecosystem is harmed by pollution or overuse, the effects can be widespread. Today, the demands of an increasing human population are adding stress and causing damage to ecosystems around the world.

Environmental Problems

At the end of the Roman Empire in the year 550, there were only about 250 million people living on Earth. Now there are more people than that just in the United States. As of 2009, the world population was more than 6.7 billion.

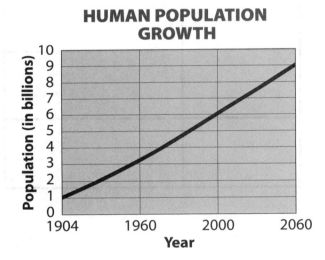

HUMAN POPULATION GROWTH

In the past, famine or disease killed people at an early age. Many children died young. The death rate was high, and the birthrate was low. With today's advances in health care, agriculture, and sanitation, human life spans have increased. The death rate has decreased, while birth rates have increased. The result has been an enormous increase in population.

Many people have become concerned about the human impact on ecosystems. Humans depend on the environment to supply everything they need to stay alive. Unfortunately, people have not always recognized this dependence. Today, there is a greater need for food, water, land, and resources than ever before. These increased needs have put a serious strain on our environment. People must learn what they can do to protect the planet and its resources, the materials that people use, for the future.

DISTINGUISH BETWEEN FACTS AND SPECULATION

It is important to distinguish between **facts** and **speculation** when you read. A **fact** is something that is known to be true. **Speculation** is an opinion or guess about something based on incomplete evidence.

Facts are usually presented in a very straightforward manner. They can sometimes be distinguished by their mathematical statistics or references to trusted sources.

Speculation is sometimes more difficult to distinguish. The author may present his or her speculation in a straightforward way as if it were fact. To identify speculation in a passage, ask yourself: *Is there a lack of examples, statistics, or references? Does the material seem to contain generalizations rather than specific information? Does the author share an opinion or a specific point of view?* If the answer to any of these is yes, it is possible that the material is presenting speculation instead of facts.

Read the following passages. Then, identify whether the passage presents facts or speculation.

1. Human population is increasing. As the population increases, the demand for resources increases as well. Some resources are considered renewable. Others are nonrenewable. Conservation is a way that we can help protect the resources we need for life.

2. Most people are far too wasteful with resources. There is not enough awareness about environmental issues. At the rate at which humans are using up resources, there won't be any left for the next generation.

The first passage presents

_____.

The second passage presents

_____.

Limited Natural Resources

Natural resources are those things in the environment that we use to survive. The air we breathe, the water we drink, and the food we eat are natural resources. The products we use to build our homes, stay warm, and make all the things we use every day are also natural resources. All these resources are in limited supply and must be used by everyone on Earth. Many scientists fear that because of our rapidly growing population and wasteful habits, we are using Earth's resources too quickly.

Scientists divide resources into two categories. **Renewable resources** are those that can be replaced within an average lifetime. Trees, animals, and crops are renewable resources.

Nonrenewable resources cannot be replaced easily. Topsoil and fuels, such as coal and oil, are nonrenewable resources. It took millions of years for fuels like coal and oil to form. When they are gone, they are gone forever.

Conservation is a method of using nonrenewable resources in ways that do not waste them. Forests, water, soil, and any other resources can be conserved by recycling, reusing, or reducing our use of products containing these resources.

Pollution

Human activities not only use up our natural resources, they also pollute, or contaminate, the environment. **Pollution**, man-made waste that contaminates the environment, takes many forms. Solutions can be as simple as picking up litter or as complex as reducing the output of pollutants from industrial plants.

Garbage

Solid waste is all the garbage that comes from homes, businesses, mines, farms, and even schools and hospitals. The chart below shows that the amount of trash we produce has greatly increased. In the 1960s, we produced trash at the rate of about 2.5 pounds of trash per person each day. Now we produce about 4.6 pounds of waste per person per day. Most of that waste ends up in landfills. Some waste is burned. But helpfully, more waste than ever before is recycled into other products.

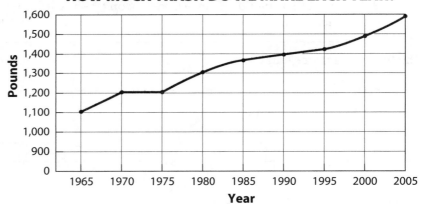

HOW MUCH TRASH DO WE MAKE EACH YEAR?

Hazardous Wastes

Hazardous wastes come from the hundreds of chemicals that we produce and use in the United States. These chemicals include house paints and automotive oils and fluids, as well as many poisons used by farmers. Safely disposing of these wastes is an enormous problem. More and more states are trying to control how these hazardous wastes are handled.

Air Pollution

Some air pollution is invisible. Compounds called chlorofluorocarbons (CFCs) cannot be seen, but they have caused a lot of damage. CFCs are now banned, but they once were released from aerosol spray cans. They damaged the ozone layer in the upper atmosphere. Ozone protects Earth from harmful radiation from the Sun.

Smog is a visible form of air pollution. It forms when water droplets in the air mix with compounds of nitrogen and sulfur, both of which are released when fossil fuels are burned.

Today, the US government tries to limit air pollution. Laws set air-quality standards for cars, factories, and power plants.

Water Pollution

Earth's water collects in ponds, lakes, rivers, and oceans. Water also collects between soil particles and cracks in rocks underground, where it is called groundwater. Any of these supplies of water can become polluted. Moreover, because water flows from place to place, pollutants can spread many miles from their original source.

Water pollution can come from mines, factories, and power plants. It can come from sewage that is improperly treated. Fertilizers from farms and lawns can also pollute the water. When fertilizers wash into a lake, they help algae and other plants grow too quickly. This can begin a process that kills fish and slowly drains the lake.

Pollutants can also travel from the air into the water. Gases from burning coal can mix with water droplets in the air. The droplets fall to Earth's surface as acid rain. Acid rain can kill plants when it mixes into the soil. It kills fish when it mixes into lakes and rivers.

Reading Skill
Distinguish Between Facts and Speculation

Many people have strong feelings about environmental issues, such as conservation, pollution, endangered species, and climate change. When reading about these topics, look for clues to tell you whether the author is presenting facts or speculation. Are statistics provided? Are specific examples given? If so, the material is likely factual. Does the text contain opinions or generalizations? If so, then the author is likely presenting his or her speculations.

THINK ABOUT SCIENCE

Directions: Answer the questions below.

1. What are two reasons for increased population growth?

    _____

2. Compare and contrast nonrenewable and renewable resources.

    _____

    _____

Uses of Land and Water

Like all other living things, humans need food, water, air, and a place to live. As the human population grows, so does its need for these resources. Three hundred years ago, for example, North America was home to only a tiny fraction of its human population today. Vast forests and open grasslands covered much of the land. Today, most of these tracts of forest and grasslands have been replaced by cities, suburbs, farms, and ranches. Water has been channeled out of lakes and rivers for human use.

Changes like these help humans live and grow. But they have unwanted effects, too. Sometimes these effects take many years to observe. Cutting down trees from a hillside makes room for new houses, but the soil may eventually wash away and carry the houses with it.

In the United States, many laws affect how people use land and water. Often these laws and policies are controversial. Should a farmer be allowed to sell his land for a new housing development? Where can one build new factories or power plants? The opinions and interests of businesses, landowners, and community members can all be different.

Endangered Species

In the mid-1800s, about sixty million bison lived on the grasslands of North America. By the year 1900, there were fewer than one thousand. Hunters with rifles were the cause. The bison were overhunted, meaning they were killed much faster than their numbers could increase.

Overhunting is one way that a species can become endangered. Many species are endangered because their habitats, or natural homes, are being destroyed. An **endangered species** has very few individuals left alive, and the species could die out completely. Species with no members still alive are described as **extinct**.

The current list of endangered species is very long. It includes animals of all shapes and sizes, from tiny snails to large elephants, tigers, and gorillas. Some of these species live only in zoos. Others are on the verge of extinction, meaning they could be lost forever.

Yet the American bison has recovered, as has the bald eagle and other species that were once endangered. They have been helped by nature preserves, conservation programs, and the work of scientists and citizens.

THINK ABOUT SCIENCE

Directions: Match the effects in the column on the left with their cause listed on the right.

_____ 1. damage to the ozone layer

_____ 2. acid rain

_____ 3. bison loss during the 1800s

_____ 4. loss of forests and grasslands

A. building of cities and farms

B. over-hunting

C. CFCs and burning fuels

D. pollutants released by industry

Global Climate Change

According to many scientists, the most significant environmental issue is **global climate change**. This issue is sometimes called *global warming* because it involves rising temperatures across Earth.

Climate is the average weather conditions from year to year. In a stable climate, the weather changes in a predictable way every year. Living things depend on these changes. The weather often determines the growing season for plants and the breeding season for animals.

Yet in recent years, scientists have been observing evidence of warming temperatures and climate change. In places across Earth, record high temperatures are being recorded more frequently. Traditional winter weather is lasting a shorter time and is not as cold.

The most dramatic changes are occurring near the North and South Poles. Glaciers are melting and ice sheets are thinning. Yet important changes are happening everywhere. In the western United States, wildfires resulting from dry summer weather are more common. In the southern Atlantic Ocean, hurricanes are more numerous than in the past.

What is causing climate change? Some scientists argue that the cause is the burning of coal, oil, and other fossil fuels. These fuels release a gas called carbon dioxide. Carbon dioxide is a natural part of Earth's atmosphere. It helps trap heat in the atmosphere. Carbon dioxide levels appear to be rising, causing Earth's temperatures to rise as well.

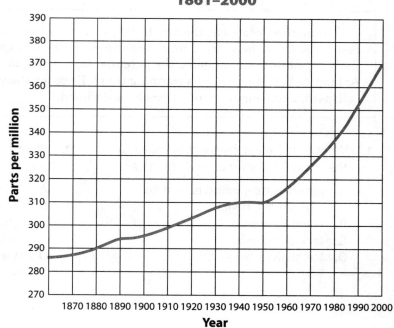

LEVELS OF ATMOSPHERIC CARBON DIOXIDE, 1861–2000

THINK ABOUT **SCIENCE**

Directions: Answer the questions below.

1. What is climate and how is it different from daily weather?

   _____

2. What is global climate change?

   _____

Vocabulary Review

Directions: Complete the sentences using one of the following words:

conservation   natural resources   pollution   speculation

1. Air, water, and other substances that make up the environment are _____.

2. Any contamination of the environment is _____.

3. An opinion or guess based on incomplete evidence is _____.

4. Using nonrenewable resources in a way that does not waste them is _____.

Skill Review

Directions: Read the passage below and answer the questions.

> Antarctica, located at the South Pole, is covered by ice sheets that are thousands of feet thick. Over the years, different countries have sent expeditions to explore this area. Early expeditions investigated the potential for whaling in the region. Since 1945, Antarctica has experienced a warming trend. Huge ice shelves have disintegrated or collapsed, and penguin populations have declined.

1. What is the most reasonable cause for the loss of ice shelves in Antarctica?

    A. numerous expeditions sent to explore the region
    B. warming trends
    C. lack of snowfall
    D. ice shelves break off

2. What is the most reasonable cause for the decline of penguin populations?

    A. They have been over-hunted.
    B. They are prey for whales.
    C. Whaling has reduced their food supply.
    D. Warmer temperatures have caused loss of habitats.

Directions: Answer the questions below.

3. Can a human activity cause multiple environmental problems at the same time? Give an example to support your answer.

_____

_____

4. How can a lake become polluted even if humans never visit it?

_____

_____

Skill Practice

Directions: Read the passage below and choose the correct answer to each question that follows.

> In an international study, scientists researched whether dangerous chemicals pass from farm crops to the people who eat them. The chemicals they searched for—DDT, HCH, and PCB—are all used in farming. In humans, these chemicals are stored in fat cells and are seldom shed. However, nursing mothers shed fat cells in breast milk. The scientists found that the mothers' milk contained chemicals that were commonly used on farms of their region. In China, India, and Mexico, mothers pass on DDT and HCH to their babies. In West Germany, the United States, and Japan, they pass on PCBs.
>
> Women who were exposed to any of these chemicals at their jobs were excluded from the study. This allowed the researchers to assume that the chemical contamination came from diet, not the work environment.
>
> The study shows that toxic chemicals can be passed from mother to child. However, the scientists do not want to discourage breastfeeding. They say that the infant's exposure exceeds daily intake guidelines. However, the exposure is only for a short time and the daily intakes are still within acceptable limits for a lifetime.

1. How do DDT, HCH, and PCB enter the environment?

    A. They are used by farmers to grow food.
    B. Women use them to prevent nausea while pregnant.
    C. They are by-products of the petrochemical industry.
    D. Farmers use them to store their crops.

2. Why were women who may have been exposed to the chemicals on their jobs excluded from this study?

    A. The goal of the research was to show diet as a cause.
    B. The women were sick from the high concentrations of the chemicals.
    C. The women's food sources contained the chemicals.
    D. The women were too busy for the study.

3. Why were PCB levels higher in the breast milk of mothers from the United States and Japan than they were in the milk of mothers from China and India?

    A. Mothers in China and India breast-feed more often than mothers in the United States and Japan.
    B. PCBs have been used more in China and India.
    C. PCBs were used only in meat production in China and India, and these women had little or no meat in their diets.
    D. PCBs are used more in the United States and Japan.

Directions: Choose the best answer to each question.

1. Which question would an ecologist most likely investigate?

   A. What is the maximum length of human life?
   B. How are Emerald ash borers, a type of beetle from Asia, changing forests in North America?
   C. What caused dinosaurs to become extinct?
   D. Why are some tomatoes red, while others are yellow or orange?

2. Which of these biomes can be thought of as a "cold desert?"

   A. grassland
   B. tundra
   C. tropical forest
   D. temperate forest

3. The number of organisms within a population that a habitat can support without losing its resources or damaging them severely is the habitat's

   A. tolerance.
   B. limiting factor.
   C. carrying capacity.
   D. population.

4. Which of these is an example of mutualism?

   A. a bee pollinating a flower
   B. a tick sucking blood from a dog
   C. a remora eating debris from a shark's mouth
   D. an orchid attaching itself to a tree

5. In which of these types of symbiosis is the host neither harmed nor helped?

   A. predation
   B. parasitism
   C. mutualism
   D. commensalism

6. Which of these is an abiotic factor?

   A. a bacterium
   B. sunlight
   C. a dead tree
   D. an insect

7. Which outcome best describes the future of an endangered species?

   A. The species will soon become extinct.
   B. The species might become extinct, or might recover.
   C. The species will thrive again someday.
   D. The species will live only in zoos, not the wild.

Questions 8–10 refer to the following passage.

Hardly anyone in Minamata, Japan, noticed when fish were found floating in the water and shellfish frequently died. Two years later, seagulls were dropping out of the sky. Yet, fishermen continued to fish. People continued to eat the fish. Then a six-year-old girl was admitted to the hospital with brain damage. Within a year, fifty-two people were affected. Twenty-one died.

Before long, 103 people had died, and 700 more had been seriously and permanently damaged. Severe mental retardation, tremors, and limb deformities were among the symptoms.

This is the story of mercury poisoning through man-made environmental pollution. Mercury in industrial wastes was dumped into the ocean and polluted the feeding grounds of shellfish and fish. The poisoned fish were eaten by animals and humans.

Mercury must accumulate in cells before there is enough to become toxic. As a result, smaller animals were affected first. (They weighed less, so they needed less mercury accumulation in their bodies to display symptoms.) Eventually, enough mercury accumulated in the brain cells of humans to cause damage. The symptoms of this tragedy were first noticed among small children.

8. What caused the tragedy in Minamata?
   A. poisoned seagulls
   B. mercury poisoning
   C. climate changes
   D. poor diet

9. Why were people affected by the industrial wastes dumped in the ocean?
   A. They swam in the polluted waters.
   B. The mercury got into their drinking water.
   C. They ate the fish and shellfish that were poisoned by mercury.
   D. They lived near the polluted waters.

10. Why were seagulls "dropping out of the sky?"
    A. They died from eating poisoned fish.
    B. They died from drinking contaminated water.
    C. The fishermen decided to hunt seagulls instead of catching fish.
    D. The seagulls died from a virus.

11. Differentiate between the seven different types of biomes: deserts, tundras, grasslands, tropical forests, temperate forests, oceans, and freshwater areas.

_____

_____

_____

_____

_____

_____

_____

_____

_____

12. Compare the relationship between predator and prey to the relationship between parasite and host.

_____

_____

_____

_____

_____

_____

_____

_____

_____

13. Explain how limiting factors influence the carrying capacity of a habitat.

_____

_____

_____

_____

_____

_____

_____

_____

14. Draw an example of one of the types of biomes, including at least three biotic and three abiotic factors. Label the biotic and abiotic factors. Then describe why each factor is biotic or abiotic.

_____

_____

_____

_____

_____

_____

15. Classify each of these as either renewable or nonrenewable resources. Place each resource under the correct label at right.

cattle
coal
crops
minerals
natural gas
oil
sunlight
topsoil
trees
wind

Renewable

_____

_____

_____

_____

_____

Nonrenewable

_____

_____

_____

_____

_____

Check Your Understanding

On the following chart, circle the number of any item you answered incorrectly. Next to each group of item numbers, you will see the pages you can review to learn how to answer the items correctly. Pay particular attention to reviewing those lessons in which you missed half or more of the questions.

Chapter 3 Review

Lesson	Item Number	Review Pages
Ecosystems	1, 2, 11, 14	88–95
Carrying Capacity	3, 12, 13	96–101
Symbiosis	4, 5, 12	102–109
Disruption	6, 14	110–117
Environmental Issues	7, 8, 9, 10, 15	118–125

Application of Science Practices

CHAPTER 3: ECOSYSTEMS

Question

What are the relationships between the components of an ecosystem?

Background Concepts

An ecosystem is a group of living organisms that interact with each other and the nonliving components in their environment. The living components of an ecosystem can include plants, animals, and microbial life. The nonliving components of an ecosystem can include water, soil, rocks, air, and sunlight.

An ecosystem is in dynamic equilibrium, meaning that conditions within the system stay more or less balanced over time. This balance is achieved by a self-correcting negative feedback system. For example, say there is plenty of plant material to feed a herd of deer. A drought occurs, reducing the food supply. Now there is too little food to support the herd. Some deer migrate to find food elsewhere, or they die. Equilibrium is restored.

Investigation

1. Do some research to learn about different ecosystems and select one that you find interesting. Or select an ecosystem in or near where you live. Remember, an ecosystem is a system formed by the interactions among organisms, and also between organisms and their environment. An ecosystem can be as vast as an ocean or as small as a compost pile.

2. Draw and label the parts of the ecosystem you chose. Draw and label examples of producers, consumers, and detritivores, or microorganisms that feed on or break down organic waste. Also draw and label examples of nonliving components in the ecosystem.

Application of Science Practices

Critical Thinking

1. Imagine that you removed one of the living components from the ecosystem you chose. Predict and describe how this would affect the other organisms in the ecosystem.

    _____

    _____

    _____

    _____

    _____

    _____

2. Now imagine that you removed one of the nonliving components of the ecosystem you chose. Predict and describe how this would affect the living components in the ecosystem.

    _____

    _____

    _____

    _____

    _____

    _____

Evidence

Do research or contact a local expert to find evidence to test your predictions. What evidence did you find?

_____

_____

_____

_____

_____

_____

_____

_____

Foundations of Life

How many different kinds of living things do you think exist on Earth? Estimates have varied from about three million to nearly 100 million. In a recent study, a team of scientists put forth the number 8.7 million, a total that does not include the many species of bacteria that inhabit the planet. Of the 8.7 million species, almost 6.5 million live on land. The remaining species live in water.

Living things are found in all corners of Earth—from the deepest oceans, to the driest deserts, to the coldest ice caps—and every place in between. The planet is truly alive with living things.

In this chapter, you will study the variety of living things, from single-celled microbes to multicellular organisms.

In this chapter you will learn about:

Lesson 4.1: The Cell
Cells are the building blocks of life. Plants are made of cells. So are animals and all other life forms. Some scientists estimate that the adult human body has from 75 to 100 trillion cells. Cells contain internal structures that carry out specific jobs.

Learn how plant cells and animal cells are alike and how they are different. Also find out how cells work.

Lesson 4.2: Simple Organisms
A microbe is a simple organism made from a single cell. Bacteria are microbes. Some scientists say that the majority of cells in the human body are bacterial cells, most of which live in the digestive system.

Learn about the earliest and simplest organisms that lived on Earth—the prokaryotes. Then read about eukaryotes and more complex microbes and multicellular organisms, such as mold, mushrooms, and ferns.

Lesson 4.3: Invertebrates
What distinguishes invertebrates from vertebrates? Learn about the basic characteristics of invertebrate animals, including sponges, worms, mollusks, and insects. Also read about the process of metamorphosis.

Lesson 4.4: Vertebrates
You, a dog, and a whale are all vertebrates. Discover what characteristics are common to all vertebrates and investigate the differences between cold-blooded and warm-blooded animals.

Goal Setting

You encounter different species of organisms every day. They live in your home, in your classroom, and outdoors. Which of these organisms interests you most? What would you like to know about this organism?

Choose an organism and draw its picture.

Write the name of the organism:

_____

Draw a picture of your organism.

Use the chart to write questions you have about the organism. As you read, return to the chart to write the answers to your questions.

Questions	Answers

The Cell

KEY CONCEPT: Cells are the basic units of structure and function in living things.

Have you ever watched or played in a musical group? If so, you know that even though members of a group play together, each person has a special role. One person may sing, while others play instruments. When each individual musician performs well, the group plays successfully, and the audience enjoys the music.

Every cell in a living thing works the same way. It has individual parts that must work together in order to function successfully.

The Structure of Cells

In the mid-1600s, Robert Hooke was using a microscope to look at a piece of cork. He noticed that the cork seemed to be made of "little rooms," which he called **cells**.

It was not long before scientists realized that all living things were made of cells. Now we know that cells are the building blocks of life.

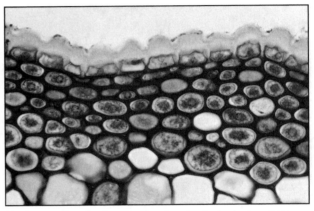

Hooke saw hundreds of cells when he looked at a piece of cork under a microscope.

Cytoplasm

Cells are basic to life, but they are not solid bits of matter. Inside each cell is a jellylike substance known as **cytoplasm**. Some scientists call this the "stuff of life."

Cytoplasm is hard to describe because it is constantly changing. However, it always contains some of the same ingredients:

- About 70 percent of the cell's interior is made of water.

- Cytoplasm contains about 20 different **amino acids**. These acids combine and recombine to form thousands of different kinds of proteins. The proteins are used to build and repair the cell.

- **Carbohydrates** provide the cell with energy.

- **Fats** are the cell's storage tanks. Fats contain all the energy the organism does not use immediately.

- **Nucleic acids** control the cell's activities.

DETERMINE CONCLUSIONS

A conclusion is a kind of summary. Authors normally use the last paragraph or two in a text to present a conclusion. In it, they restate the primary message in an argument or the central idea in an explanatory text. They also include a few key points to support the primary message or central idea.

You can form your own conclusions after reading. Begin by revealing a text's central idea. Then, determine which information the author provided that helps you build support for the main idea you described.

In some cases, it may be necessary for you to consider your personal knowledge and experience before coming to a conclusion. Sometimes, authors don't include all of the information you need to form a conclusion. You can use what you know to fill in the missing pieces. For example, you may not know the names or roles of all of the parts within a cell. If you have a general understanding of what a cell is, however, you can draw conclusions about how the parts of a cell work together as you read the text.

Reread the text on the previous page about how members of a musical group work together. Which of these is a conclusion you can draw about the text?

> (1) One member of the group is more important than the others.
> (2) The members of the group don't have to work together.
> (3) The success of the group depends on each member's participation.

Sentence 3 is the best conclusion that can be drawn from the text.

Core Skill
Support Conclusions

A conclusion is a kind of summary. It states a text's main idea and also provides important details that support, or prove the logic or reasonableness of, the main idea.

It is helpful when reading a science text to pause after paragraphs or sections to form a conclusion. Reread the section "Cytoplasm." Write a conclusion for the section. Underline each detail you use to support your conclusion.

THINK ABOUT **SCIENCE**

Directions: Match the words on the left with their descriptions on the right.

_____ **1.** cytoplasm   **A.** building blocks of protein

_____ **2.** carbohydrates   **B.** the cell's storage tanks

_____ **3.** nucleic acids   **C.** substance inside every cell

_____ **4.** fats   **D.** the cell's energy source

_____ **5.** amino acids   **E.** controllers of cell activity

Cell Structures

Cytoplasm contains a variety of structures that help the cell to **function**, or work, properly. As you read about cell structures, refer to the diagrams of plant and animal cells below.

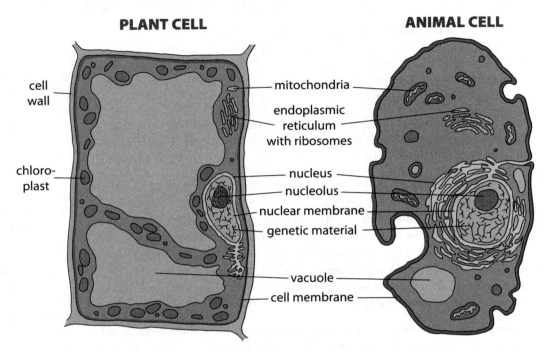

PLANT CELL

ANIMAL CELL

- cell wall
- chloro- plast
- mitochondria
- endoplasmic reticulum with ribosomes
- nucleus
- nucleolus
- nuclear membrane
- genetic material
- vacuole
- cell membrane

Cell Membrane

Beginning at the outer edge of the cell, notice the **cell membrane**. This structure is soft and flexible, but it holds the cell together. It also serves as a gate. Substances needed by the cell can pass through the membrane and into the cell. Other substances are shut out. Waste products pass out of the cell through this membrane.

Nucleus

The **nucleus** is one of the larger structures of the cell. It controls the activities of the cell. Like the cell, the nucleus is made of a substance much like cytoplasm. It is also surrounded by a membrane. Long, thin strands of genetic material float in the nucleus. Inside the nucleus is the **nucleolus**, a dense, round body that makes specialized cell structures called ribosomes.

THINK ABOUT **SCIENCE**

Directions: Answer the question in the space provided.

Are the cell membrane and nucleus part of plant cells, animal cells, or both kinds of cells?

_____

_____

Specialized Cell Structures

Cells may differ in small ways, but all cells contain a membrane, cytoplasm, and nuclear material. In addition, most also contain some specialized structures that help the cell to function. Just a few of these structures are described below.

Most cells contain **mitochondria**, which are shown near the top of both diagrams on the previous page. These sausage-shaped structures trap the energy from food and release it to the cell.

Throughout the cell is a network of canal-like structures known as the **endoplasmic reticulum** (ER). The ER leads from the cell membrane to the nucleus. It transports and stores substances needed by the cell.

Ribosomes are the tiny dots seen on the ER. They combine amino acids into proteins. The **vacuoles** have a variety of functions. Some digest food. Others store or dispose of waste. In plant cells, the vacuole may be filled with water, which helps to support the stem and leaves.

Specialized Cell Structures in Plants

Examine the diagrams again. Notice that plant cells have several structures that are not found in animal cells. The **chloroplasts** contain the chlorophyll that plants use to trap energy from the Sun for the process of photosynthesis.

The outermost structure just outside the plant's cell membrane is the **cell wall**. This rigid wall provides support for the cell. When Robert Hooke found cells in a slice of cork, he was looking at the cork's cell walls.

How a Cell Works

To stay alive, most cells need to take in food, water, and oxygen and to eliminate waste. This means cells must let materials pass through the cell membrane without losing their cytoplasm. Transporting materials across the cell membrane takes place one molecule at a time. Like a filter, the cell membrane has tiny holes that allow some small molecules to pass through it. Larger molecules are filtered out.

Core Skill
Determine Conclusions

As you read the first section of this page, ask yourself: *What central idea links the text in these paragraphs?* When you answer that question, you have found the main idea, the basis of your conclusion. In this section, the main idea is that the cell has many specialized cell structures. Ask yourself: *Why does a cell have specialized parts?* Combine what your read with what you already knew to conclude why a cell has specialized parts.

THINK ABOUT SCIENCE

Directions: Answer the questions in the space provided.

1. What do mitochondria do?

   _____

2. How is the outer edge of a plant cell different from that of an animal cell?

   _____

3. When is chlorophyll kept in the cell? Is it useful in both plant cells and animal cells?

   _____

Diffusion

Material, such as water or simple carbohydrates, tends to seek a balance on either side of the cell membrane. If more of a specific kind of molecule, such as water, is on the outside of a cell, the tendency is for those molecules to spread evenly into the cell. Scientists describe this process of spreading molecules evenly through an area as **diffusion**. Diffusion moves molecules in and out of cells without using any of the cell's energy.

Active Transport

Cells can also move molecules from less crowded areas outside the cell to more crowded areas inside the cell. This movement against the normal flow of diffusion is called **active transport**. The process uses energy and works something like a revolving door. Special molecules in the membrane pick up materials outside the cell. They then turn to the inside of the cell and release the material.

Vocabulary Review

Directions: Complete the sentences below using one of the following words:

cells   diffusion   function   nucleus

1. Several structures are necessary for a cell to _____.

2. A cell's activities are controlled by the _____.

3. Cells use a process called _____ to maintain a balance on either side of the cell membrane.

4. All living things contain _____, the basic building blocks of life.

Skill Review

Directions: Read the text below and choose the best answer to each question that follows.

(1) The shape of a cell is often related to its function. (2) Nerve cells, for example, are long and thin. (3) This enables them to send information from one part of the body to another. (4) Red blood cells carry oxygen through the body. (5) These cells are in the form of a flattened disk that can flow through thin blood vessels. (6) A sperm cell is a cell that travels to an egg cell. (7) Each sperm cell has a tail that allows it to travel quickly.

1. What is the best title for the passage?
   A. "Red Blood Cells"
   B. "What Cells Do"
   C. "The Shape and Function of Cells"
   D. "How Nerve Cells Send Information"

2. Which sentence in the passage states the main idea?
   A. Sentence 1
   B. Sentence 2
   C. Sentence 4
   D. Sentence 7

Skill Review (continued)

Directions: Study the diagram. Then, use the lines provided to describe what is happening in each stage. (HINT: Refer to the section "Diffusion" on the previous page.)

FACILITATED DIFFUSION

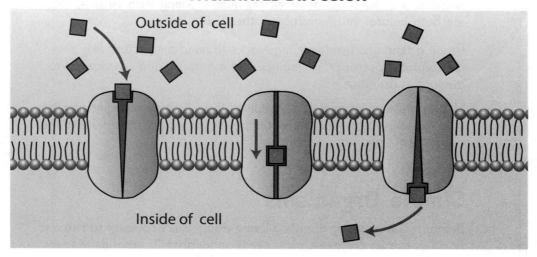

3. _____

_____

_____

_____

_____

Skill Practice

Directions: Choose the best answer to each question.

1. Which of the following is found in plant cells but not in animal cells?

   A. cell membrane
   B. chloroplast
   C. mitochondria
   D. nucleus

2. What is the purpose of the cell wall in plant cells?

   A. to release energy to the cell
   B. to trap energy from the Sun
   C. to separate the cells from the rest of the plant
   D. to give strength to the cells of the plant

3. Which cell structure is responsible for the creation of proteins?

   A. endoplasmic reticulum
   B. vacuole
   C. nucleolus
   D. ribosomes

4. Sea plants often contain more molecules of iodine than does the seawater around them. Iodine moves from the seawater into the plants. What process is described by this example?

   A. active transport
   B. diffusion
   C. interdependence
   D. diversity

Simple Organisms

KEY CONCEPT: Single-celled organisms, including bacteria and protists, are the simplest of all organisms. Along with viruses, they are both helpful and harmful to other living things.

Recall a time you used building blocks to build a structure like a castle or a skyscraper. Each block had a purpose and contributed to the larger structure.

You can think of living cells as building blocks. Some organisms consist of only a single cell. All of the activities within the cell contribute to the cell's survival and reproduction.

Lesson Objectives

You will be able to

• Identify basic characteristics of microbes

• Recognize different types of microbes

• Understand the role of microbes in the environment

Skills

• **Core Skill:** Compare and Contrast Information

• **Core Skill:** Cite Textual Evidence

Vocabulary

compare
decomposer
microbe
organs
thrive

Simple Organisms

Before it is possible to discuss a living thing, it is necessary to name it. In other words, every organism needs an identity. Thousands of years ago, for example, a Chinese emperor studied, tasted, and named hundreds of herbs used to make medicine. These names made communication about the medicinal plants possible.

Later, a Greek philosopher organized animals into two large groups, those with and without red blood. Today, those groups represent the vertebrates and invertebrates.

As technology improved, scientists gathered more information, which influenced the classification systems they devised. Today, scientists generally agree upon three broad domains of life. These domains are based on cell types, and they include the Archaea, or Archaebacteria, Bacteria, and Eukarya. The last group, the Eukarya, are divided into four kingdoms—Protista, Fungi, Plantae, and Animalia.

Microbes

The greatest number of living organisms on Earth are also the smallest. Trillions and trillions of different kinds of bacteria populate the planet. These organisms are **microbes**, meaning they are too small to be seen without the help of a microscope. Most microbes are less than $\frac{4}{1,000}$ inch (0.1 millimeter) across.

Although you cannot see microbes, you may be familiar with some of their effects. Some bacteria, for example, cause decay. Some bacteria and viruses cause disease. Others microbes are essential ingredients in making alcohol, bread, and cheese. Yeast is a microbe that causes bread dough to rise.

Microbiology is the study of microbes. **Biotechnology** is the use of microbes to make useful materials for human consumption.

CITE TEXTUAL EVIDENCE

As you read, you probably question what you read, or you use what you read to form ideas of your own. In either case, you look for evidence in the text to support your ideas. This evidence may be in the form of printed words, or it may be numerical. Both words and numbers can represent data, or pieces of information, that are helpful in supporting or justifying your thoughts.

For example, read the following text. Then, answer the question that follows and use text evidence to support your claim.

Yellow Fever

A bite from an infected mosquito can transmit the virus that causes the disease yellow fever. Within days, the virus, normally found in tropical and subtropical parts of South America and Africa, can result in a variety of symptoms over a period of several days.

At first, an infected person develops a fever and headache. Skin and the whites of the eyes turn yellow, a condition called jaundice. Symptoms disappear after three to four days, and many people recover. For those who don't, the heart, liver, and kidneys begin to fail. Internal bleeding occurs, and patients may have seizures. Eventually, the disease can be fatal.

There is no cure for yellow fever, so people living or traveling in affected areas rely on insect repellant, protective clothing, and vaccinations to prevent contracting the disease.

What conclusion can you draw about how the yellow fever virus got its name?

_____

_____

_____

21st Century Skill
Productivity and Accountability

Scientists collaborate with other scientists to investigate answers to their questions. They are productive individuals, meaning they apply their creativity toward specific goals. They design and implement investigations, focusing on what's important and overcoming obstacles that may prevent them from moving forward.

Scientists hold themselves accountable for their work. This means that they pay close attention to detail and to the precision of their observations and measurements. They also publish the results, sharing their findings with other members of the scientific community. In what way does publishing their results help scientists to be accountable?

THINK ABOUT SCIENCE

Directions: Answer the question below.

What is a microbe?

_____

_____

Bacterial Shape	
bacilli (rod-shaped)	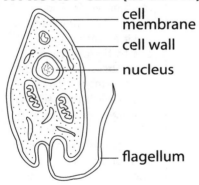 **Bacilli**
cocci (round-shaped)	**Cocci**
spirilla (spiral-shaped)	**Spirilla**

A PROTIST CELL (*EUGLENA*)

- cell membrane
- cell wall
- nucleus
- flagellum

Archaea

Archaea, which were once called archaebacteria, are single-celled organisms that are thought to have been the first organisms to have lived on Earth. They thrive in extremely harsh conditions, such as hot springs or in waters with high salt content. Archaea are also found in oxygen-free and highly acidic environments, such as the digestive tracts of animals.

Bacteria

Bacteria grow well in warm, moist places, but they also inhabit harsh environments, such as deserts, the arctic tundra, and ocean waters. They also live inside the digestive tracts of humans and other animals.

Some bacteria are valuable **decomposers**. They break down the bodies of once-living things, adding nutrients to the soil that plants need to **thrive**, or grow. Decomposers also help prevent the accumulation of dead matter on Earth's surface and in its soil and water.

Bacteria can cause food to spoil, a process that can be slowed by freezing, boiling, or drying food. Some bacteria cause disease, such as strep throat and tuberculosis. Bacterial diseases can be treated with antibiotics, such as penicillin, an antibacterial substance produced by mold, a kind of fungus.

Three basic shapes help identify bacteria. They are bacilli (rod-shaped), cocci (round-shaped), and spirilla (spiral-shaped).

Eukarya

Eukarya include all organisms with eukaryotic cells, or cells containing membrane-bound organelles, or cell structures. Organisms in Kingdom Protista, or the protists, are single-celled eukaryotes. Some have plant-like traits. Others, like *Euglena*, have traits of both plants and animals. *Euglena* contain chlorophyll and make their own food. They also swim and capture food.

Slime molds are also protists. When food is scarce, the cells of one kind of slime mold join to create colonies that look like blobs of petroleum jelly. These blobs migrate to reproduce elsewhere.

WRITE TO LEARN

When you write to compare and contrast two topics, you describe their similarities and differences. Write a brief text to compare and contrast *Euglena* and colonizing slime mold.

THINK ABOUT SCIENCE

Directions: Identify each organism by writing a *B* for bacteria, an *A* for archaebacteria, or a *P* for protists.

_____ 1. most familiar form of single-celled organism

_____ 2. found in very extreme conditions

_____ 3. may have traits of both plants and animals

_____ 4. can be identified by shape

Viruses

Viruses are microbes, but not cells. Many scientists hesitate to call them living things because they do not grow or reproduce. They have no cytoplasm, no nucleus, and no cell membrane. They do, however, have a core of genetic material surrounded by a protein shell.

Viruses depend on living cells for their survival. Once they invade a living cell, they control the cell and use it to produce more viruses. They are responsible for a host of diseases including flu, the common cold, chicken pox, mumps, measles, and AIDS. AIDS is believed to be caused by the HIV (human immunodeficiency virus). This virus harms the body's natural defenses so that other viruses or bacteria can strike. Unable to fend off disease-causing organisms that take advantage of an already weakened system, the body weakens further, and the victim dies.

A VIRUS

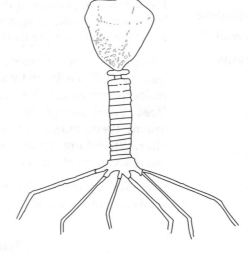

Multicellular Organisms

All organisms begin life as a single cell, but some do not stay single-celled. Cells in multicellular organisms specialize. Their shapes tell you what special functions they perform.

Tissues are groups of specialized cells. Different kinds of tissues may form an **organ**, a part of the body that performs one or more functions. For example, the heart contains muscle tissue, blood tissue, and nerve tissue.

Groups of organs that work together to perform a life process are known as an **organ system**. For example, the tongue, teeth, stomach, and intestines are all part of the digestive system. They all work together to remove nutrients from food. Examples of simple multicelled organisms include fungi and simple plants, such as algae, mosses, and ferns.

THINK ABOUT SCIENCE

Directions: Answer the question below in the space provided.

1. Why aren't viruses classified as living things?

   _____

   _____

2. Write the following terms in order from the simplest to the most complex structure: *organ, tissue, organ system.*

   _____

A FUNGUS CELL

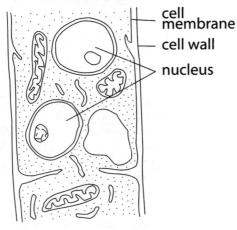

cell membrane
cell wall
nucleus

Fungi

Mold, mildew, yeast, and mushrooms are all **fungi**. With the exception of yeast, fungi are multicellular organisms. They live in environments that are continuously moist or wet.

Like plants, fungi are rooted in one place. Their cytoplasm is at least partially enclosed by a rigid cell wall. However, fungi cannot make their own food because they do not contain chlorophyll. They "feed" by releasing a chemical that digests the organic matter around them. Fungi feed on plants and animals that are dead or alive. When fungi attach themselves to living organisms, they can cause a variety of diseases, one of which is athlete's foot.

Some fungi reproduce by growing a new cell, called a *bud*. Others reproduce by releasing tiny reproductive cells, called *spores*.

FUNGUS

Core Skill
Cite Textual Evidence

Red, dry skin on the bottom of a person's feet can be a sign of athlete's foot. So can cracked, peeling skin between the toes. The fungus thrives in wet, warm places, like the showers and locker rooms that athletes use. This led to the name of the disease, but anyone exposed to the fungus can be infected, and infection can occur anywhere on the body that tends to be damp and warm. Cite text evidence that explains why the fungus infects such areas of the body.

Algae

Algae are the simplest nonflowering plants (a category that also includes mosses and ferns). Although all algae contain chlorophyll, not all algae are green. Some are red or brown. Algae cannot survive out of water. Because algae live in water and absorb their food from the water, they do not need true roots. The water supports the algae, so they do not need true stems.

ALGAE

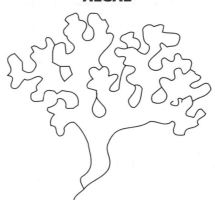

Mosses

Mosses live in wet places on land. Like algae, mosses do not have a group of specialized cells to transport water or to carry food. Instead, mosses depend on diffusion to move water into their cells. Because diffusion is a slow process, mosses do not grow to be very large.

Mosses have two stages in their life cycle. The first stage produces both male and female cells. These cells meet to produce the second stage, a spore-bearing plant. Spores are tiny reproductive cells. Upon ripening, the capsules that contain spores open and scatter the spores. Spores may be scattered by the wind or by animals brushing up against the ripened capsules. If the spores land on fertile ground, they develop into new plants.

Ferns

Some plants have true roots, stems, and leaves but no true seeds. Instead, these plants reproduce by forming spores on the undersides of their leaves. The most familiar plant in this category is the **fern**. Most ferns live in damp, shady places. Instead of stretching up to the Sun, fern stems grow horizontally underground.

Years ago seedless plants, such as ferns, were abundant. As they died, they were buried, and the earth pressed them. Today, we dig up this decayed ancient plant matter and burn it as coal.

MOSS

FERN

THINK ABOUT SCIENCE

Directions: Complete the following chart by filling in the blanks below with the correct words.

Organism	Structure	Chloroplast	Mobile	Cell Covering	Genetic Material
1. ___	incomplete cell	no	no	none	surrounded by protein
Archaea	2. ___	no	no	cell wall	floats in cell
Bacteria	usually single cell	some	some	3. ___	floats in cell
Protist	single cell	some	4. ___	cell wall	in nucleus
Fungus	5. ___	no	no	incomplete cell wall	in several nuclei
Plant	multicell	6. ___	7. ___	cell wall	in nucleus
Animal	multicell	8. ___	yes	cell membrane	9. ___

Vocabulary Review

Directions: Complete the sentences below using one of the following words:

decomposers   microbe   organs   thrive

1. The role of _____ is to break down dead organisms, returning nutrients to the soil.

2. If you need a microscope to see an organism, it is considered a(n) _____.

3. A group of tissues working together to perform a particular function is a(n) _____.

4. A plant will _____ if it gets enough water, nutrients, and sunlight.

Skill Review

Directions: As you read the passage below look for important information. Then answer the questions that follow.

Polio Research

The American researcher **Jonas Salk** knew that three types of infectious **viruses** cause a disease called polio. They are the Brunhilde, the Lansing, and the Leon viruses. He also knew that **immunity** to one type of these viruses did not ensure immunity to the others.

Dr. John Enders found a method of growing viruses in cultures outside the human body. Salk used this discovery in his research. He developed a **vaccine** to protect people from the disease. The vaccine contained **inactivated** poliomyelitis viruses from each of the three types.

---

1. What text features did you see that gave you information about the topic of the passage?

   _____

2. What words might you look up to help you better understand the passage?

   _____

3. What do you predict you will learn about in a lesson that uses this passage?

   _____

Skill Review (continued)

Directions: Look at the diagram. Then answer the questions.

4. What information does this Venn diagram provide?

_____

5. Can viruses be used to make yogurt? How do you know?

_____

6. What does the center section of the diagram show?

_____

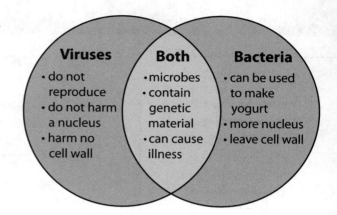

Viruses
- do not reproduce
- do not harm a nucleus
- harm no cell wall

Both
- microbes
- contain genetic material
- can cause illness

Bacteria
- can be used to make yogurt
- more nucleus
- leave cell wall

Skill Practice

Directions: Study the chart. Then choose the best answer to each question.

INFECTIOUS DISEASES THAT OCCUR DURING CHILDHOOD

Disease	Cause	Symptoms	How It's Spread	Prevention/Treatment
mumps	virus	fever; swelling of salivary glands; sometimes a headache and pain when swallowing	air; direct and indirect contact	vaccine
measles (rubeola or red measles)	virus	cough, runny nose, and fever followed by red rash beginning on face	air; direct and indirect contact	vaccine
German measles (rubella or three-day measles)	virus	slight fever and runny nose; red rash beginning on face; swelling of lymph nodes in neck	air; direct and indirect contact	vaccine
chicken pox	virus	slight fever; red bumps beginning on chest and back; severe itching of bumps	air; direct and indirect contact	not currently preventable except with a special vaccine used for children and adults who have certain diseases
cold	virus	runny nose and sneezing; sometimes cough and sore throat; general weakness	air; direct and indirect contact	wash hands and cover mouth and nose; treat the symptoms
influenza	virus	fever; chills; cough and sore throat; general aches and pains and weakness	air; direct and indirect contact	vaccine for older adults and people with certain chronic illnesses
strep throat	bacteria	sore throat; swelling of lymph nodes in neck; sometimes fever and headache	air; direct contact	treated by taking an antibiotic, usually penicillin, for ten days

1. How can most viral diseases of childhood be prevented?

    A. through vaccination
    B. by washing hands
    C. through regular doctor visits
    D. through antibiotics

2. A red rash might be a symptom of which diseases?

    A. measles or chicken pox
    B. flu or a cold
    C. mumps or measles
    D. flu or chicken pox

3. Which disease is caused by bacteria?

    A. measles   C. cold
    B. strep throat   D. flu

Invertebrates

KEY CONCEPT: Animals are made of many cells and can be classified as invertebrates or vertebrates.

You may have twisted and squeezed a kitchen sponge. Sponges like these are usually human-made. But you may have also used the skeleton of a real sponge for bathing. You can twist and squeeze these sponges, too.

Sponges are invertebrates, meaning they have no backbone. If they did, you would not be able to bend, twist, and squeeze them.

Invertebrates

Among animals are those with backbones, or **vertebrates**, and those without backbones, or **invertebrates**. Invertebrates share common traits, such as being multicellular, or made of more than one cell. And unlike plant cells, animal cells have no cell walls. Other than sponges, cells within invertebrates are organized into tissues, or cells that share a common function.

Most invertebrates are mobile, meaning they are able to move freely. But some, such as clams and oysters, spend their adult lives anchored, or attached to something solid, like rocks or even the bodies of other animals. An invertebrate's body shows symmetry. Bilateral symmetry, or two-sided symmetry, means that if you were to draw a line down the middle of the animal's body, the two sides, or halves, would look identical.

Most invertebrates reproduce sexually, meaning male and female invertebrates produce sex cells. The cells join to form a new organism.

Invertebrates depend on other living things for food. Some animals filter food from water. Others hunt for and capture food.

There are different kinds of invertebrates, including sponges, cnidarians, mollusks, arthropods, insects, worms, and echinoderms. The simplest of all of these animals are the sponges.

HYDRA

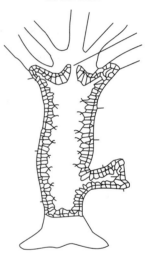

Sponges and Cnidarians

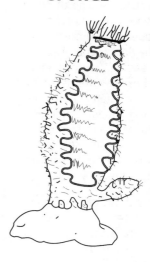

Sponges look like a bag full of holes. They lack a head, arms, and legs, and once a young sponge has settled in one spot, it never moves again. Sponges live on the floor of the ocean. They take in tiny bits of food that they filter from passing water.

Cnidarians include corals, hydras, and jellyfish. These animals have tentacles that they use to sting or trap prey. The tentacles push food into a body cavity where it is digested.

Corals often form large colonies. The colonies secrete a stony skeleton, called a reef. Coral reefs provide homes for a wide variety of animals in warm, ocean waters.

FOLLOW A MULTISTEP PROCEDURE

Investigation is the foundation of science. Scientists ask questions and design experiments to find answers to their questions. Their experiments are often multistep procedures, or a specific sequence of steps. Each step is carried out completely and carefully. Such caution leads to results that can be replicated, or repeated by other scientists.

Imagine that a scientist is curious about how *Euglena*, microorganisms commonly found in freshwater ponds, respond to light other than white light. Read the following multistep procedure for observing *Euglena* behavior.

1. Observe *Euglena* beneath a microscope.

2. Focus white light over the organisms and observe their behavior.

3. Attach a red filter to the light. Turn on the light and observe the organisms' behavior.

4. Attach a green filter to the light and observe their behavior.

5. Attach a blue filter to the light and observe their behavior.

Explain why this procedure begins by studying the effects of white, or natural light, on *Euglena*.

THINK ABOUT SCIENCE

Directions: Answer the question below.

Describe the difference between invertebrates and vertebrates.

_____

_____

To improve your understanding of science text, it is helpful to ask questions before you begin to read. After reading, you can answer those questions, or form conclusions about the texts' most important ideas. After forming conclusions, return to the text to find evidence to support your ideas.

As practice, read again the text on worms and write a conclusion about the roles of worms in the environment. Then highlight the evidence in the text that supports your conclusion.

Worms

Worms is a term that describes a huge variety of animals. Different kinds of worms live in the soil, in the ocean, and as parasites.

WORM

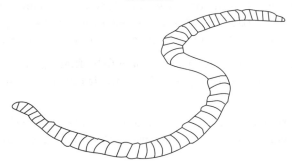

Just one kind of worm, the roundworm, may have as many as half a million species. Roundworms are small and can be found everywhere. Scientists counted 90,000 tiny roundworms in just one rotting apple. An acre of ground may contain up to 10 million.

Earthworms are among the most advanced worms. They have a mouth at the head end, an anus for waste removal at the tail end, and a digestive tract in between. Short, stubby bristles help them move. Although earthworms have no eyes or ears, they have a nervous system and respond to both light and sound.

Every earthworm has both male and female reproductive organs. They mate in pairs, exchanging sperm so that each individual fertilizes the other's eggs. Sometimes an earthworm can fertilize its own eggs.

Role of Worms in the Environment

Many roundworms are **parasites**, meaning they live inside the body of a larger animal and feed off it. Examples include hookworms and eyeworms. They cause diseases in livestock and in humans.

In contrast, earthworms are beneficial. By burrowing through the soil, they help break up soil particles and mix oxygen into the soil. They also return nutrients to the soil through their wastes.

THINK ABOUT SCIENCE

Directions: Check the statements below that apply to earthworms.

_____ 1. Short bristles help them move.

_____ 2. They have a simple digestive system.

_____ 3. They have both male and female reproductive organs.

_____ 4. They have no nervous system.

_____ 5. They cause devastating diseases in humans.

Mollusks

Mollusks are generally made up of a soft body and shell. Slugs, squids, and octopuses also fit into this group, although they don't have shells like the others. Mollusks have no skeletal system, but they have all the other basic organ systems (muscular, digestive, circulatory, nervous, excretory, respiratory, and reproductive). Their bodies include a foot, a mantle, a shell, and a mass of body organs.

Role of Mollusks in the Environment

Many mollusks feed on bits of decaying plants and animals, which helps clean up their environment. Scientists sometimes use certain mollusks to measure the level of pollutants in their sea environments.

Some mollusks, such as snails and slugs, eat plants, including many farm crops. But mollusks help humans, too. Squid, clams, and oysters are mollusks that are important food sources for humans and other animals. The shells and pearls produced by some mollusks are used for jewelry.

SNAIL

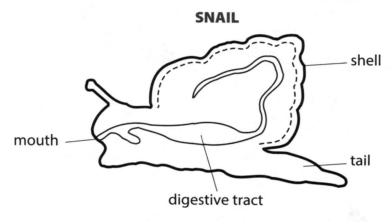

Arthropods

Arthropods are animals with legs that bend. Spiders, ticks, lobsters, and insects are common arthropods. They also have eyes and antennae for sensing the world around them. They have a hard outer covering known as an **exoskeleton**.

The material that forms the exoskeleton is not living tissue and cannot grow with the rest of the animal. As a result, arthropods must shed their exoskeletons to grow. The process is called **molting**. Before molting, the animal produces a soft, new exoskeleton under the old one. During molting, the old exoskeleton splits open and the animal crawls out. After a short wait, the new exoskeleton swells and hardens into the animal's new size. Although exoskeletons work well for small animals, they would be too heavy for large animals.

Insects, the largest group of arthropods, not only respond to the world around them, they **adapt**, or change, and so quickly that they have become among the most successful animals on Earth. This success is related to their ability to live on land, in air, and in water. Scientists have identified more than 700,000 species of insects, and no one thinks that the list is complete. Insects also reproduce at an amazing rate. A housefly may lay one hundred eggs at a time. The young will hatch, develop, and lay their own eggs in as few as ten days.

Insect Body Structure

All insects have six legs and three body regions: a head, a thorax, and an abdomen. A mouth and one pair of antennae are located on the head. They may or may not have wings.

Insect mouths are specialized for eating. Some, such as the grasshopper, have mouth structures designed for chewing. Others, such as the mosquito, have structures designed for piercing and sucking. The butterfly has a long tube for sipping nectar from flowers. Insects' legs may also be specialized. They may be suited for walking, jumping, swimming, or clinging to other animals.

Some insects have tiny, simple eyes. They see poorly and make up for it by their sense of touch. Other insects see more clearly with compound eyes, which are eyes made up of many small parts. Insects have no noses, and instead smell with their antennae, or feelers.

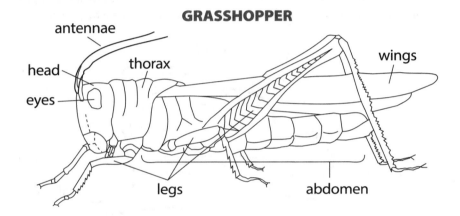

GRASSHOPPER

THINK ABOUT **SCIENCE**

Directions: Answer the questions below.

1. What organisms are included in the group of animals called mollusks?

_____

2. How does shedding an exoskeleton help an arthropod?

_____

3. What are two things that all insects have in common?

_____

_____

Insect Life Stages

Like jellyfish and corals, insects live their lives in stages. They go through either three or four distinct stages as they develop. The process is called **metamorphosis**. All insects begin as eggs. Insects that live in four stages hatch into wormlike creatures called larvae. In this stage, the larvae eat constantly and grow quickly. They then go into a resting stage (the pupa), in which they wrap themselves in a cocoon. Inside the cocoon, the tissues of the larva change into the tissues of an adult insect. When the insect comes out of the cocoon, it is an adult.

METAMORPHOSIS

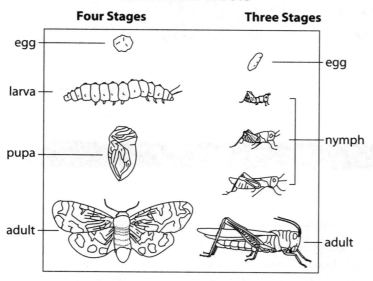

Insects that go through only three stages of metamorphosis skip both the larva and pupa stages. Instead they hatch into nymphs. The nymphs are much smaller than the adults. They have no wings, and their reproductive structures are incomplete. With each molting, however, a nymph is more and more like an adult.

Role of Insects in the Environment

Many farm crops depend on insects to pollinate them. Insects are also an important part of the food chain, providing food for many animals, and even humans in some cultures. Yet insects are also pests, and much worse. Locusts, beetles, and termites destroy crops and lumber. The bites of mosquitoes, ticks, and lice spread many serious diseases.

THINK ABOUT **SCIENCE**

Directions: Answer the questions below.

1. Write the four stages of metamorphosis—pupa, egg, adult, and larva—in the correct sequence, beginning with the earliest stage.

   _____

2. Compare and contrast three-stage metamorphosis and four-stage metamorphosis. Describe how they are the same and different.

   _____

21st Century Skill
Communication and Collaboration

Scientists often work collaboratively, or together, to design and carry out investigations. Scientists work as team members, collecting data, such as facts, solutions, and other information. They then record these data for later examination. Team members also communicate their data and conclusions with other scientists. They may communicate by writing articles, putting information online, or speaking at conferences. Communicating is an essential part of science practice because it unites scientists in the pursuit of new knowledge and understanding. List and describe some practical advantages of scientists sharing data.

WRITE TO LEARN

When you use a set of steps to describe a procedure, you tell how to do something in one action at a time. Think of a meal or dish that you like to prepare. Write instructions describing how to prepare it. Write a clear sequence of steps that someone with little cooking experience could follow successfully.

Directions: Match each definition on the left with the correct term on the right. Write the correct letter on the line.

1. _____ metamorphosis
2. _____ adapt
3. _____ invertebrate
4. _____ parasite
5. _____ vertebrate

A. a change to fit a new situation

B. an animal with a backbone

C. the three or four stages of insect development

D. an animal without a backbone

E. an animal that harms another animal by living on or inside it

Skill Review

Directions: Read the passage below and then answer the questions that follow.

> Because sponges remain in one place and grow slowly, people once thought they were not even alive. Sponges live by filtering food out of the water. Inside their hollow bodies are tail-like structures attached to special cells. These structures move to create a current inside the sponge. As water moves through its body, the sponge filters out tiny food particles.

1. Which statement best summarizes the passage?

   A. People once thought sponges were not alive.
   B. Sponges have tail-like structures inside their bodies.
   C. Sponges live by filtering food out of the water.
   D. Sponges cannot move and therefore they cannot find food.

2. Which statement would be LEAST important to include in a summary of the passage?

   A. People once thought sponges were not even living things.
   B. Inside their hollow bodies are tail-like structures attached to special cells.
   C. These structures move to create a current inside the sponge.
   D. As the water moves through its body, the sponge filters out tiny food particles.

Directions: Arrange the steps below into a time line for the life of a grasshopper.

- The adult grasshopper continues to grow and molt.
- Grasshopper eggs are deposited in the soil.
- The nymph develops into an adult.
- A nymph hatches from the egg.

_____

_____

_____

_____

Skill Practice

Directions: Choose the best answer to each question.

> Most insects are solitary creatures that meet others of their kind only to mate. Some, however, live in large, organized colonies with an obvious division of labor. Termites, wasps, ants, and bees are among these social insects. Bees, for example, live in large hives. Each nest has one queen bee that spends its entire life laying eggs. Drone bees have but one job—to mate with the queen. They can do little else and cannot even feed themselves. Most of the bees in the nest are worker bees. As the name suggests, they do all the work in the colony. They build and clean the hive, collect food, care for the young, guard the nest, and feed the queen and the drones.
>
> In recent years, scientists have learned that worker bees communicate with one another through dance. When a bee finds a good source of nectar, it returns to the nest and does a special abdomen wiggling dance. The speed and direction of the wiggles give other bees information about the location of the nectar.

Questions 1 and 2 refer to the passage above.

1. Based on the passage, which statement describes the social behavior of insects?
   A. Social insects survive by dividing labors and cooperating.
   B. Bees are the only insects that live in colonies.
   C. Insects are solitary organisms.
   D. Worker bees communicate through a kind of song.

2. Why is communication among worker bees important?
   A. It helps the colony survive by increasing the efficiency of the worker bees.
   B. It reduces the risk of worker bees getting lost.
   C. It helps the queen bee lay more eggs.
   D. It increases the size of the hive.

Questions 3 through 5 refer to Lesson 4.3.

3. What is the simplest animal with a brain?
   A. sponge
   B. insect
   C. jellyfish
   D. earthworm

4. Why do arthropods molt?
   A. Their old exoskeleton wears out.
   B. They cannot move with an exoskeleton.
   C. They outgrow their old exoskeleton.
   D. Only pupa have exoskeletons.

5. If you found an organism with three body regions, six legs, one pair of antennae, a mouth, and no wings, how would you identify it?
   A. a mollusk
   B. an insect
   C. a sponge
   D. a spider

Vertebrates

Lesson Objectives

You will be able to
- Give characteristics of vertebrates
- Explain the difference between warm- and cold-blooded animals

Skills

- **Core Skill:** Follow a Multistep Procedure
- **Reading Skill:** Analyze Author's Purpose

Vocabulary

amphibians
instinct
mammals
reflex
respond

KEY CONCEPT: Vertebrates are animals with backbones. They have more developed systems than invertebrates.

A bird and a dinosaur may seem to be very unlikely relatives. Yet, a bird skeleton that is millions of years old closely resembles the skeleton of a small dinosaur. Like all vertebrates, this "early bird" had two pairs of limbs and a backbone. If not for its feathers, it might have been grouped with the dinosaurs.

Vertebrates are animals with backbones. Although there is huge variety among vertebrates, they have many characteristics in common.

Vertebrates

Vertebrates have an internal skeleton made of bone, a type of tissue. Bones grow as an animal grows. Backbones, also called vertebral columns or spines, can support large and heavy bodies, such as those of elephants, giraffes, and whales. Yet they can also bend to allow motion. Skeletal joints, like hinge joints and ball-and-socket joints, allow bone movement.

Most vertebrates have similar body organization. A head equipped with sensory organs sits at the top of the vertebral column. There may also be a tail at the base. Vertebrates have two pairs of limbs that may be specialized as legs, arms, wings, fins, or flippers. Systems of internal organs tend to be organized in similar ways and positions.

Vertebrates **respond**, or react, to the world around them in many ways. Some behaviors, such as reflexes and instincts, are inherited from the parents. A **reflex** is a rapid, automatic response that usually protects an animal from harm. One example is the rapid closing of an eyelid when an object moves too closely toward an animal's eye.

An **instinct** is an automatic behavior that is usually more complicated than a reflex. The way a bird builds a nest or follows its mother, for example, is an instinct. A mother acting to protect her young is another instinct. So are certain reactions to danger or threat, such as fleeing, fighting, or hiding. In each of these examples, an animal acts without having been taught why or how to act.

Most vertebrates can also learn new behaviors. If you repeatedly place food in one spot of an aquarium, a fish will look there for its dinner. A parrot learns to imitate certain words. A human learns to speak, read, and write.

Intelligent behavior, such as problem solving and decision making, is more complex. Among vertebrates, birds and mammals are more likely to show intelligent behavior.

ANALYZE AUTHOR'S PURPOSE

Authors always have a purpose for writing. They may write to entertain you or to inform you. They may write to describe something or to explain how to do something.

Authors who explain how to do something often include procedures, or steps for you to follow. The procedures are often numbered in the order you are expected to complete them. It is important to complete each step and to complete steps in the order they are given. Otherwise, the results are unpredictable.

Consider a scientist who wants to explain how to prepare a wet mount for viewing a specimen under a microscope. The scientist lists and numbers several steps similar to these:

1. Use a dropper to place a drop of water on a clean slide.

2. Use a toothpick to pick up your specimen.

3. Touch the toothpick to the water to release the specimen. If the specimen does not come off automatically, turn the toothpick gently in the water.

4. Start on one side of the drop of water and lower a cover slip carefully.

Explain the author's purpose in writing and numbering simple directions in this procedure.

_____

_____

Core Skill
Follow a Multistep Procedure

In science, investigations often include multistep procedures, or tasks that have several steps that must be completed in the order in which they are listed.

Review the steps in the procedure for preparing a wet mount. Note the special direction in the last step. Explain why the author directs readers to start at one edge of the drop of water to lower the cover slip.

THINK ABOUT SCIENCE

Directions: Write a short response.

What is the difference between an instinctive behavior and a learned behavior?

_____

_____

Cold-blooded Animals

The term *cold-blooded* does not mean that an animal has cold blood. Instead, it means that the animal's body temperature changes with its surroundings. In a cold environment, the animal's temperature cools. In a warm environment, the temperature rises.

Fish

There are three main groups of fish. All have an internal skeleton and an outer covering of scales.

Some fish, like sharks and rays, have internal skeletons made of **cartilage**, a tough, elastic tissue. These are cartilaginous fish. Most fish, however, have bony skeletons. Trout, tuna, and salmon are examples of bony fish. Jawless fish like lampreys and eels form a third group of fish.

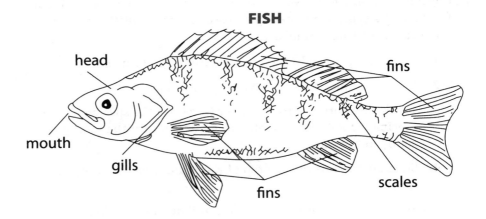

FISH

head — mouth — gills — fins — fins — scales

All fish live in water. They have fins and a tail for swimming. Gills help them take in oxygen from the water.

Fish reproduce by laying eggs. The males spread sperm over the eggs. Most fish leave their eggs, and the young hatch on their own. But some fish guard the eggs and even the young offspring.

A variety of fish can be found in almost every water habitat. Some fish can live only in salt water, while others can live only in freshwater.

Large fish have long been an important food source for humans. Now, overfishing and pollution is causing the numbers of these fish to dwindle.

THINK ABOUT SCIENCE

Directions: Write a short response to each question.

1. What happens to the body of a cold-blooded animal in warm surroundings?

    _____

2. Describe how human activities can affect fish populations.

    _____

Amphibians

Amphibians include frogs, toads, and salamanders. Like insects, amphibians undergo metamorphosis as they grow and develop. A frog's life cycle begins when it hatches from an egg in water. The young frog is called a tadpole, and it looks and lives much like a small fish. Like fish, young frogs use gills to breathe.

During metamorphosis, a tadpole develops lungs for breathing air and legs for moving on land. As adults, frogs live on land. When they are ready to reproduce, they return to the water to lay their eggs.

METAMORPHOSIS IN THE FROG

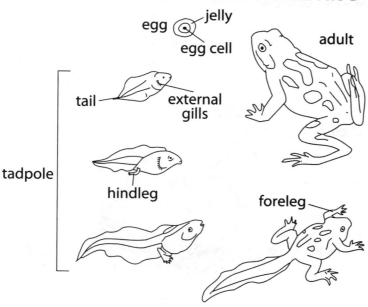

Because they are cold-blooded, the activity level of some amphibians decreases when the temperature drops. In cold areas, they may bury themselves in mud and hibernate. When amphibians are inactive, they live off body fat until they can feed again in the spring. Some amphibians also bury themselves during very hot, dry weather and become inactive.

In addition to breathing through their lungs, adult amphibians may also breathe through their skin. The skin must be thin and supple to let gases, such as oxygen and carbon dioxide, pass through it. This means that many pollutants can pass through the skin as well. Often the pollutants are in the water. Because pollutants can accumulate in the bodies of frogs and other amphibians, these species are useful monitors of water quality and overall environmental health.

THINK ABOUT SCIENCE

Directions: Choose the three terms in each set of words that describe amphibians.

1. egg, nymph, tadpole, adult

2. lungs, skin, respiration, scales

3. warm-blooded, eggs, water, reproduce

Reptiles

Familiar **reptiles** include snakes, turtles, alligators, and lizards. Dinosaurs were reptiles too, but they have been extinct for a long time. No one knows for sure why the dinosaurs became extinct, but many scientists think that the process began when a huge asteroid hit Earth.

The bodies of modern reptiles are dry and covered with scales. The scales are made of a hard material that is similar to human fingernails. The scales help to protect the animal from enemies and from drying out.

Reptiles have well-developed lungs for breathing. Their skin is not used for respiration and does not need to remain moist. As a result, they can live their entire lives on land.

REPTILE EGG

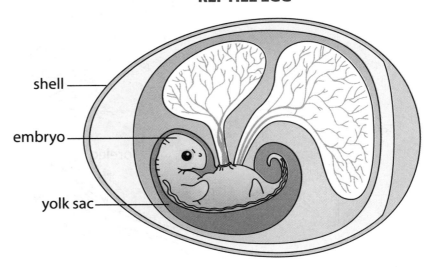

Unlike amphibians, reptiles do not go through metamorphosis. They never live in water like tadpoles. Instead, young reptiles are small versions of the adults. Reptiles do not even need water to lay their eggs. Instead, the eggs have a shell that keeps the moisture inside the egg. The male sperm must reach the egg before the shell forms, so reptile fertilization is internal. A few reptiles give birth to live young.

The American alligator is a reptile that was once very close to becoming extinct. Its habitat, the swamps and wetlands of the southeastern United States, was disappearing rapidly. There was a big demand for products made of alligator skin, such as belts, pocketbooks, and shoes. To keep up with this demand, hunters killed many alligators.

Today, the American alligator is one of the greatest conservation success stories. Populations of American alligators are now thriving, thanks to laws that preserve their habitats and protect them from being hunted.

Warm-blooded Animals

Birds and mammals are warm-blooded. This means that their internal temperature does not depend on the weather outside. They use the energy released from food to keep their bodies warm. Feathers and fur trap air, which provides insulation from the cold. As a result, the body temperature of birds and mammals remains nearly constant.

Birds

Birds are more similar to other animals than they may appear to be. Wings are specialized front limbs. Feathers are modified versions of the reptile's scales. Bird's bones are light and hollow, which helps many birds fly. Other birds, such as the ostrich, only walk and run. Still other birds are excellent swimmers.

BIRD

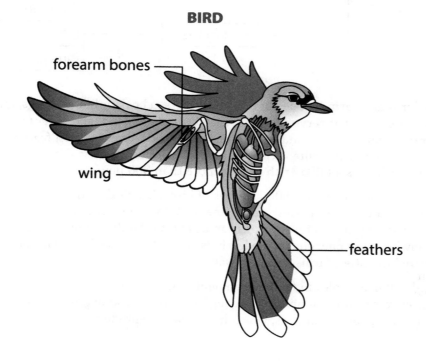

forearm bones

wing

feathers

THINK ABOUT **SCIENCE**

Directions: Match the words with their description on the right. Write the correct letter on the line.

_____ **1.** reptile   **A.** covers the bodies of reptiles

_____ **2.** bird   **B.** keeps moisture inside the egg

_____ **3.** scales   **C.** vertebrates with dry, scaly skin

_____ **4.** shell   **D.** warm-blooded animal

BIRD SPECIES

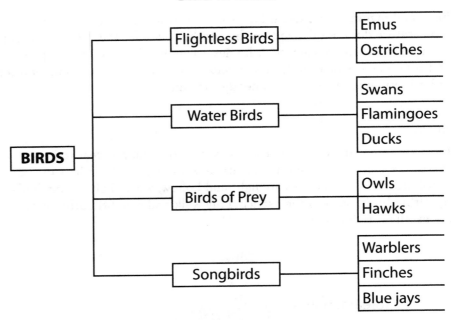

All birds are warm-blooded and have feathers, a beak or bill, two legs, and two forearms that serve as wings. They lay eggs on land, and parents typically care for eggs and young chicks. Beyond these common features however, birds are quite varied. A hummingbird weighs less than an ounce, whereas a tall ostrich weighs more than 300 pounds.

Earth is home to about 9,000 species of birds. Tropical lands are home to about half of these species, among them parrots, toucans, and parakeets. Other species live in forests, grasslands, and wetlands. Unfortunately, many birds are now endangered. Lands where they once lived have been cleared for cities, farms, and highways.

Birds fill many roles in the environment. Some eat insects, others are scavengers that clean up the decaying remains of dead organisms. A few birds also pollinate flowers, which helps plants reproduce.

THINK ABOUT SCIENCE

Directions: Write a short response to the question.

What characteristics are common to all birds?

_____

_____

Mammals

The fossil record shows that the first mammal appeared about two million years ago. A **mammal** is a vertebrate that has hair and that makes milk for its young.

Early mammals were small shrewlike creatures. They probably ate insects and parts of plants. They were warm-blooded and had large eyes, which suggests they were active in the cool of the night when the great dinosaurs were resting. After dinosaurs died out, the mammals spread throughout the world. Some mammals, such as bats, fly. Some, such as moles and shrews, live underground. Whales and dolphins live in the ocean. Hundreds of species of mammals now share the land.

Although some mammals lay eggs (the platypus), most mammals are born alive. Some mammals, such as kangaroos and opossums, nourish their newborn young in pouches. These animals are called **marsupials**. Some mammals get together only for mating. Others live in pairs or family groups. Still others live in large herds or colonies. Mammals are further divided into groups according to their foot and tooth structures.

WRITE TO LEARN

A diagram is often the best way to present information about how to use or set up equipment. Locate a diagram that explains how to use an electronic device, such as a phone or camera. Write a paragraph interpreting the instructions that are presented in the diagram.

MAMMALS

As a group, mammals are the most intelligent creatures on Earth. Their ability to adapt to new situations, to learn from past experience, and to communicate with one another has enabled them to thrive in many habitats. Many mammals use tools and build simple shelters. They live in complex social groups that require cooperation and communication.

THINK ABOUT SCIENCE

Directions: Write a short response to the question.

What are three characteristics that distinguish mammals from other vertebrates?

1. _____

2. _____

3. _____

Vocabulary Review

Directions: Complete the sentences below using one of the following words:

amphibians   instincts   mammals   reflex   respond

1. _____ have hair and feed their young milk produced from special glands.

2. Frogs and salamanders are examples of _____.

3. An animal's _____ can help protect it from danger.

4. Animals _____ to their environment using reflexes and learned behaviors.

5. An automatic response that protects an animal from harm is a(n) _____.

Skill Review

Directions: Answer the following questions by interpreting the diagram shown below.

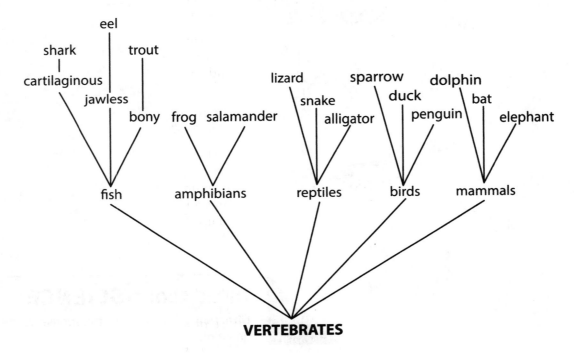

1. Lizards belong to the group of vertebrates called _____.

2. A shark is an example of a _____ fish.

3. Bats belong to the group of vertebrates called _____.

4. Fish, amphibians, reptiles, birds, and mammals belong to the larger _____ group.

5. Frogs and salamanders belong to the group of vertebrates called _____.

Skill Practice

Directions: Read the following passage. Choose the best answer to each question.

> **Penny:** WHAT DID YOU DO TO PENNY?
> **Koko:** BITE.
> **Penny:** YOU ADMIT IT?
> **Koko:** SORRY BITE SCRATCH. WRONG BITE.
> **Penny:** WHY BITE?
> **Koko:** BECAUSE MAD.
> **Penny:** WHY MAD?
> **Koko:** DON'T KNOW.
>
> The conversation printed above may seem odd. Part of the oddity may be that the speakers were using sign language and that Koko is a gorilla.
>
> Koko is the subject of an ongoing scientific experiment conducted by Dr. Penny Patterson. In the 1970s, Patterson taught Koko ASL—the hand sign language used by deaf people. As of 2013, Koko is still signing!
>
> Koko talks about being happy, sad, or angry. She talks about things she did in the past, and she sometimes lies to avoid getting into trouble. She asks people for hugs and makes up insults about those she doesn't like. Although Koko can't read, she enjoys looking at picture books and signs to herself when she recognizes a picture.
>
> One Christmas, Koko signed that she wanted a cat for a present. Her teachers gave her a stuffed cat. Koko responded angrily to the stuffed cat and covered it with a blanket. Finally, the scientists realized that Koko wanted a real cat. Six months later, they brought her a real kitten. At first sight, she signed, "Love that," and she did.
>
> Koko once lived with a male gorilla, Michael, who also learned sign language. Koko and Michael often signed to one another. Unfortunately, Michael died in 2000.
>
> Nonhuman vocal cords do not allow speech like ours, so scientists look for other ways to communicate with gorillas and chimpanzees. In addition to sign language methods, these animals have learned to use elaborate computer programs to imitate language. The chimpanzee Kanzi learned to respond immediately to spoken language. When the teacher said, "I hid the surprise by my foot," Kanzi lifted up the teacher's foot to look for the prize.

1. Why can't scientists communicate with gorillas and chimpanzees the same way they communicate with humans?

   A. Gorillas and chimpanzees are too intelligent to speak to humans.
   B. The vocal cords of gorillas and chimpanzees are not designed for speech.
   C. Gorillas and chimpanzees respond only to voice commands.
   D. Gorillas and chimpanzees are not smart enough to remember words.

2. Based on the passage and your own knowledge, what statement could you make about Koko's use of language?

   A. Gorillas have no emotions.
   B. Gorillas are not capable of planning.
   C. Gorillas are capable of intelligent thought.
   D. Gorillas will one day use speech to communicate with humans.

Directions: Choose the best answer to each question.

1. Why do you think scientists refer to cytoplasm as the "stuff of life"?

   A. The nature of cytoplasm varies among organisms.
   B. It contains all of the materials a cell needs to function.
   C. About 70 percent of cytoplasm is composed of water.
   D. Cytoplasm is the building block of all matter.

2. Which toy best models the role of cells in living things?

   A. a toy car that moves along a track
   B. a doll that runs on a battery
   C. small blocks that snap together
   D. a set of marbles

3. Which set of living things is largest?

   A. mammals and reptiles
   B. insects, fish, and amphibians
   C. Archaea, bacteria, protists, fungi, plants, and animals
   D. invertebrates and vertebrates

4. Which structure is present in plant cells but not animal cells?

   A. cell wall
   B. cell membrane
   C. mitochondria
   D. ribosomes

5. Why are viruses classified as nonliving things?

   A. Viruses need a host to help them reproduce.
   B. Viruses cause disease.
   C. Viruses do not contain DNA.
   D. Viruses are not made of cells.

6. Which organism is the simplest of all invertebrates?

   A. sponges
   B. cnidarians
   C. cartilaginous fish
   D. jawless fish

Review

7. What characteristic distinguishes diffusion from active transport?

   A. Diffusion moves against the normal flow inside a cell, but active transport moves with it.
   B. Diffusion delivers molecules to a cell and active transport carries them out.
   C. Active transport transports water, and diffusion transports nutrients.
   D. Active transport requires energy, but diffusion does not.

8. Which group or groups include animals that undergo metamorphosis?

   A. insects and amphibians
   B. insects and fish
   C. mollusks and reptiles
   D. mammals only

9. In order to survive, arthropods must go through the molting process. What is the function of molting? Molting allows an arthropod to

   A. breathe.
   B. reproduce.
   C. grow.
   D. eat.

10. Mosses lack many of the parts found in larger, more advanced plants. Which part or parts would mosses need to be able to grow?

    A. leaves
    B. roots and stems
    C. flowers
    D. fruits and seeds

11. Which characteristic do birds share with mammals?

   A. Their forearms are adapted for flight.
   B. They have bills or beaks.
   C. They represent the most intelligent creatures on Earth.
   D. They are warm-blooded.

13. What does the nature of an amphibian's skin tell you about where amphibians live?

   A. They must live in or near water, where their skin can remain moist.
   B. They can live where there is reduced oxygen.
   C. They are more likely to withstand periods of drought.
   D. They adapt quickly to rapid changes in weather conditions.

12. How do birds contribute to plant survival?

   A. Birds chase away animals that would eat the plants.
   B. Birds assist in plant reproduction.
   C. Birds deliver nutrients plants need for growth.
   D. Potential pollinators follow birds.

14. Which kinds of animals are most often filter feeders?

   A. vertebrates that undergo metamorphosis
   B. vertebrates that inhabit cave environments
   C. invertebrates that remain attached to solid structures throughout their adult lives
   D. invertebrates that live on the surface of bodies of water

Review

Directions: Questions 15–17 refer to the following passage.

The discovery of microbes was a new beginning for scientists who were searching for the causes of disease. Louis Pasteur, a French chemist, was one of those scientists. In the mid-nineteenth century, he began to explore the idea that some diseases might be caused by microbes.

Pasteur discovered how to grow bacteria in the laboratory. He then experimented with several different kinds of bacteria. He found that different types of bacteria cause different diseases. One bacteria—bacillus—caused a deadly sheep disease, called anthrax. Pasteur discovered that he could make sheep immune to this disease by giving them a very weak form of the bacillus bacteria. His method was the first vaccine. The vaccination protected the sheep because their bodies developed antibodies to fight off the weak bacillus. When the stronger bacillus invaded, the antibodies were ready to destroy them.

A disease called rabies is caused by a virus. Although viruses were too small for him to see, Pasteur applied his knowledge of immunity from microbes to develop the first effective vaccine against rabies.

Pasteur discovered that food spoiled because of the presence of microbes. He developed a process for protecting perishable foods, such as milk, from bacteria. The process, which is called pasteurization in his honor, destroys harmful microbes with controlled heat.

15. Which task were scientists unable to do before the invention of the microscope?

   A. treat any illness
   B. conduct experiments on spoiled milk
   C. find the causes of diseases
   D. discover the role of bacteria in disease

16. Giving a dog a rabies shot prevents it from contracting the rabies disease. Which process is used when giving rabies shots?

   A. sterilization
   B. pasteurization
   C. vaccination
   D. antibiotic

17. Which process controls the spread of bacteria in perishable foods?

   A. immunization
   B. sanitation
   C. pasteurization
   D. vaccination

18. What makes your digestive system a suitable environment for Archaebacteria, or Archaea?

_____

_____

_____

_____

20. Describe the characteristics of protists that distinguish them from other microbes.

_____

_____

_____

_____

19. Explain why mosses depend on diffusion to get the water and nutrients their cells need.

_____

_____

_____

_____

_____

_____

_____

_____

_____

_____

_____

_____

21. Draw and label a diagram of a _Euglena_. Then explain what makes the _Euglena_ unique among microbes.

_____

_____

_____

22. Use the Venn diagram to compare fungi to plants.

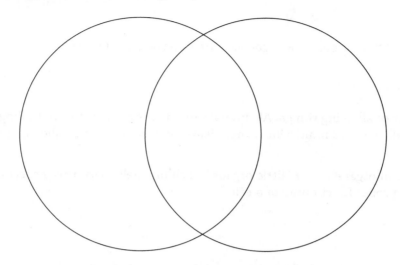

Check Your Understanding

On the following chart, circle the number of any item you answered incorrectly. Next to each group of item numbers, you will see the pages you can review to learn how to answer the items correctly. Pay particular attention to reviewing those lessons in which you missed half or more of the questions.

Chapter 4 Review

Lesson	Item Number	Review Pages
The Cell	1, 2, 4, 7	134–139
Simple Organisms	3, 5, 10, 15, 16, 17, 18, 19, 20, 21, 22	140–147
Invertebrates	6, 8, 9, 13, 14	148–155
Vertebrates	8, 11, 12	156–165

CHAPTER 4: FOUNDATIONS OF LIFE

Question

What is the purpose of the different parts, or organelles, of an animal or plant cell?

Background Concepts

Cells are the building blocks of all living things. All animals and plants are made of cells. Single-celled organisms have only one cell, whereas multicellular organisms can have millions, billions, or even trillions of cells.

Plant and animal cells contain organelles, or "little organs." Each organelle performs a specific function that supports the proper functioning of a cell.

Investigation

1. Use library or online resources to find illustrations of plant and animal cells.
2. Draw, build, or use a software tool to model a plant cell. Label and explain the purpose of each organelle. Be sure to include the cell wall, cell membrane, and nucleus.

3. Draw, build, or use a software tool to model an animal cell. Label and explain the purpose of each organelle. Be sure to include the cell membrane and nucleus.

Application of Science Practices

Interpretation

Compare the models. List the organelles that appear in plant cells but not in animal cells.

_____

_____

Answer

What is the purpose of the organelles of the animal and plant cells you included in your models? Why don't plant and animal cells have identical organelles?

_____

_____

_____

_____

_____

_____

_____

_____

Evidence

Plants are producers, and animals are consumers. How are the differences in their organelles related to these different roles in an ecosystem?

_____

_____

_____

_____

_____

_____

_____

Heredity

Smell the scent of a flower. Watch a pet at play. Listen to actors in a movie. The things you smell, see, and hear are observable traits. Those traits were inherited from the parents of the plants, pets, and people you observed.

The traits you observe in an organism are the organism's phenotype. The code, or instructions that result in that phenotype, are called the organism's genotype.

A genotype also includes information that is not expressed. It is the complete set of genetic information encoded in genes that reside on chromosomes inside the nucleus of each cell. You can use two parents' genotypes to determine the possible genotypes among their offspring.

In this chapter you will learn about:

Lesson 5.1: Genetics
DNA is hereditary material that occurs in almost every cell in a living thing's body, including yours. Most of this DNA is found in a cell's nucleus. The information in DNA is stored in units called genes, and each gene is made of a sequence, or order, of base pairs. Those sequences of pairs are a kind of code that guides an organism's development, growth, and functioning.

Lesson 5.2: Genotypes and Phenotypes
All of the genetic information carried in an organism's cells represents its genotype. All of the observable characteristics in a living thing, such as the shape of a leaf and the scent of a flower, are the organism's phenotype. Being familiar with the genotypes of two parent organisms makes it possible to predict the possible combinations of alleles, or variations of genes, that will occur in the offspring.

Goal Setting

What are some things about genetics that you find most interesting? What would you like to know about genetics?

A KWL chart is a graphic organizer that you can use to identify what you know (K), want to know (W), and have learned (L). List what you already know about genetics in the KWL chart. Then list what you want to know. As you read and discover answers to your questions, return to the chart to record what you have learned. Return to the KWL chart as often as you like throughout your reading.

What I Know	What I Want to Know	What I Learned

Genetics

KEY CONCEPT: Genes carry the codes for human traits. They are located on chromosomes within the nucleus of every living cell.

When Leonardo da Vinci painted his famous portrait of Mona Lisa, he dipped his brush into pots of different colored paints to create the colors of her eyes, hair, and skin. His brushstrokes formed the waves of her hair, the shape of her hands, and her famous smile.

The paints and brushstrokes of biology are the genes that determine each individual's hair and eye color, hair texture, or face shape. The differences among genes from person to person make each individual "portrait" unique.

Lesson Objectives

You will be able to

• Relate genes to chromosomes

• Identify how traits are passed from parents to offspring

• Explain the structure and processes of DNA

Skills

• **Core Skill:** Make Predictions

• **Reading Skill:** Summarize Text

Vocabulary

chromosome
dominant
genes
genetics
prediction
recessive
trait

Genetics

Genetics is the study of how **traits**, or characteristics, are inherited, or passed from parents to offspring. The Greek word *genesis* means "born of" or "produced by." Genes are units of heredity, located on chromosomes inside a cell's nucleus. For species that reproduce sexually, offspring receive half of their genetic information from each parent.

Gregor Mendel

Gregor Mendel (1822–1884) worked in a garden in a monastery. He was not content raising vegetables. He was interested in the variations, or slight differences, that he observed in plants.

Mendel observed that some of his pea plants grew yellow peas, while others grew green peas. Some pea plants grew tall, while others grew short. Mendel conducted careful experiments on thousands of pea plants. Then he proposed a model that explained his evidence.

Today, Mendel is recognized as the father of modern genetics. Mendel's work in genetics applies not only to pea plants but also to all plants and animals, including humans.

Purebred and Hybrid

Mendel investigated the seven traits of pea plants shown in the chart. Each trait has two forms. The color of the pea flower, for example, is either purple or white. The shape of peas is either round or wrinkled.

To begin his experiments, Mendel obtained several **purebred strains** of pea plants. Pea plants that are purebred tall always produce tall plants. Pea plants that are purebred short will produce short offspring.

SUMMARIZE ACCURATELY

Science texts often contain complex information and many details. To communicate efficiently, people in many professional careers must prepare summaries of texts. A summary is a short version of a text that includes only the main points.

To prepare a summary, first skim the text. As you skim, circle any unfamiliar words you find. Individuals working in science often use technical vocabulary to discuss ideas related to their fields. Use a dictionary or other resources to determine the meaning of each unfamiliar word. Define each term in your own words before you attempt to read.

Then, read the text. As you read, pause in your reading to ask yourself questions, such as *What is this about? What is the main idea?* Look for important information and try to find the main points of the passage.

Read the following passage and underline the main points.

> Today, scientists have identified the genes that cause certain diseases, including cystic fibrosis and hemophilia. They also are working on methods to treat these diseases by altering, or making changes to, affected DNA. These treatments are examples of genetic engineering. Genetic engineering is the subject of much debate among politicians, patients, and the scientific community. People have different opinions about how and when this new technology should be used.

When you have finished, state the main idea of the passage clearly in two or three summarizing statements. Use your own words. Remember that a summary should be accurate. It must be free of opinion and should not contain information that is not present in the original text.

SEVEN TRAITS OF PEA PLANTS

Character	Dominant Trait	Recessive Trait
Flower color	Purple	White
Flower position	Axial	Terminal
Seed color	Yellow	Green
Seed shape	Round	Wrinkled
Pod shape	Inflated	Constricted
Pod color	Green	Yellow
Stem length	Tall	Dwarf

THINK ABOUT SCIENCE

Directions: Answer the questions below.

1. What is genetics?
2. What is one example of a question about pea plants that Mendel investigated?

Mendel crossed, or mated, a purebred tall pea plant and a purebred short pea plant. The seeds formed by this cross are called hybrids. A **hybrid** has parents from two different purebred strains.

Mendel planted the hybrid seeds. He expected that they would grow into plants of medium height. To his surprise, all the hybrid seeds grew into tall plants. It was as if the trait for shortness had disappeared!

Mendel then continued his experiment. He crossed the hybrid pea plants with one another, then planted their seeds. This time, the trait for shortness returned. One-fourth of the new pea plants were short. The remaining plants were tall.

Genes and Alleles

Mendel used the term *factors* to explain how plants control traits. Today, Mendel's factors are called **genes**.

As Mendel concluded, a pea plant receives two copies of each factor, or gene. One copy comes with each parent. There are two forms of each gene. The forms are called **alleles**. For the trait of height in pea plants, one allele codes for tallness. Another allele codes for shortness. A plant that is hybrid for height has one allele of each kind.

Recall that Mendel observed only tall pea plants in his first round of hybrids. He explained this observation by proposing that for each trait, one allele is **dominant** over the other allele. For the trait of plant height, the allele for tallness is dominant. The allele for shortness is masked, so it is called **recessive**.

Human Traits

Not all traits, especially in humans, are expressed as Mendel described. Nevertheless, the idea of dominant and recessive alleles helps explain much about human inheritance.

For example, to determine eye color in humans, the allele for brown eyes (B) is dominant over the allele for blue eyes (b). This means that a person that has both alleles (Bb) has brown eyes. But this person could pass the allele for blue eyes to a son or daughter.

Chromosomes and DNA

The nucleus of every living cell carries genetic material—the **chromosomes**—that determines the traits of that organism. Chromosomes are made of many different proteins and a substance called deoxyribonucleic acid, or **DNA**. Each molecule of DNA consists of millions, even billions of tiny units linked together in a chain. The chain resembles a twisted ladder. Each small section of DNA, which we call a gene, forms part of the rung and rail of that ladder.

The gene is like an instruction manual for assembling specific molecules of the body. Some genes determine eye or hair color. Each trait is determined by genetic information.

The Genetic Code

How is DNA able to code for all the traits of the human body? The answer lies with the tiny units, called **nucleotides**.

DNA is made of only four nucleotides. They are adenine (A), guanine (G), cytosine (C), and thymine (T). Each "rung" of the DNA ladder consists of pairs of these four nucleotides repeated over and over again in a specific order. The order of the nucleotides is the way that DNA codes for genes.

Surprisingly, almost all DNA is identical from person to person. Only small changes in the DNA of certain genes account for the differences that make each person unique.

In 2003, scientists completed the **Human Genome Project**. This project identified all of the genes in human DNA. It also identified the order of all the nucleotides in these genes. The number amounted to about three billion.

Reading Skill
Summarize Text

Read this page and prepare a summary. When writing a summary, don't let the presence of unfamiliar words stop you. Some examples of unfamiliar words on this page include *chromosomes, DNA, nucleotides,* and *Human Genome Project.* Use a dictionary to learn the meaning of these terms.

A summary should not include the writer's own ideas. Does your summary express your opinion? If it does, you will need to edit your summary so that it presents only the key details from the text.

THINK ABOUT SCIENCE

Directions: Match each term on the left with the best description on the right.

_____ 1. purebred

_____ 2. chromosome

_____ 3. DNA

_____ 4. nucleotides

_____ 5. gene

A. an organism with two of the same alleles for a particular trait

B. substance that makes up chromosomes

C. segment of DNA that controls specific traits

D. structure that contains genetic information

E. adenine, guanine, cytosine, and thymine

WRITE TO LEARN

One way to get more from what you read is to predict, or think about, what you expect will happen or what the writer will say. Locate an editorial in a newspaper or magazine. After you read the first few paragraphs, make a prediction about what you think the writer will say next. Write your prediction in a notebook. Make new predictions as you continue reading.

You can think of
initiative as a first step
toward a goal. You set
a goal and make a plan
to reach it. Your plan
includes a schedule of
checkpoints, or tasks
you want to complete
at certain times. The
checkpoints help you
make adjustments, solve
problems, and stay on
the path leading to your
goal.

Throughout the process,
you may be working
independently. If that's
the case, self-direction
is essential. You do
what is necessary to
reach your goal, solving
problems along the way,
and pausing occasionally
to reflect upon your
progress.

What goal would you
like to set for yourself?
Think about the steps
you would need to take
to reach your goal. List
the steps and create a
schedule. Then, applying
both initiative and self-
direction, work toward
success.

DNA Replication

The human body is made of trillions of cells. With a few exceptions, each of these cells contains exactly the same DNA. How is this possible? When a cell divides to make new cells, its DNA undergoes a copying process. This process is called **DNA replication**.

Recall that the structure of DNA is like a twisted ladder. The two halves of the ladder are joined by their nucleotides, also called bases. Notice that the bases always pair in the same combination. Adenine (A) always pairs with thymine (T). Cytosine (C) always pairs with guanine (G).

During DNA replication, the two halves of the ladder "unzip," or separate. The exposed bases then pick up and bind to new bases from their surroundings. In this way, each half of the unzipped DNA serves as a template, or model, to build a new half. The result is two new DNA molecules, each an exact copy of the original.

DNA "UNZIPPING"

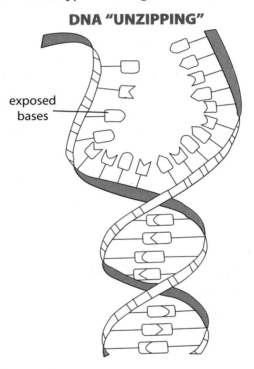

exposed
bases

Forty-Six Chromosomes

In most human cells, DNA and genes are divided among 46 chromosomes. DNA replication is part of the process that copies chromosomes when new cells are made.

Two of the 46 chromosomes are different from the others. These are the sex chromosomes, also called the X and Y chromosomes. The sex chromosomes determine gender. Cells in men have one X and one Y chromosome. Cells in women have two X chromosomes.

All 46 chromosomes are present in the first cell of a developing human baby. Half of the chromosomes, including either an X or a Y, come from the father. These 23 chromosomes were carried by a sperm cell. Sperm are male reproductive cells. The other 23 chromosomes, including one X chromosome, come from the mother. They are carried by an egg cell, the female reproductive cell. A sperm and egg cell combine to form a new individual.

Genetic Errors

Usually, the replication of DNA is perfect, and two identical chromosomes result. Sometimes, however, a mistake occurs in the instructions. We say the gene has **mutated**—or changed its instructions. This change can affect the way the organism develops. If the organism survives the mutation, that mutated gene may be inherited by future generations. Many inherited mutations are harmful. Inherited conditions include muscular dystrophy, hemophilia, and sickle-cell anemia.

DNA Transcription and Translation

How is DNA decoded? The process involves two important steps. The first step is called **transcription**. In this step, DNA serves as a template, or model, to make a molecule of RNA. RNA is short for ribonucleic acid.

Unlike DNA, a molecule of RNA can leave the nucleus and travel to other cell parts. Away from the nucleus, RNA acts as a model to make proteins. This process is called **translation**. The cell uses proteins to help control all of its chemical processes.

In any cell, only certain genes are ever transcribed and translated. Active genes differ among body cells. In certain stomach cells, for example, some of the active genes help make stomach acid. In nerve cells, active genes help make the chemicals that nerves use to send messages.

DNA and Living Things

Scientists have identified and studied DNA in a huge number of organisms. As they have discovered, DNA controls traits in plants, animals, and even fungi and bacteria. Viruses contain either DNA or RNA.

With only a few exceptions, the DNA in every living thing functions just as it does in humans. Whether you are investigating the DNA of a parrot or a pineapple, you will find that it uses the same four nucleotides and the same genetic code. It also uses the same processes of transcription and translation.

THINK ABOUT SCIENCE

Directions: Circle the words that make each sentence true.

1. In a DNA molecule, adenine binds with (thymine, cytosine).

2. Most human cells have (46 chromosomes, 23 chromosomes).

3. DNA replication depends on the halves of a DNA molecule (separating, dissolving).

4. A "mistake" in DNA replication is called a genetic (mutation, expression).

Vocabulary Review

Directions: Complete the sentences using one of the following words.

chromosome **dominant** **genes** **genetics** **recessive** **trait**

1. The study of how traits are inherited is ______________.

2. A ______________ allele can be expressed only when there are two of the same allele.

3. ______________ are located on chromosomes.

4. Alleles for a ______________ trait will mask those for a recessive trait.

5. An example of a human ______________ is the presence or absence of freckles.

6. A ______________ is made up of proteins and DNA.

Skill Review

Directions: Choose the best answer to each question. Questions 1 and 2 refer to the diagram below.

1. What are the chances that an offspring of Parents 1 and 2 will have brown eyes?

   A. 2 in 4, or 50%
   B. 3 in 4, or 75%
   C. 4 in 4, or 100% (the offspring will definitely have brown eyes)
   D. 0 in 4, or 0% (the offspring will definitely not have brown eyes)

2. The allele for brown eyes is ________, while the allele for blue eyes is ________.

   A. dominant, recessive
   B. recessive, dominant
   C. purebred, hybrid
   D. hybrid, purebred

POSSIBLE OFFSPRING OF TWO BROWN-EYED PARENTS WITH RECESSIVE GENE FOR BLUE EYES

		Parent 2	
		B	**b**
Parent 1	**B**	**BB** brown	**Bb** brown
	b	**Bb** brown	**bb** blue

B = brown
b = blue

Directions: Answer the questions below based on the following passage.

> Over many generations, some family members may show a form of a particular trait, either dominant or recessive. Other individuals may inherit an allele for a particular trait, but not show it. A pedigree shows both individuals who show the trait, as well as those who are carriers.

3. Based on the passage, what is a carrier?

____

4. Based on the passage, what is a pedigree?

____

Skill Practice

Directions: Choose the best answer to each question. Questions 1 and 2 refer to the diagram below.

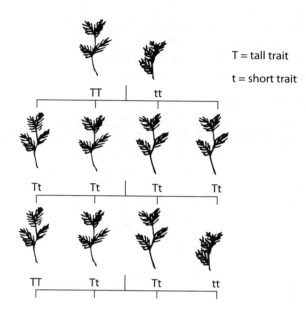

T = tall trait

t = short trait

TT   tt

Tt   Tt   Tt   Tt

TT   Tt   Tt   tt

1. The diagram shows a cross between a tall plant and a short plant. Each row shows a possible generation. What can you conclude about the trait for short size (t)?

   A. The trait is dominant.
   B. The trait is recessive.
   C. The trait is a mutation
   D. The trait is a blend.

2. If a plant grows tall, what can be concluded about its genes?

   A. It has two alleles for tall height (TT).
   B. It has only one allele for tall height (Tt).
   C. It has two alleles for short height (tt).
   D. It has either one or two alleles for tall height (TT or Tt).

Directions: Questions 3 and 4 refer to the following passage.

Scientists have discovered thousands of human diseases that have a genetic link. One such disease is PKU, or phenylketonuria. Most people have at least one copy of the dominant allele (P) that provides the code for the body to produce a special enzyme. This enzyme changes a harmful acid in milk into a harmless acid that humans can digest.

A child with two recessive PKU genes (p) cannot make this acid-changing enzyme. If a PKU child drinks milk, the harmful acids will build up and damage several body organs—especially the brain.

Hospitals now test all newborns for PKU. Those with the recessive trait are given special diets that prevent the buildup of harmful acids. These children will not get sick from PKU if they continue eating this diet. They still carry the recessive gene, however, and pass it on to their offspring.

3. What is the cause of PKU?

   A. an enzyme that makes a harmful acid
   B. a recessive gene that prevents the body from making an important enzyme
   C. a special enzyme
   D. a damaged brain

4. Why do hospitals now test all newborns for PKU?

   A. PKU is highly contagious.
   B. PKU children need a special diet.
   C. PKU children will be mentally disabled.
   D. The harmful gene can be fixed.

Genotypes and Phenotypes

KEY CONCEPT: Traits, or characteristics, are transmitted from one generation to the next. This transmission is called heredity. The young, also called offspring, resemble their parents. However, there are also differences, or variations, between them. The traits we observe in an organism represent its phenotype. The genetic information underlying the phenotype is called the genotype.

Have you ever looked at photographs of celebrities and their children and noticed a strong resemblance? Or perhaps you resemble someone in your family. An organism's traits, or characteristics, are passed from parents to their offspring on chromosomes. Chromosomes are threads of proteins and a substance called DNA, or deoxyribonucleic acid, which contains small sections called genes. Each gene lives in a specific spot on a chromosome. This spot is called a gene's locus. Different genes live at different loci. You can think of genes as computer code for specific traits.

Traits can be dominant or recessive. Dominant traits are expressed. They are the traits that you observe. Recessive traits may or may not be expressed, depending on other factors.

The Science of Heredity

When your smart phone doesn't work as you expect it to, you try to figure out the problem. Sometimes the solution is as simple as turning on a power button. However, sometimes the problem is more complex. The problem may be internal, making it impossible to solve without further investigation into the phone's computer code.

Heredity is similar to a smart phone in that it has visible features and invisible programming. Heredity is the transmission of traits from parents to their young, or **offspring**. Transmission is the process of spreading or passing along from one to another. The traits you observe in the offspring represent its **phenotype**. The offspring's **genotype**, however, is its set of genes, which reside on the organism's chromosomes.

Two organisms may share the same phenotype, that is, they may look or act in similar ways. However, they can share the same genotype only if they share the same genes.

Geneticists are scientists who study heredity and genetic codes. To understand an organism's genotype and phenotype, it is necessary to understand alleles.

Heredity and the Allele

A gene can have different forms, or **alleles**, for a specific trait. You inherit one allele for a specific trait from each parent. There are usually two kinds of alleles, with one being stronger than the other. These alleles are called dominant and recessive.

More than one pair of alleles controls many of the traits you see in yourself and in other organisms. However, a single pair of alleles does determine some traits, such as the length and color of a cat's hair.

There are special terms used to describe the pairs of alleles that appear on chromosomes. To understand those terms, it is helpful to know the meanings of several word parts.

The prefix *homo* means "the same." The prefix *hetero* means "different." The base *zyg* means "union," as in *zygote*, which is the fertilized egg that forms after the union of an egg and sperm cell.

When two alleles are identical, that is, when they are both dominant or they are both recessive, they form a **homozygous** pair. When two alleles are different, meaning one is dominant and the other is recessive, they form a **heterozygous** pair.

When a dominant allele exists within an organism's genetic code, it is always expressed. A recessive trait is expressed only when both alleles are recessive. The allele for short hair in a cat is dominant. Conversely, the allele for long hair is recessive. A cat with homozygous dominant alleles will have short hair. A cat with heterozygous dominant alleles will also have short hair. Only a cat with homozygous recessive alleles has long hair.

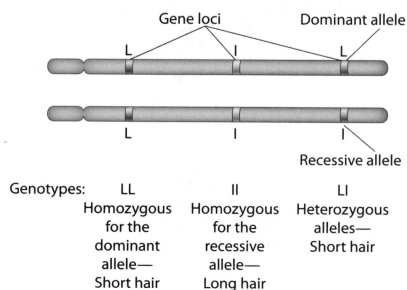

Many of the learning materials used to teach genetics rely on several examples of dominant and recessive alleles. Over time, scientists have come to different conclusions based on evidence in hand.

For example, a cleft chin takes a variety of forms. It may be a hollow dimple or a vertical or Y-shaped depression. Some scientists concluded that two alleles controlled the presence or absence of a cleft chin. In 1941 scientists concluded the trait was recessive. Other scientists later claimed it was dominant but affected by a person's environment. Further studies didn't support the conclusion that the trait was either dominant or recessive. A strong genetic influence exists, but the appearance of the trait among offspring doesn't conform to mathematical models of dominant and recessive traits.

Scientists make reasoned judgments based on evidence. But the evidence may vary, resulting in different judgments. Explain why it is important to **distinguish**, or tell apart, the differences among multiple judgments before accepting a scientific conclusion.

A **Punnett square** is a tool you can use to determine possible combinations of alleles among offspring. Recall Gregor Mendel's work with pea plants. He learned through experimentation that when purple-flowered plants were crossed with white-flowered plants, the offspring expressed the same phenotype. This first generation of offspring all had purple flowers. Then Mendel studied the results of self-pollination among the offspring. The results in the second generation included plants with both purple and white flowers in a ratio of 3:1.

Among pea plants, pairs of genes control the expression of several traits. Flower petal color is one trait. Others include the color of the pea pod and stem length. Another trait is the appearance of the pea pod, which may be inflated or pinched.

	P	p
P	PP	Pp
p	Pp	pp

Ratio of purple flowers to white flowers is 3:1

A single pair of genes does not determine all phenotypes. Multiple genes control some traits, called polygenic traits. Among humans, for example, eye color is a polygenic trait influenced by variations and interactions among several genes.

Some pairs of genes do control some human phenotypes, however. One pair determines the nature of human earwax. The dominant allele results in wet, sticky, yellowish-brown earwax. The recessive allele results in dry, crumbly, gray-to-tan earwax.

You can use Punnett squares to show the results of different parental crosses. In a **monohybrid** cross, each member of the parental generation, or **P generation**, is different for a particular trait. Suppose that in the case of earwax, for example, one parent is WW for wet earwax and the other is ww for dry earwax.

The offspring of the P generation are called the filial generation, or **F generation**. If you use a Punnett square to model the possible allele combinations, you see that all of the first filial generation, or F1, are Ww, or heterozygous for wet, sticky, yellowish-brown earwax.

	W	W
w	Ww	Ww
w	Ww	Ww

All offspring are heterozygous for wet, sticky, yellowish-brown earwax.

In a **dihybrid** cross, the parents in the P generation are different for two traits. So consider another human trait, freckles. Freckles, or dark spots on the skin, are caused by the uneven distribution of skin's natural pigment, or **melanin**. Although the quantity and darkness of freckles is influenced by exposure to the sun, there is also a genetic factor at work. People with freckles have at least one dominant allele for a protein that controls melanin production.

When you build a Punnett square to model the possible combinations among the F1 generation, you begin by writing all of the possible allele pairs for each parent. In this example, you can use W and w to represent alleles for earwax and F and f to represent alleles for freckles.

	WF	Wf	wF	wf
WF	WWFF	WWFf	WwFF	WwFf
Wf	WWFf	WWff	WwFf	Wwff
wF	WwFF	WwFf	wwFF	wwFf
wf	WwFf	Wwff	wwFf	wwff

wet earwax, freckled: 9; wet earwax, not freckled: 3; dry earwax, freckled: 3; dry earwax, not freckled: 1. Ratio: 9:3:3:1

When you list the possible allele combinations among the offspring, you see that there is a greater probability that some combinations will occur before others. The production of sex cells in each parent yields four possible combinations of alleles: WF, Wf, wF, and wf. When the parents are crossed, the allele combinations form the ratio 9:3:3:1.

THINK ABOUT **SCIENCE**

Directions: Describe the possible phenotypes of a dihybrid cross between parents that are heterozygous for earwax and freckles.

_____

_____

_____

X-Linked Inheritance

The sexes of each species differ in their sex chromosomes. In humans, for example, females have two X chromosomes. Males have one X and one Y chromosome. The Y chromosome is smaller than the X chromosome and carries less genetic information.

Because females inherit an X chromosome from each parent and males inherit one X chromosome from the mother, genes that appear on an inherited X chromosome are called sex-linked or X-linked. Scientists have found hundreds of genes on the X chromosome. Almost all of the genes carried on this X have no corresponding genes on the Y. Consequently, a male expresses whatever allele he inherits on the X chromosome, even if it is a recessive allele.

Among X-linked traits are genes for red-green colorblindness, male pattern balding, and hemophilia, a blood disorder. Red-green colorblindness is a recessive trait. A female may carry the recessive trait on one of her X chromosomes, but because the trait is recessive, it does not express itself, and the woman has normal vision. However, as a carrier of the trait, the woman can pass the trait to her offspring.

Female carrier

		X^C	X^c
Male with normal vision	X^C	$X^C X^C$	$X^C X^c$
	Y	$X^C Y$	$X^c Y$

Probability is described mathematically as the number of desired outcomes out of all possible outcomes. There are four possible outcomes in the offspring of a cross between a female carrier for red-green colorblindness and a male with normal vision. The probability of the couple having a red-green colorblind son is 1 out of 4, which can also be expressed as $\frac{1}{4}$, or 25 percent. The probability of having a female carrier is also 1 out of 4, or $\frac{1}{4}$, or 25 percent. Both of the remaining offspring, or 2 out of 4 ($\frac{1}{2}$, or 50 percent), will have normal vision.

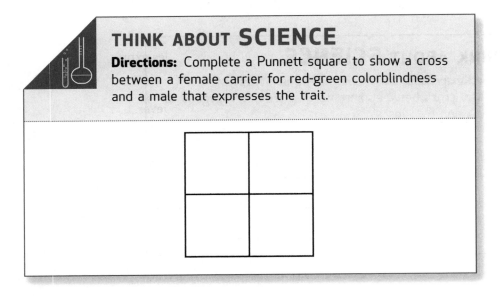

THINK ABOUT SCIENCE

Directions: Complete a Punnett square to show a cross between a female carrier for red-green colorblindness and a male that expresses the trait.

Exceptions to Dominance

There are two exceptions to the rules of dominance among alleles. One is incomplete dominance and the other is codominance.

Incomplete Dominance In examples of incomplete dominance, one allele in a pair is unable to express its phenotype fully. The result is a combined phenotype in which the dominant allele isn't fully dominant. Take, for example, the color of flowers called snapdragons. Use the Punnett squares below to examine the results of two crosses. The first is a monohybrid cross of homozygous red and homozygous white snapdragons. The second is a cross between two members of the F_1 generation.

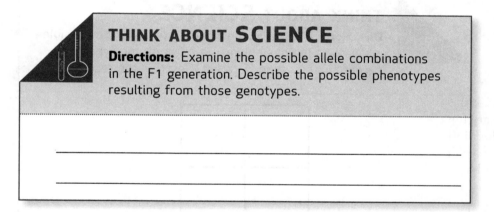

THINK ABOUT SCIENCE

Directions: Examine the possible allele combinations in the F1 generation. Describe the possible phenotypes resulting from those genotypes.

_____

_____

Codominance The prefix *co-* means "together." Two codominant alleles express their corresponding phenotypes as "together."

A red horse is homozygous RR. A white horse is homozygous WW. However, a cross between a red and a white horse results in heterozygous offspring that have hair of both colors. These offspring have the genotype RW, and both alleles express themselves.

Another example of codominance occurs among humans. Codominant alleles determine human blood type. The three alleles for human blood type and the symbols used to represent them are listed in the following table. The allele for Type O blood is recessive and Type A and Type B are dominant. However, when Type A and Type B alleles appear in offspring, they share dominance, resulting in Type AB blood.

Blood Types	Symbols
I^A	A
I^B	B
i	O

Examine the results of a cross between two parents with the same blood type AB.

There is a 25 percent chance that one of their offspring will be either Type A or Type B. There is a 50 percent chance that the offspring will be Type AB.

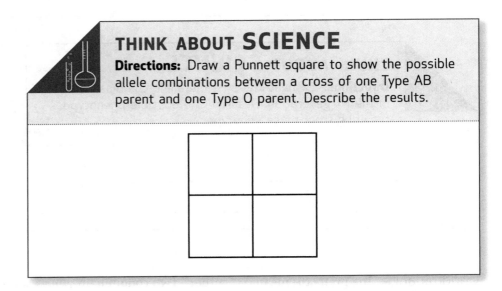

	I^A	I^B
I^A	$I^A I^A$	$I^A I^B$
I^B	$I^A I^B$	$I^B I^B$

THINK ABOUT SCIENCE

Directions: Draw a Punnett square to show the possible allele combinations between a cross of one Type AB parent and one Type O parent. Describe the results.

Extract DNA

There are a number of procedures you can follow to extract DNA, the material that carries the genetic code, from items you may find in your kitchen. Take, for example, strawberries.

You need only a few materials to extract, or remove the DNA from strawberries. Gather three strawberries, dishwashing soap, table salt, rubbing alcohol, ice, a bowl, two transparent plastic cups, a sealable plastic bag, a coffee filter, a teaspoon, and a plastic coffee stirrer.

Wear protective clothing while you work, and clean up accidental spills to avoid slipping.

Read the procedure on page 191.

WRITE TO LEARN

How could you explain genotypes and phenotypes to someone who has not studied genetics? What examples could you use to clarify the relationship between the two?

Write a paragraph defining a genotype and a phenotype. Include examples to support your definitions.

Extract Strawberry DNA

1. Put a bottle of rubbing alcohol in a bowl of ice.

2. Remove the green leaves from three strawberries.

3. Put the strawberries into a plastic bag, squeeze out the air, and seal it. Use your hands to smash the contents gently.

4. Prepare the DNA extraction liquid in a plastic cup: mix 2 teaspoons of liquid dish soap, 1 teaspoon of salt, and ½ cup of tap water.

5. Remove 2 teaspoons of the DNA extraction liquid from the cup and put it into the bag with the smashed strawberries.

6. Reseal the bag and gently squeeze the mixture again for one minute. Avoid making soap bubbles as you squeeze.

7. Place a coffee filter inside the second plastic cup.

8. Pour the smashed strawberries from the plastic bag into the filter. The strawberry liquid will empty into the cup, and the strawberry pulp will remain in the filter.

9. After the strawberry liquid has finished draining, remove the coffee filter with strawberry pulp and throw it away.

10. Measure the amount of strawberry liquid in the cup. Then return the liquid to the cup.

11. Measure an equal amount of cold rubbing alcohol and pour it into the cup with the strawberry liquid. Do not stir or mix.

12. Watch for a white cloudy substance to form above the strawberry liquid. This is strawberry DNA!

13. Use a plastic coffee stirrer to twirl or spool the DNA. It will be a goopy glob of material that under the right conditions can remain stable for many years.

Directions: Write the word that matches the description.

allele   genotype   heredity   offspring   phenotype   **Punnett square**

1. _____ a diagram used to show possible allele combinations

2. _____ one form of a gene

3. _____ the passing on of traits from parents to their young

4. _____ an organism's observable traits or characteristics

5. _____ what is born from or grows from an organism, such as the young of humans, plants, or animals

6. _____ the genetic code for a given trait

Skill Review

Directions: Read the text. Then complete the activity.

1. A carnation's color is determined by the alleles for red (RR) and white (WW) petals. A carnation that is heterozygous RW has pink petals. Explain why a heterozygous carnation is pink.

   _____

   _____

2. Draw a Punnett square to determine the possible outcomes of a cross between two heterozygous carnations. Describe the results.

   _____

   _____

3. A pink snapdragon (RW) is crossed with a white snapdragon (WW). Draw a Punnett square to determine the possible outcomes of the cross. Describe the results.

   _____

   _____

4. A person has Type O blood, but the parents have different blood types. List the genotypes of the parents.

   _____

   _____

Skill Practice

Directions: Read the text. Then complete the activity.

1. A homozygous red four-o'clock flower is crossed with a homozygous white four-o'clock flower. Neither allele is completely dominant over the other. Draw a Punnett square to show the cross. List the possible genotypes and describe the phenotypes of the offspring.

   _____

   _____

2. A simple cut can cause bleeding, but the bleeding eventually stops. That's because your blood contains special proteins called clotting factors that form a plug over the wound. Your blood carries several clotting factors, but sometimes one or more of these factors is missing, and abnormal bleeding can occur. This disorder, called hemophilia, is recessive and most often affects males. Where is the gene for this condition carried? Justify your answer.

   _____

   _____

3. Imagine you are a doctor who treats a man with hemophilia. This man would like to be a parent. What advice would you give him?

   _____

   _____

4. Some cats have hair that has more than one color along each strand. This kind of color-banded hair is called agouti, and it is determined by a dominant agouti allele (A). Cats without the agouti dominance are a single color.

   Recall that the allele for short hair (L) in cats is dominant to the allele for long hair (l).

   Choose a method for determining the possible genotypes resulting from a dihybrid cross between cats with the genotypes AaLl and aall.

5. List the possible genotypes in the F_1 generation and describe the chance of each type occurring.

   _____

   _____

6. Describe the possible phenotypes of the offspring.

   _____

   _____

Review

Directions: Choose the best answer to each question.

1. Assume the characteristic for tallness is dominant. The genotype for a pea plant that is homozygous dominant for tallness would be

   A. Tt
   B. TT
   C. tt
   D. TT x tt

2. How many chromosomes are in the nucleus of most human cells?

   A. 12
   B. 23
   C. 24
   D. 46

3. What is an allele?

   A. a recessive trait
   B. a dominant trait
   C. a form of a gene
   D. a trait that is always expressed

4. Which factor distinguishes parrots from pineapples?

   A. the number of chromosomes in their cells
   B. the specific nucleotides that form their DNA
   C. the method of transcription
   D. the method of translation

5. What do cells use to control their chemical processes?

   A. proteins
   B. DNA
   C. RNA
   D. alleles

Directions: Read the text and then write a response.

6. Explain the difference between a genotype and a phenotype.

   _____

   _____

7. Explain the difference between a homozygous and heterozygous pair of alleles.

   _____

   _____

8. A plant that is monohybrid for a dominant trait R is crossed with a plant that is monohybrid for the recessive allele of the same gene. Draw a Punnett square to show the possible allele combinations in the first filial generation.

Review

9. A plant that is dihybrid for a pair of traits WwPp is crossed with a plant that is monohybrid for both traits. Draw a Punnett square to show the possible allele combinations in the first filial generation.

10. Explain X-linked inheritance to someone who is hearing the term for the first time.

_____

_____

_____

_____

_____

_____

_____

_____

_____

_____

_____

_____

Check Your Understanding

On the following chart, circle the number of any item you answered incorrectly. Next to each group of item numbers, you will see the pages you can review to learn how to answer the items correctly. Pay particular attention to reviewing those lessons in which you missed half or more of the questions.

Chapter 5 Review

Lesson	Item Number	Review Pages
Genetics	1, 2, 3, 4, 5	176–183
Genotypes and Phenotypes	6, 7, 8, 9, 10	184–193

CHAPTER 5: HEREDITY

Question

How do phenotypes provide clues to underlying genotypes?

Background Concepts

An organism's phenotype is its set of observable characteristics. That is, it is the expression of its genotype, or the genetic information it inherited from its parents. Offspring receive half of their genetic information from their mother and the other half from their father.

An allele is a form a gene can take. For example, a single pair of genes determines whether someone has freckles. The dominant allele (F) results in freckles, while the recessive allele (f) results in no freckles. If a person inherits the genotype Ff, the dominant allele expresses itself, even though it is paired with a recessive allele. So, the offspring's phenotype will express skin with freckles.

Investigation

1. Create an imaginary superhero. List two or three traits that are observable in your superhero.

2. Assume that a single pair of alleles determines each observable trait.

3. Also assume that your superhero inherited one dominant allele and one recessive allele for each trait.

4. Now work backward two generations to your superhero's parents and grandparents to determine the possible sources of the alleles that your superhero could have inherited for each trait.

5. Use a diagram like the one below to show the possible traits in the members of each generation.

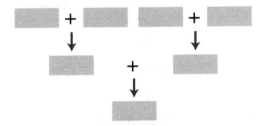

Application of Science Practices

Interpretation

Describe your superhero's phenotype.

_____

_____

Use letters to write the genotype for each expressed trait. For example, if your superhero can tolerate extreme heat, you might use the genotype Hh, with H representing the dominant trait of tolerance and h representing the recessive trait of intolerance.

_____

_____

Answer

How do phenotypes provide clues to underlying genotypes?

Assume that your superhero has a child with another superhero with the identical genotype. What possible genotypes might the child inherit?

_____

_____

Evidence

Draw a Punnett square to show how your superhero's child's inherited phenotype could result from the alleles that the child could have inherited. (See page 186 for an example of a Punnett square.) Explain the genotype, including which allele is dominant and which is recessive.

Evolution

Do you enjoy visiting zoos or watching videos about animals in faraway places? The enormous variety of living creatures on Earth has fascinated human beings for thousands of years. Each year scientists exploring new places continue to find new kinds, or species, of living things.

Over the centuries, scientists have wondered how so many kinds of living things developed. This chapter describes how living things share common ancestors, and discusses scientists' explanations for how new species form.

In this chapter you will learn about:

Lesson 6.1: Biological Evolution
In the 1830s, English naturalist Charles Darwin studied living things found in different parts of the world. Darwin concluded that those organisms that are best suited for their environment are most likely to reproduce. In this lesson, you will learn how Darwin's observations led him to develop his theory of evolution.

Lesson 6.2: Common Ancestry and Cladograms
Scientists group, or classify, living things based on how closely they are related. Learn more about how scientists study evolutionary relationships among living things.

Lesson 6.3: Speciation
New species can appear over time. Scientists studying the fossil record have learned about the past history of living species. In this lesson, you will learn how a species can accumulate enough differences to become a new species over time.

Goal Setting

There are many reasons to study evolution. You might want to

- learn more about the variety of living things found on Earth.

- determine how species form and change.

- understand the role of environment in the evolution of a species.

Think about your reasons for wanting to study evolution. What do you hope to accomplish by reading this chapter? Write your goals down. As you read through this chapter, focus on accomplishing the goals you set for yourself.

_____

_____

_____

_____

_____

_____

_____

_____

_____

_____

_____

_____

Biological Evolution

KEY CONCEPT: Fossils indicate that organisms have changed over time. The theory of evolution is scientists' best explanation for how those changes occur.

Earth is constantly changing. Earth's living things are constantly changing, too. The theory of evolution explains how Earth's living things change in response to a changing environment.

Sometimes Earth's community of living things changes relatively quickly. For example, soon after an asteroid struck Earth sixty-five million years ago, all the dinosaurs died. Yet many changes to living things occur more slowly and gradually. Both fast and slow changes led to the community of living things alive today.

The History of Life

Many scientists, along with men and women in other disciplines, have asked questions about the origin of life on Earth. English naturalist Charles Darwin (1809–1882) was neither the first nor the last scientist to propose answers to such questions. Yet his answers were powerful, and remain so, because he supported them with logic and evidence. **Evidence** is something that provides proof.

In 1859 Darwin's *The Origin of Species* was published. In this book he presented his theory of evolution. The term ***evolution***, simply stated, means "change over time." Darwin's concept of evolution was that older species of living things gave rise to newer species. He also proposed an explanation for how this occurred.

Darwin was not the only naturalist to suggest that life evolved. He was the first, however, to organize his ideas into a theory. A scientific **theory** is a logical explanation of events and evidence from the natural world. A theory also can be used to predict future events.

Since Darwin's time, scientists have discovered an overwhelming amount of evidence in support of the theory of evolution. Although certain parts of Darwin's original theory have been modified, or changed, the core of Darwin's work remains accepted today.

What Darwin Knew

On the remote Galápagos Islands of South America, Darwin observed many different kinds of finches. He noted that the beaks of the finches, which are small seed-eating birds, varied in size and shape. The different types of beaks were suited to the different kinds of food available on the islands. Darwin's observations of the finches eventually gave rise to later research suggesting that the similarities among the finches indicated a common ancestor in the distant past.

CITE TEXTUAL EVIDENCE

In the science texts you read, each paragraph or section has a main idea. Those ideas are supported by details. In science, those details are often data, or bits of information in word or numerical form. In addition to providing supporting information, data provide evidence, or support, for claims writers make in their work.

When you read, it is important to identify an author's claims. It is also important to determine how well the evidence supports those claims. Too little data or unrelated data should make you question the reliability of a writer's claims.

As you seek evidence to support either a writer's claims or your own, ask yourself: *What claim or claims is the writer making? What evidence supports those claims? Is there sufficient evidence, or is it necessary to find more?*

Read the following paragraph. Identify the writer's claim and then underline textual evidence that supports the claim.

> Adding broccoli and spinach to your diet could boost memory and overall brain power. Both of these vegetables contain magnesium, a chemical that studies show improved learning and memory in young and old rats. Increased magnesium levels cause physical changes in the part of the brain associated with memory. In laboratory studies, higher levels of magnesium were linked to an increased number of synapses among neurons, or brain cells, and to greater speed of communication across the brain.

THINK ABOUT **SCIENCE**

Directions: Answer the questions below.

1. What is evolution?

   _____

2. What is the theory of evolution?

   _____

Core Skill
Identify Hypotheses

A **hypothesis** is a proposed explanation based on limited evidence. That evidence is based on existing knowledge and observations. A hypothesis is an idea that will be either supported or rejected after careful experimentation and investigation. Although a hypothesis incorporates facts, it is not a statement of fact itself, and it is testable by any number of scientists.

After extensive and repeated testing, evidence may support a hypothesis. Eventually, a **theory** summarizes the hypothesis or group of hypotheses.

As you read about Darwin's theory on this page, identify the hypotheses that form his theory.

The similarity of embryos also pointed Darwin toward the idea of evolution. The embryos of most vertebrates look very similar. At one point, the human embryo has structures in the neck area that resemble the gill slits and gill arches of fish at a similar point in development. Other organisms have equally puzzling structures. Snakes, for example, have tiny, useless leg bones.

Darwin's Theory

Darwin's theory consisted of three major ideas. First, he proposed that all living things—some two million different species—developed from just a few primitive, simple organisms. Over a period of millions of years, gradual changes occurred in these first species.

Sometimes small groups of organisms became isolated from others. In time, each succeeding generation was less and less like the original group. Eventually, the offspring differed so markedly that they could no longer breed with the original population. They had become a unique species. Through this concept of evolution, Darwin explained how some species of amphibians developed into reptiles, and how some species of reptiles developed into birds and mammals.

Second, Darwin had read studies that reported that populations multiply faster than their available food supply. As a result, organisms compete for food. Darwin concluded that nature acts as a selective force in which the least fit, or the least able to live to reproduce, die and the most fit survive. Darwin called this idea "survival of the fittest," or **natural selection**.

Finally, Darwin proposed that in a changing environment, some species adapt to different conditions. Individuals of species that have adapted successfully live long enough to reproduce. The offspring of the survivors are better suited to their changing environment, and they, too, will survive to reproduce. In this way, the evolution of a species is driven by the benefits of adapting, or changing to adjust, to its environment.

What Darwin Did Not Know

It seems amazing that Darwin was able to develop his theory simply from his observations of many species and their structural similarities and differences. In his era, scientists did not know how traits were passed from parents to offspring. They had only begun to unearth the tremendous variety of fossils that now helps scientists trace the course of evolution. Darwin's theory was an incredible burst of understanding.

Since Darwin's time, scientists have found additional evidence that all life is related. For example, all organisms use the same classes of molecules, including sugars, proteins, and DNA. DNA is the molecule that carries genes, the structures that code for all the traits of an organism. Not only do all species use DNA, they all decode it in the same way.

Darwin, Modified

Darwin suggested that evolution happened like a slow, smooth march across time. But today, scientists argue that this march was more uneven. There were long periods without change interrupted by rapid change. This idea, known as **punctuated equilibrium**, suggests that evolution is much like a reading passage. Instead of one long stream of words, history provides sentences that stop and start—and stop and start again. The punctuation can be provided by any event, such as the impact of a comet that rapidly changes the environment.

Darwin also pictured evolution as if all the different species were on one ladder, starting with the simplest organisms on the bottom rung and climbing to humans at the top. Today, scientists describe evolution as more like a bush. Some branches have broken off the bush. Other branches have hundreds of twigs. Species at the bottom of the bush may be equally complex as those further up.

How Evolution Works

An **adaptation** is a change in a species that improves its chances of survival. For evolution to progress, populations of living things develop adaptations to their environments. One of the most famous examples of adaptation involves the wing color of moths

About 200 years ago, most of one species of moth living near Manchester, England, had light-colored wings. The light wing color blended in well with the natural color of the local trees. Dark moths were fewer in number, however. One factor causing the lower numbers of dark moths was the fact that moths with dark wings were easier for birds to see and eat.

Then coal-burning factories came to Manchester. Soon the tree trunks were dark with soot. Now the dark-colored moths were hard to see against the soot. The numbers of moths with light wings began to decrease. Scientists still need to learn more about how bird behavior and moth migration patterns influence the changing size of the moth population.

THINK ABOUT SCIENCE

Directions: Match the words in the column on the left with their description on the right.

1. _____ species
2. _____ natural selection
3. _____ punctuated equilibrium
4. _____ adaptation

A. times of stability interrupted by times with rapid changes

B. survival of the fittest

C. smallest group of classification system

D. change in a species that improves its chances of survival

Mutations

One way that adaptations develop is because of mistakes or errors in the genes that are passed from parent to offspring. This kind of "mistake" is called a **mutation**. Many mutations cause harm. Others have little impact. Occasionally, however, a mutation benefits an organism. Because of mutations, offspring can develop at least slightly different traits from their parents. According to the theory of evolution, these differences may lead to a new species over many generations.

Darwin thought that mutations occurred both rarely and completely at random. But evidence today suggests this may not always be so. In one experiment, for example, a population of bacteria was reduced to very low numbers. It began developing mutations at a very fast rate, including several mutations that helped it survive.

The effects of mutations are sometimes easy to observe. Eighty years ago, for example, doctors began to prescribe the drug penicillin. Penicillin helped the body fight bacterial infections, and it was very effective. Today, however, mutations have helped bacteria become resistant to penicillin. The penicillin of eighty years ago would be ineffective today.

Unfortunately, bacteria gradually develop resistance to every antibiotic—or bacteria-killing drug—that they are exposed to. Doctors are concerned that their ability to invent new antibiotics will not keep pace with the need for them.

THINK ABOUT SCIENCE

Directions: Answer the questions below.

1. What is a mutation?

   _____

   _____

2. Are all mutations harmful? Explain.

   _____

   _____

3. Explain why doctors must continue to develop new antibiotics.

   _____

   _____

Fossils

A **fossil** is the preserved remains of an organism. A fossil can form from a small scrap of a plant, a bone, a shell, or an entire animal. Fossils have formed in different ways. One way is when a dead organism is covered quickly with sand or silt. The covering prevents bacteria and other organisms from beginning the process of decay. If the sand or silt is squeezed into rock, the organism's body may be preserved inside. Scientists have found many thousands of fossils that were formed in this way.

Other fossils were formed when insects or other small animals were trapped in tree sap. The sap hardened with the dead animal inside, preserving it nearly perfectly. Still other types of fossils were trapped in ice. Whole bodies of wooly mammoths were preserved this way, including the hair and blood. One mammoth even had food in its mouth.

One fossil proved to be more than just another ancient animal. The coelacanth (SEE-la-canth) is a lizardlike fish. The fossil of a coelacanth showed that it had fins that moved like legs and that allowed it to crawl on the beach. The fish also had a bony skeleton and joints in its skull. The fish could even twist its head.

About fifty years ago, scientists found a live coelacanth off the coast of Africa. Fourteen years later, another was found.

The Ages of Fossils

Scientists use several techniques to determine the age of fossils. One method is to compare the ages of the layers of the rocks where a fossil is found. Generally, layers of rock deep in Earth are older than layers closer to the surface. A fossil found in a rock layer close to the surface is usually younger than a fossil found in a deeper layer.

Scientists can also determine a fossil's age by measuring the amount of certain radioactive isotopes, which are special forms of an element. One such isotope is carbon-14. Carbon-14 is a rare but ever-present form of carbon, an element that all living things take in and use. When a plant or animal dies, it stops taking in carbon. As a result, the amount of carbon-14 decreases at a certain rate, known as a half-life. By measuring the amount of carbon-14 or other radioactive isotopes, the approximate age of the fossil can be calculated.

Eras of Life on Earth

By examining fossils of different ages, scientists have pieced together the history of life on Earth. The table below shows the major eras of Earth's history.

THE FOSSIL RECORD

Era	Period	Epoch	Outstanding events	Millions of years ago
Cenozoic	Quaternary	Recent	Major human civilizations	Less than 1
		Pleistocene	Homo sapiens	2?
	Tertiary	Pliocene	Later hominids	6
		Miocene	Increase in mammal populations	22
		Oligocene	Early hominids; grasses, grazing mammals	36
		Eocene	Primitive horses	58
		Paleocene	Mammals, dinosaurs become extinct	63
Mesozoic	Cretaceous		Flowering plants	145
	Jurassic		Birds; dinosaurs climax	210
	Triassic		Early mammals; conifers, cycads, dinosaurs	255
Paleozoic	Permian		Mammal-like reptiles; trilobites extinct	280
	Pennsylvanian		Deserts, coal forests	320
	Mississippian		Club mosses, horsetails	360
	Devonian		Insects, amphibians	415
	Silurian		Land plants and animals	465
	Ordovician		Early fishes	520
	Cambrian		Early trilobites; marine animals	580
Pre-cambrian			Early marine animals	1 billion
			Green algae	2 billion
			Bacteria, blue-green algae	3 billion

Notice in the chart that just as new species have evolved over the years, others have become extinct, or disappeared completely. For example, dinosaurs evolved about 225 million years ago and became extinct all at once about 65 million years ago. An event that kills a huge number of species all at once is called a mass extinction. Earth has suffered many mass extinctions in its long history.

One of the largest mass extinctions occurred at the end of the Paleozoic era, about 280 million years ago. Before that time, small sea creatures called trilobites dominated Earth for about 240 million years. Then they all disappeared at once. The cause of their extinction is a mystery.

Vocabulary Review

Directions: Complete the sentences using one of the following words:

adaptation   **evidence**   **evolution**   **fossil**   **mutation**

1. The preserved remains of a living organism is a(n) _____.

2. A(n) _____ is a change in a species that improves its chances of survival.

3. A scientific theory is based on logical reasoning and _____.

4. Change over time is called _____.

5. A(n) _____ is a change in the genetic information that is passed from parent to offspring.

Directions: Read the following passage and answer the questions that follow.

> Evidence shows that living things have changed over time. Scientists have collected a huge number and variety of fossils, which together show changes that different species have undergone. Darwin's theory of evolution explains how these changes occurred. As a young man, Darwin traveled the world. He observed and recorded the differences in the organisms he encountered. These observations helped provide evidence for his theory. Today, scientists apply knowledge not available to Darwin to provide more evidence that evolution has occurred.

1. What is the main idea of this passage?

    A. Charles Darwin proposed the theory of evolution.
    B. Gene mutation proves evolution.
    C. Living things have changed over time.
    D. Fossil evidence proves the theory of evolution.

2. Write two details that support the main idea.

    _____

    _____

Directions: Read the statements in the paragraphs below. In the first paragraph, select the statement that best describes a benefit of *adaptation*. In the second paragraph, select the statement that best describes a benefit of *mutation*.

3.   (1) Charles Darwin observed many different species of living things. (2) He saw variation among the individuals of each species. (3) He noted that individuals competed with one another for limited resources, such as food. (4) By chance, variations in some individuals reduce the likelihood that they will live long enough to reproduce. (5) Darwin concluded that a species is successful when it develops variations that improve its chances of survival.

4.   (1) A mutation is a mistake or an error in a gene. (2) Some kinds of mutations can be passed from parent to offspring. (3) Many mutations cause harm because they produce an abnormal protein that cannot function. (4) Other mutations have little impact. (5) Mutations are a source of slightly different traits among some individuals of a particular species.

Skill Practice

Directions: Choose the best answer to each question.

1. Which statement is a major idea of Darwin's theory of evolution?

   A. All organisms have DNA as their genetic material.
   B. Species have remained the same over time.
   C. Complex forms of life evolved from simpler forms.
   D. Rapid changes in organisms occur after periods of stability.

2. Which fact did Darwin learn from studies by another individual?

   A. On the Galápagos Islands, only finches with large beaks could fit the seeds in their mouths.
   B. Populations multiply faster than the food supply.
   C. Populations multiply at the same rate as food supplies.
   D. Different species can interbreed.

3. According to the idea of punctuated equilibrium, at what rate does evolution occur?

   A. only in slow, smooth stages
   B. in one rapid burst after another
   C. in rapid bursts after periods of long stability
   D. at a constant rate throughout time

4. What is one modern modification to Darwin's theory of evolution?

   A. Species develop adaptations that help them survive in their environment.
   B. Parents pass on traits to their offspring.
   C. All organisms descended from a common ancestor.
   D. Evolution develops in a pattern like a bush, with species developing like branches and twigs.

Common Ancestry and Cladograms

KEY CONCEPT: Cladistics is an analytical method scientists use to hypothesize about the relationships among existing organisms. The foundation of the method is an agreement that members within any clade, or group, share a common evolutionary past.

In previous lessons, you learned that offspring inherit genetic information from their parents. Genetic changes occur within a population, and those changes are inherited by the offspring of subsequent generations. Over many generations, these changes can lead to new kinds of organisms. Animals that are now extinct may be ancestors of common species. For example, scientists generally believe that birds have common ancestors in dinosaurs. The process of change among descendants is called biological evolution.

Lesson Objectives

You will be able to

- Describe the purpose of cladistics
- Interpret a cladogram
- Identify assumptions behind cladistics

Skills

- **Core Skill:** Determine Meaning
- **Core Skill:** Integrate Explanations with Visual Representations

Vocabulary

ancestry
assumptions
cladistics
cladogram
diverge
homologous
phylogeny
systematics
taxonomy

A Family Tree

Perhaps someone in your family has remarked that you share a characteristic with an ancestor—a family member who lived many generations ago. Perhaps while looking at old family photographs, you have found that you share similar physical traits with a grandparent or great-grandparent. You share things you don't see, too. That is one reason a new doctor asks so many questions about your family's health history.

Today, people can use online research tools and services to learn more about their ancestors. They can use this biological line of descent, known as their ancestry, to build diagrams called family trees. These diagrams take a variety of forms, from trees with limbs to connected boxes, but their purpose is the same. They show the connections among past and present family members.

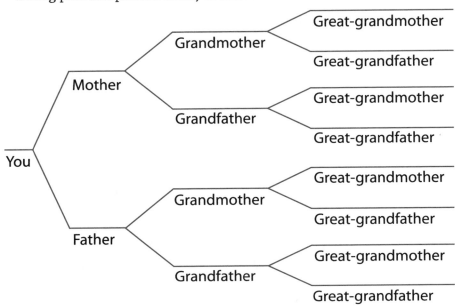

You could research your family history to draw a family tree like this one.

What Is Cladistics?

Scientists examine relationships and construct patterns of events in Earth's living history. These events have led to the study of **biodiversity**, the variety of living organisms, their biological makeup, their ecosystems, and their distribution around the world. With so many living things past and present in so many places, scientists continue to classify, or group together, living things based on shared characteristics. This science of biological classification, called **taxonomy**, makes it possible for scientists to communicate effectively about specific organisms.

Swedish scientist Carolus Linnaeus is often called the "Father of Taxonomy." In 1735 Linnaeus published an eleven-page book presenting a new classification system. Fellow scientists were intrigued by Linnaeus' system, which grouped organisms by shared traits.

In addition to creating a shared-trait taxonomic system for identifying organisms, Linnaeus opened the study of science to systematics, which includes methods for classifying organisms by their shared and evolutionary traits. **Systematics** is the study and classification of organisms according to their biological diversity and evolution.

In the 1950s, German scientist Willi Hennig proposed that systematics should reflect evolutionary history, or **phylogeny**. Hennig called this approach to classification **phylogenetic systematics**. His system focused on **monophyletic** groups, the groups of one ancestor and all of the descendants of that ancestor. This grouping, also called a **clade**, forms the structure for cladistics. **Cladistics** is a systematic method for making and testing predictions about evolutionary relationships among living things.

The Main Ideas Behind Cladistics

The foundation of cladistics is the accepted understanding that members of a group, or clade, share the same evolutionary history. This shared history means that animals within the same group have more in common with each other than they do with organisms outside of their group. It also means that within any group, members share unique characteristics that did not exist among their distant ancestors.

Cladistics is based on several **assumptions**, or beliefs that are thought to be true, but without proof. The first is that life appeared on Earth once, and only once. You can conclude from this assumption that all organisms alive today are related in some way.

Core Skill
Determine Meaning

Recognizing Greek and Latin word parts can help you figure out the meaning of many science vocabulary terms, including terms used in the study of cladistics. Consider the word *systematics*. The term comes from the Latin word *systema*, meaning "an arrangement."

When the Greek prefix *phyl–* appears before a vowel, as it does in the word *phylogenetics*, it means "a family or tribe." The word part *gen* comes from the Latin word *genus*, which means "type or kind." Therefore, you might interpret the word *phylogenetics* to mean "a kind of family or tribe."

Use the meanings of the following word parts to help you understand and remember some of the technical words that appear in this lesson.

-plesio, from the Greek word *plesios*, meaning "near"

-apo, from the Greek word *apo*, meaning "from, after, in descent from"

-morph, from the Greek word *morphe*, meaning "form, shape, outward appearance"

The Greek word *klado* means "branch or twig." In what way might cladistics be associated with branches of a tree?

Scientists analyzing giraffe DNA, or genetic information, have evidence of at least six different lines of descent from a common ancestor. Five of those six lines also contain genetically unique populations. The result is at least eleven genetically distinct populations. So, people who believe all giraffes are the same should reconsider their opinions based on evidence.

People need good information to make sound decisions and revise opinions. Unfortunately, not all information is accurate or complete, and often, people are reluctant to change long-standing opinions. The willingness to seek and confirm valid evidence and the mental flexibility required to use that evidence to change an opinion are critical in the workplace, at home, and in society.

Imagine you own your own company. You make short-term and long-term decisions that affect your company's success. How is evidence more valuable than opinion to your continued success?

The second assumption is related to how new species come into being. New species arise when existing species **diverge,** or split into two groups. The rise of two new species from a parent species is called a "splitting event."

Finally, the third assumption emphasizes the constancy of change. The characteristics of living things continue to change over time. Scientists use these changes to identify different ancestral groups. They call an ancestor's original characteristics **plesiomorphic**, or primitive. The characteristics they change into are **apomorphic**, or divergent.

One example of a plesiomorphic trait, or plesiomorph, is the trait for four legs among mammals. All mammals inherited this trait from a common ancestor long ago. Consequently, such a trait is not helpful in determining closer evolutionary relationships within groups, or clades.

An example of an apomorphic trait, or apomorph, is a giraffe's long neck. Long ago, its ancestor had a short neck. The divergent trait appears in living giraffes but did not appear in its ancestor.

Scientists describe evolutionary relationships in diagrams similar to family trees. These diagrams, called cladograms, show where traits diverge over time. Here is an example of a simple cladogram.

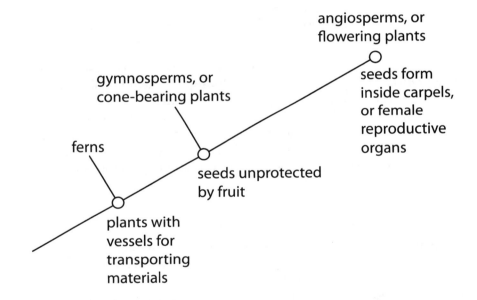

THINK ABOUT SCIENCE

Directions: Look closely at the simple cladogram. Notice the circles, which are called nodes. Each node represents a splitting event that gives rise to species different from the ancestor. What splitting event occurred that made gymnosperms different from their ancestor?

_____

_____

How Do You Make a Cladogram?

The process of building cladograms is difficult and relies on extensive research and scientific evidence. However, to understand the process, you can follow some simple steps.

Begin by choosing a group of organisms that interests you. Let's say you choose a squid, frog, carp, eagle, whale, and cow.

Next, draw a table. List your organisms across the top of your table. Then, think of ways the animals are alike and how they are different. List your ideas in your table. Then use the letter X to identify which animals have which traits. Look at the following example.

	Squid	Frog	Carp	Eagle	Whale	Cow
Has a backbone		X	X	X	X	X
Warm-blooded				X	X	X
Produces eggs	X	X	X	X	X	X
Gives birth to young					X	X

Now, that you have completed a data table, draw circles around the pictures of the organisms that share the same traits.

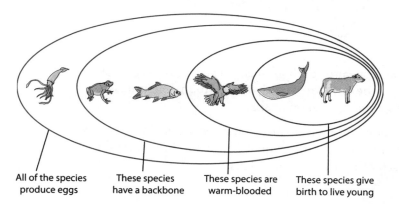

All of the species produce eggs   These species have a backbone   These species are warm-blooded   These species give birth to live young

Finally, arrange the relationships among your chosen animals in a cladogram. The example below is one of a variety of cladogram styles.

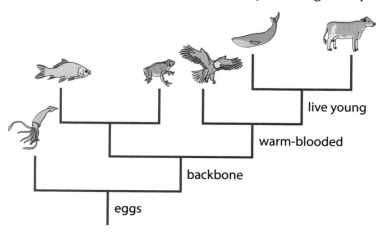

live young

warm-blooded

backbone

eggs

Core Skill
Integrate Explanations with Visual Representations

In print and online materials, you often find visual representations of the text you are reading. Those representations include graphs, tables, maps, diagrams, models, photographs, illustrations, and videos. Their purpose is to help explain and support the text. Consider this example:

Paleontologists, scientists who study the fossil record, have assembled the skeleton of the modern kangaroo's distant ancestor. This ancestor, called a nambaroo, was a marsupial about the size of a dog, walked on four legs, had fangs, and ate rainforest fruits and fungi. No nambaroos exist today, but there are four species of descendant kangaroos, the largest of all marsupials.

Nambaroo

size of a small dog; walked on four legs; had fangs; ate fruit and fungi; had strong forearms that helped it lope, or run; existed 25 million years ago; no living descendants

marsupial; share a distant ancestor

hops on two legs; eats grasses; four living species; can reach 200 pounds

Kangaroo

Use the text and the diagram to identify two features that the nambaroo and kangaroo share.

Ingroups and Outgroups

Imagine that a scientist is studying the character traits of some of the Australian marsupials—kangaroos, wallabies, koalas, and wombats. These organisms share common character traits, such as giving birth to small live young, that complete their development within a mother's pouch. Now add a **taxon**, or group, that is not among the marsupials being classified—the Australian crocodile. The crocodile and marsupials share a **homologous** feature, meaning a feature that is similar in form and structure. They all have five digits at the end of each limb.

A homologous feature suggests a common ancestor. Scientists consider the shared feature a plesiomorph, a primitive trait. The crocodile, however, does not belong with the other organisms that are being classified. It represents an **outgroup**, and the marsupials being classified are the **ingroup**.

Why would scientists use ingroups and outgroups to explain evolutionary history? Because the outgroup helps scientists distinguish between plesiomorphs and apomorphs.

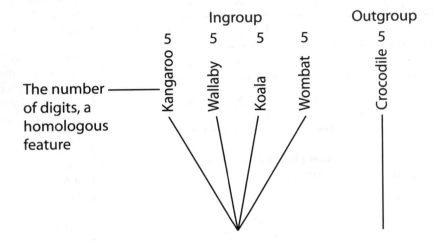

Think about how a homologous feature can appear in both groups. Its appearance suggests that there is a common ancestor older than the ancestor of only the ingroup. The trait is a plesiomorph. It came before later splitting events, or apomorphs.

Although the ingroup and outgroup are related, there is great distance between them in evolutionary history. The organisms with the most recent common ancestor, in this case, the marsupials, are more closely related to each other than to crocodiles.

The Principle of Parsimony

Scientists prefer to seek the simplest explanations to explain branches on evolutionary trees. This preference is called the **principle of parsimony**. Scientists assume that a tree with the fewest branches, or splitting events, is most likely to be true. In other words, the fewer the changes, the more likely it is that a cladogram shows an accurate picture of phylogeny, or evolutionary history.

Consider the two hypotheses below. Notice that each group has an outgroup, a taxon that is not being classified, but shares a trait with the taxa (plural of *taxon*) that are. The question is, which trait is primitive and which is divergent?

In the first hypothesis, trait "b" is at the bottom of the tree. So, it represents a primitive trait that evolved once. Trait "a" is the divergent trait. It evolved twice, resulting in its appearance in the outgroup and taxon C.

In the second hypothesis, trait "a" is at the bottom of the tree. So, it represents the primitive trait that evolved once. Now look for divergent trait "b," and you will see that it also evolved only once.

Next, count the number of evolutionary changes in each hypothesis. The first hypothesis has two, while the second has one. If you apply the principle of parsimony, the second hypothesis is more likely to be true. Trait "a" is the primitive trait, or plesiomorph. Trait "b" is the divergent trait, or apomorph.

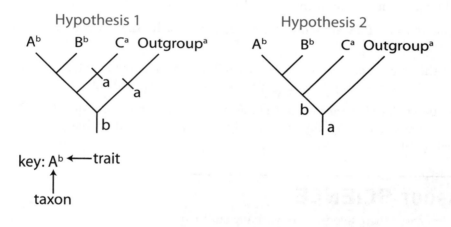

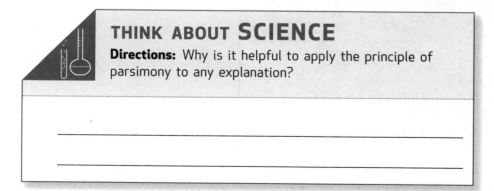

THINK ABOUT **SCIENCE**

Directions: Why is it helpful to apply the principle of parsimony to any explanation?

_____

_____

The Parts of a Cladogram

Scientists enter specific traits as computer data to generate cladograms. At first, they entered only morphological data, such as having one or more cells, being endothermic, having a backbone, and having hair. Today, more scientists are using molecular data, meaning genetic information. Regardless of the input, or kind of data that goes into a computer, the output is the same. Computers generate cladograms with specific features that allow scientists to communicate about the results.

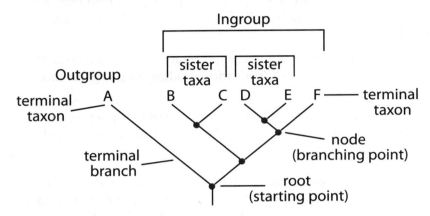

The cladogram indicates that Taxon A is the outgroup. It is used to determine the trait that came first in evolutionary history—the plesiomorph.

Taxons B through F form the ingroup. They are more closely related to each other than they are to the outgroup.

The plesiomorph appears at the tree's root. The branches that follow are "splitting events." They indicate the appearance of new species possessing apomorphs, or divergent traits.

In this cladogram, there are two pairs of sister taxa in the ingroup—taxa B and C and taxa D and E. The organisms within each sister group are more closely related to each other than they are to the other taxa in the ingroup.

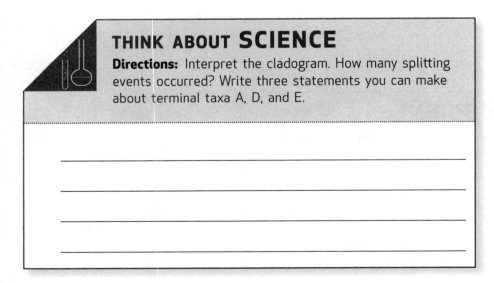

THINK ABOUT SCIENCE

Directions: Interpret the cladogram. How many splitting events occurred? Write three statements you can make about terminal taxa A, D, and E.

Reasons for Cladistics

Cladistics is the most commonly used method for classifying organisms. The method allows scientists of all kinds to discuss an immense variety of organisms, including those that are living and those whose histories are recorded in fossil evidence.

There is another reason cladistics is so popular. The method of systematics helps scientists predict evolutionary relationships that have not yet been observed. In other words, there may be no existing evidence of a relationship, but the analysis of existing evidence makes it likely that new evidence will be found.

Cladistics also gives scientists opportunities to test hypotheses. Consider, for example, the trait of web design among spiders. Orb-weaving spiders produce webs of precision and order. Cobwebs are less orderly. So scientists hypothesized that the trait for orb-webs came after the trait for spinning cobwebs. Cladistics has turned that hypothesis on its head. Evidence now suggests that orb webs came before cobwebs.

Applying a systematic approach to the study of Earth's biodiversity gives scientists a more thorough understanding of events within life's evolutionary history.

WRITE TO LEARN

Mollusks include a variety of organisms such as shelled clams and wormlike slugs. A long-standing hypothesis suggested that soft, wormlike mollusks, which have no shells, came before their shelled relatives. Genetic evidence proves otherwise. Describe the position of each species on a cladogram and justify your description.

Vocabulary Review

Directions: Match each word to its definition.

1. _____ ancestry
2. _____ assumptions
3. _____ cladistics
4. _____ cladogram
5. _____ diverge
6. _____ homologous
7. _____ phylogeny
8. _____ systematics
9. _____ taxonomy

A. a method for making and testing predictions about evolutionary history

B. the science of biological classification

C. a trait that is similar in shape and location

D. split into two groups

E. methods for classifying organisms

F. unproven beliefs that are thought to be true

G. a diagram reflecting evolutionary relationships

H. a line of descent

I. the study of evolutionary history

Directions: Read and complete each activity.

1. Explain how phylogeny and systematics are related to cladistics.

   _____

   _____

2. The horse's oldest ancestor was about the size of a fox but looked more similar to a dog. Its feet had pads, and its toes ended in tiny hooves, instead of claws. The modern horse has only one toe on each foot, and this toe is a hoof. The remaining toes form tiny bumps higher on the horse's legs. Use the terms *plesiomorph* and *apomorph* to describe the modern horse's hoofed feet.

   _____

   _____

   _____

3. Explain the purpose of an outgroup in a cladistical analysis.

   _____

   _____

   _____

Directions: Examine the cladogram and answer the questions that follow.

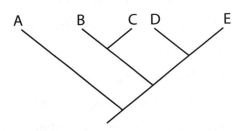

4. Which taxa represent the ingroup? _____

5. Which taxa show the closest evolutionary relationship? _____

6. How many apomorphs appear on the cladogram? _____

7. What is the terminal taxon that identifies the outgroup? _____

Skill Practice

Directions: Read the text. Then complete the activity.

1. Explain the method of cladistics to someone who is unfamiliar with the method.

   _____

   _____

   _____

2. Explain how cladistics can cause scientists to reject a longstanding hypothesis.

   _____

   _____

3. Draw a flow chart showing the basic steps in making a cladogram.

4. A student wants to understand the evolutionary relationships among a parrot, a donkey, and a butterfly. Suggest three traits that the student could use to determine how the organisms are alike and different.

   _____

   _____

   _____

5. Give the reason for the appearance of homologous features in a cladogram.

   _____

   _____

6. Write a persuasive statement explaining the need for cladistics.

   _____

   _____

Speciation

KEY CONCEPT: Speciation refers to the evolutionary process by which new biological species form. The pressures of a different environment, the isolation of a population, or genetic changes that result in successful adaptations may lead to a species with characteristics unlike its ancestors.

Think of everyone who lives in your community. Together, you form a population. A population is all the individuals of a species living in the same place at the same time. Individuals within a population have unique sets of genes. Together, all of the genes in a population form a gene pool. Over time, the gene pool in a population can change. The gradual accumulation of changes within a population is called evolution.

Classifying Organisms

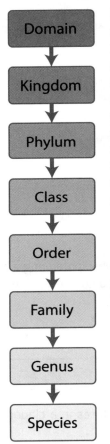

Recall that taxonomy is the classification of living things. The system developed by Carl Linnaeus in the 1700s remains in use today, although it has been modified as scientists have gained knowledge. The system is based on a **hierarchy**, meaning it is organized by levels, according to rank. The broadest group is on the top level. It leads to more and more specific groups at the lower levels. As the hierarchy moves downward, the members of each group share more and more characteristics. Species form the base of the hierarchy, and closely related species are grouped directly above, in a genus.

Every species has a scientific name. For example, Linnaeus assigned the giraffe the scientific name *Giraffa camelopardalis*. *Giraffa* represents the genus, and *camelopardali* identifies the species. Because only giraffes have this name, there is no confusion when scientists discuss members of the species.

Lamarck and Darwin

Jean-Baptiste Lamarck was one of the first scientists to explain how giraffes came to have long necks. Giraffes live on savannahs, grasslands that receive little rainfall. With so little rainfall, plant growth is limited. In his 1809 publication, Lamarck said that giraffes stretched to reach what food they could find—leaves that grow in trees. At first, giraffes stretched to reach low-growing leaves. When those were gone, the animals stretched to reach high-growing leaves. Continuous stretching led giraffes to "acquire," or gain, slightly longer legs and necks. They passed these acquired traits to their offspring. Those offspring also stretched to reach food. Over time, the giraffe's body acquired the shape you see today.

Decades later, in 1872, Charles Darwin used the process of natural selection to explain the giraffe's modern appearance. **Natural selection** is the process in which organisms best adapted to their environment tend to survive and pass on their genetic characteristics. To begin, Darwin explained that differences, or what we now call **genetic variations**, are always present in a species. So, among giraffe ancestors, there were individuals of varying height. While some were slightly shorter than most, others were slightly taller.

The second part of natural selection is related to reproduction. Whenever food became scarce, slightly taller giraffes had an advantage. They could reach food more easily than their slightly shorter relatives. So, not all giraffes survived to adulthood. Consequently, the taller giraffes lived long enough to reproduce.

Remember that offspring inherit traits from their parents, and genes carry the genetic code for height. So, offspring of taller giraffes also tended to be taller. These offspring then grew to adulthood and reproduced. The process continued over millions of years, becoming more and more common. Eventually, all of the individuals born were tall, and far different from their ancestors.

Natural selection contributes to evolution. Recall that evolution is a process of change over time, and the fossil record provides scientists with evidence of many of those changes. The **fossil record** shows physical evidence of organisms that lived over different periods of Earth's history.

THINK ABOUT SCIENCE

Directions: In what ways did Lamark and Darwin agree and disagree in their beliefs?

_____

_____

Core Skill
Determine Central Ideas

As you read, look for clues to help you identify the central or most important idea in a text. The following are some useful hints that you can use to find the central idea of a passage:

- Read the title and look for key words.

- Look for visual clues, such as graphs or illustrations.

- Find the key words that describe the most important concepts in the passage.

- Read to see if the central idea is broadly stated or if the author provides details that direct you to it.

- Check to see if the author has returned to one idea or concept throughout the selection.

- Determine the single, most important thing the author is trying to say.

What is the single most important idea of the passage "Continental Drift" on this page? Write the central idea using your own words and key words from the passage.

Evolution and the Fossil Record

The Latin word *fossilis* means "dug up." Fossils are the remains of once-living organisms. These remains are carefully "dug up" from beneath Earth's surface.

Sometimes, fossils are entire skeletons buried in ice or other places where there is no oxygen, which would cause the bones to decay. However, complete or nearly complete fossils are rare. Most often, fossils are bone fragments, teeth, and shells. Fossils may also consist of evidence of animal behavior, such as the preserved imprint of an animal's foot.

Giraffe fossils from 6 to 23 million years old have been found across Asia, Europe, and Africa. They show that at one time in Earth's history, 10 different types of giraffes lived on the planet. These giraffes were tall and had skin-covered horns on their heads but didn't possess necks as long as those of living giraffes.

Fossils from 2.5 to 6 million years ago show the first appearance of long-necked giraffes. However, not all of these groups survived. Today, many scientists think there is only one living species of giraffe, and its closest relative is *Okapia johnstoni*, or the okapi.

Continental Drift

How could giraffe fossils be found on continents where no giraffes live today? The answer may be **continental drift**, or the slow movement of continents.

Scientists hypothesize that about 200 million years ago the continents were united in a single landmass called **Pangaea**. Due to dynamic forces below the Earth's crust, Pangaea began to break apart, and a northern group of continents—Europe, North America, and Asia—split from the southern group. The southern group broke into three parts: Africa-South America, Antarctica-Australia, and India. About 140 million years ago, South America and Africa separated. Around 80 million years ago, Australia split from Antarctica. The continents drifted to their present locations, and still continue to move.

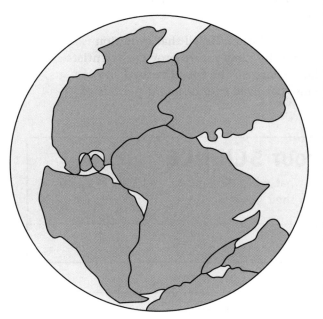

PhototAlto/PunchStock

As the continents drifted apart, organisms became isolated from one another. Fossils indicate that this isolation eventually led to the emergence of new species. In 1982, a team of American scientists in Antarctica found a fossil of a marsupial, a pouched mammal like a koala. This discovery supports the theory that marsupials migrated from South America across the Antarctica-Australia landmass before the continents separated.

Reproductive Isolation

The marsupials that reached Australia before the continents separated adapted to new habitats. New species emerged. Today, 140 species of marsupials live in Australia. Recall that *species* is the category that forms the base level of the taxonomic hierarchy. A species is a group of organisms that can interbreed with each other to produce offspring. The genetic information these organisms possess moves through a population through breeding. However, certain events, conditions, or lack of resources may prevent organisms from interbreeding. They are said to be reproductively isolated from one another. Consequently, their genes are also isolated, affecting the process of evolution.

What could isolate organisms? Moving continents can be one cause. But much smaller events can be the reason, as well. For example, a small group of mice might travel aboard a sailing ship and thrive in a new land. Or insects feeding on rotting plant matter tossed ashore might be carried to a new island by a powerful current. The mice and insects in these examples would become reproductively isolated from their original populations.

Physical barriers such as a dam on a river may separate a fish species. A jungle of thick trees may isolate a population of small mammals. A mountain range may lead to the isolation of a plant species. A meter's width of sunbaked soil can separate a population of snails.

Geographical and physical barriers aren't the only ways organisms in a population could be separated. Suppose a population of deer roams a long distance in search of food. Clusters of these deer find abundant food in a new environment along the edges of the former feeding area. When these deer mate, they mate with the deer closest to them. This takes their genes out of the larger population they left behind. The mating cluster of deer becomes an **incipient species**, meaning over a period of time, they will become a separate species from their former relatives. *Incipient* means "changing, or developing."

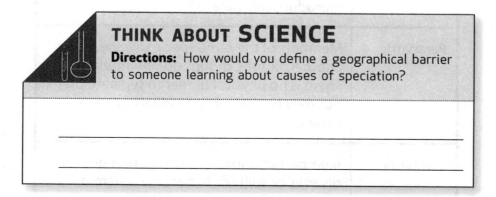

THINK ABOUT SCIENCE

Directions: How would you define a geographical barrier to someone learning about causes of speciation?

Speciation

Speciation is an event in which an organism's **lineage**, or line of descent from its ancestors, splits. Instead of one line of descent, there are two. Further splits may also occur. Examine the following diagram. Speciation led to Species A and B. Then Species B split again, leading to Species C and D.

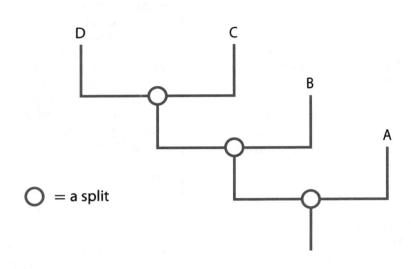

O = a split

Some scientists think that speciation occurs gradually, over long periods of time. Others think that it occurs in quick jumps and starts. Both groups of scientists use the fossil record as evidence.

The fossil record is the total collection of preserved plant and animal remains. All of the fossils that have been discovered so far suggest that life on Earth has evolved gradually. However, there are gaps in the fossil record. There are periods of time when new species seem to emerge suddenly, without the lineage splitting over a long period of time. Scientists who think speciation happens quickly point to these gaps as evidence to support their theory. However, scientists who think it happens slowly claim that the fossil record is incomplete and the fossils that support their hypothesis have not yet been found.

There are several reasons why speciation occurs. Some are described in the following chart.

The Method	How It Happens
Allopatric	A geographic barrier isolates a population.
Peripatric	For some reason, a geographical barrier, a natural disaster, or human activity, causes a very small group of individuals to become reproductively isolated.
Parapatric	Members of the same species live near one another, but the groups do not interbreed. There is no physical barrier to separate them, but they breed with individuals that are close by.
Sympatric	Members of a species live together but still diverge, or split off, to become different species.

Allopatric Speciation When a geographic barrier isolates members of a species, it interrupts the **gene flow**, or transfer of genes, within a population. For example, recall an earlier lesson that mentioned Darwin's discovery of several species of finches that live on different islands in the Galápagos archipelago, off the coast of South America. These finches share a common ancestor with the finches on the South American mainland.

A few million years ago, some finches flew approximately 650 miles across the Pacific Ocean from South or Central America. They landed on the remote Galápagos islands. Each island provided a distinctly different habitat, with different food resources. The finches, isolated on their islands, bred among themselves.

Over several centuries, changes in the birds' beaks began to appear. These changes proved successful for the birds' survival, because the adapted beak shapes allowed the finches to take advantage of the variety of seeds, insects, flowers, and leaves found on their islands.

There are now at least 13 separate species of finches living on the islands, and they all share a common ancestor. Even today, scientists observe the relationship between beak adaptations and environmental change. Climate changes have altered food supplies on the islands. Finches with the best adaptations for taking advantage of the most available food resources are surviving. Those without those adaptations are not.

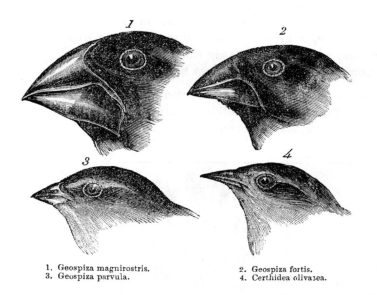

1. Geospiza magnirostris.
3. Geospiza parvula.
2. Geospiza fortis.
4. Certhidea olivacea.

THINK ABOUT SCIENCE

Directions: How do you know that the beak shapes you see in the illustration are examples of successful adaptations?

_____

_____

Core Skill
Analyze Text
Structure

Text structures are a way to organize information, or text. Common text structures include description, sequence, comparison, and cause and effect. Certain words and phrases signal the use of each structure.

Descriptions give characteristics and examples. Common signal words include _for example, such as,_ and _like._

Sequence presents content in numerical or time order. Words such as _first, next, after, when, now, finally,_ and _before_ signify this structure.

Comparisons show how things, events, or ideas are alike and different. Signal words include _however, but, although, also, alike,_ and _different._

Finally, some authors use _cause and effect_ structure to identify events, times, or facts that have consequences, or specific results. Signifying words include _consequently, as a result,_ and _therefore._

What examples of text structure can you find in this lesson? When you find an example, write a note in the margin or circle the words that provide a signal to the kind of text structure the author used.

Peripatric Speciation

Sometimes an obstacle or event causes a small group of individuals to become separated and reproductively isolated. Because there are so few individuals in the small group, any rare genes among them quickly spread across generations. The result is a new species. Consider the example of the London Underground mosquito.

To examine the bodies of *Culex pipiens* and *Culex molestus*, you might think the mosquitoes are the same species, and some scientists would agree. But others think they are separate species that formed over time, after individuals left the larger group.

Construction of the London Underground, a system of railway tunnels, began in the 1800s. Members of *Culex pipiens* flew into the tunnels and stayed. Decades later, during World War II, the underground mosquitoes bit Londoners who sought safety in the tunnel from the bombing.

Above ground, *Culex pipiens* feeds on birds. It travels and mates in large groups and remains inactive during the coldest parts of the year. However, the underground *Culex molestus* bites humans and rats, mates privately, and remains active year-round in the warmer underground temperatures. Their behavioral differences prevent them from interbreeding, meaning they are separate species.

Parapatric Speciation

Sometimes, groups of individuals live near one another but do not interbreed. There is no physical obstacle preventing gene flow. But for some reason, individuals mate with organisms closer to them. Consider the example of sweet vernal grass.

Humans mine for metals beneath the ground. The process contaminates surrounding soils, leaving behind heavy metals, such as lead, zinc, arsenic, mercury, nickel, and chromium. At some time, a group of sweet vernal grass plants began growing near a mine. Over time, this group of plants developed a tolerance for heavy metals that plants in the original population do not have. They can absorb the metals from the soil, but they are not harmed by them.

Prior to their separation, the two groups of grass plants would have interbred. But since their separation, the groups have developed different flowering times. This evolutionary change shuts off the gene flow between the two populations, leading to speciation.

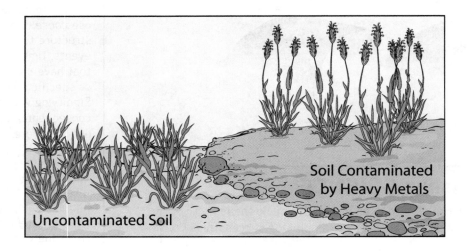

Uncontaminated Soil

Soil Contaminated by Heavy Metals

Sympatric Speciation Some members of a population live with one another, yet they do not interbreed. Because they do not interbreed, they become reproductively isolated. There may be different reasons for this isolation, but once again, the results are the same. Gene flow is interrupted. In time, one species becomes two, despite living together.

Apple maggot flies are an example of this kind of speciation. Maggots are fly larvae, meaning they hatch from fly eggs. Maggots are usually found in rotting matter, such as fruits that fall to the ground.

Centuries ago, hawthorn flies and apple flies shared a common ancestor. The ancestor flies mated and laid their eggs in the fruits of hawthorn trees. Then humans introduced a new plant to North America. Immigrants planted apple trees in their new home. Some of the hawthorn flies laid their eggs in the new fruit—apples.

When they hatch, fly larvae develop inside the fruit. As adults, male flies look for mates on the same kind of fruit they grew up in, and female flies tend to lay their eggs in the same kind of fruit they grew up in. So, offspring of hawthorn-laying flies continued laying eggs in hawthorn fruit, and apple-laying flies continued laying eggs in apples. In time, the two flies became different species.

WRITE TO LEARN

Evolutionists disagree on how speciation takes place. Explain what you would expect to observe in the fossil record if speciation occurs gradually. Then explain what you would expect to observe in the fossil record if speciation occurs in quick jumps.

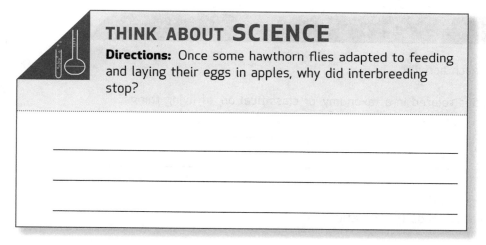

THINK ABOUT SCIENCE

Directions: Once some hawthorn flies adapted to feeding and laying their eggs in apples, why did interbreeding stop?

_____

_____

_____

Today, millions of species live on Earth. Some scientists estimate that we share the planet with about 9 million species. However, scientists continue to find new species, and some species disappear. Plus, the process of speciation is always at work. Consequently, estimates will continue to change.

Vocabulary Review

Directions: Choose the term that best matches each description.

| continental drift | fossil record | gene flow | hierarchy |
| incipient species | lineage | natural selection | speciation |

1. _____ the process by which species form

2. _____ the transfer of genes within a population

3. _____ the gradual movement of landmasses

4. _____ the total collection of preserved plant and animal remains

5. _____ line of descent; ancestry

6. _____ a group of organisms that are going to become a new species

7. _____ an order of categories or levels

8. _____ the natural process in which organisms best adapted to their environment tend to survive and pass on their genetic traits.

Skill Review

Directions: Read and complete each activity.

1. How are a domain and a species related in a taxonomy, or classification, of living things?

_____

_____

_____

2. Explain how reproductive isolation leads to speciation.

_____

_____

_____

3. A population of plants grows on a mountain. The wind and animals carry seeds far from the mountain to a different mountain. Explain what might happen that could lead to speciation.

_____

_____

4. In today's global economy, transportation vehicles and their passengers can unknowingly carry wild organisms with them. What may happen to these organisms in their new environments?

_____

_____

Skill Practice

Directions: Read and complete each activity.

1. A new species of sweet vernal grass absorbs poisonous heavy metals from soil without harm. What might happen if people planted this new species of sweet vernal grasses near abandoned mining sites?

   _____

   _____

   _____

2. Use the chart to explain the evolution of apple flies. Include the term *speciation* in your explanation.

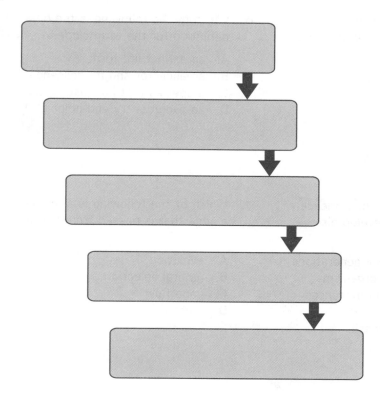

3. Explain why the fossil record is useful in understanding the process of speciation.

   _____

   _____

   _____

4. How is genetic variation related to species survival?

   _____

   _____

   _____

Review

Directions: Choose the best answer to each question.

1. Which of the following organisms was first to appear in the fossil record?

   A. insects
   B. amphibians
   C. dinosaurs
   D. fishes

2. The classification system developed by Carolus Linnaeus classifies organisms into groups based on their

   A. shared traits.
   B. range of sizes.
   C. age in the fossil record.
   D. geographic locations.

3. Which of the following forms of evidence did Charles Darwin use to develop his theory of evolution?

   A. gene mutations over many generations
   B. DNA analysis of related organisms
   C. structual similarities and differences among living things
   D. radioactive isotopes in fossils

4. The finches of the Galapagos Islands share a common ancestor with the finches on the South American mainland. The development of these different finch species is an example of which of the following methods of speciation?

   A. parapatric
   B. allopatric
   C. peripatric
   D. sympatric

5. Movement from the bottom to the top in a cladogram represents

   A. increasing population size.
   B. the passage of evolutionary time.
   C. organisms moving to new areas.
   D. the eventual dominance of one species.

6. Which of the following is the best definition for the term *clade*?

   A. an extinct life form
   B. an ancestor and its descendants
   C. a species that has just evolved
   D. an evolutionary end point

7. Which of the following is a mistake in a gene that is passed from parent to offspring?

   A. evolution
   B. natural selection
   C. gene flow
   D. mutation

8. In a cladogram, which of the following events occurs at a splitting point?

   A. A species declines in number, then becomes extinct.
   B. Species characteristics stop changing after a species forms.
   C. A parent species gives rise to two new species.
   D. Two different species breed with one another.

9. Define *natural selection* and *evolution*. Describe the relationship between these two terms.

_____

_____

_____

10. How does continental drift help to explain why marsupial species are common in Australia but relatively rare in North America?

_____

_____

_____

11. Draw a flow chart to show how the steps of evolution might have occurred to result in a longer neck length in giraffes.

12. In a cladogram, distinguish between the meanings of the terms *outgroup* and *ingroup*.

_____

_____

13. Compare the conclusions of Jean-Baptiste Lamarck to those of Charles Darwin. How were the observations made by these two scientists similar? How did their interpretations differ?

_____

_____

_____

14. Explain how taxonomy helps scientists to study living things and makes it possible for them to communicate effectively about specific organisms.

_____

_____

_____

15. Mosses are seedless plants that lack vessels for transporting water and nutrients. Show where you would you place mosses in this cladogram.

_____

_____

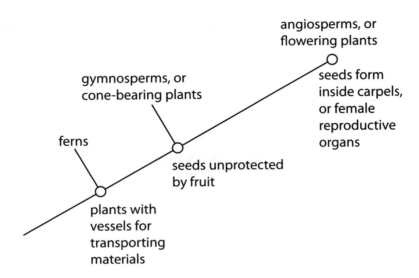

angiosperms, or
flowering plants

seeds form
inside carpels,
or female
reproductive
organs

gymnosperms, or
cone-bearing plants

ferns

seeds unprotected
by fruit

plants with
vessels for
transporting
materials

16. Two populations of the same flowering plant species live in the same region. One population flowers in late spring. The other population flowers in early summer. Explain how a difference in flowering times could lead to the development of two different species.

_____

_____

_____

_____

Check Your Understanding

On the following chart, circle the number of any item you answered incorrectly. Next to each group of item numbers, you will see the pages you can review to learn how to answer the items correctly. Pay particular attention to reviewing those lessons in which you missed half or more of the questions.

Chapter 6 Review

Lesson	Item Number	Review Pages
Biological Evolution	1, 3, 4	200–209
Common Ancestry and Cladograms	2, 6, 7, 8, 10, 12, 14, 15	210–219
Speciation	5, 9, 11, 13, 16	220–229

Application of Science Practices

CHAPTER 6: EVOLUTION

Question

How does reducing the size of a population affect its genetic diversity?

Background Concepts

Genetic drift is a mechanism of evolution, or change in a species over time. Genetic drift occurs when, by chance, some individuals produce more offspring than others. This random process leaves more genes in the population.

Simulation

Materials required

40 beads, marbles, or strips of paper, all of one color

30 beads, marbles, or strips of paper, all of a second color

20 beads, marbles, or strips of paper, all of a third color

In this activity, you will simulate genetic drift before and after a natural disaster.

1. Thoroughly mix all of the beads, marbles, or paper strips in a paper bag or box.
2. Without looking, select 40 items from the container. These items will represent the number of organisms that produce offspring. Count the items that have each color. Record the total and calculate the percentage of the total in Table 1 (number of items divided by the total number).
3. Return the counted items to the container. Mix the items again.
4. Repeat Steps 2 and 3 for a total of 3 trials.

Phenotype	Trial 1		Trial 2		Trial 3	
	Percent of Total		Percent of Total		Percent of Total	
	Color 1:	_____ /40 = _____ %	Color 1:	_____ /40 = _____ %	Color 1:	_____ /40 = _____ %
	Color 2:	_____ /40 = _____ %	Color 2:	_____ /40 = _____ %	Color 2:	_____ /40 = _____ %
	Color 3:	_____ /40 = _____ %	Color 3:	_____ /40 = _____ %	Color 3:	_____ /40 = _____ %

Application of Science Practices

Now simulate a case where the population has been severely reduced, due to a natural disaster.

1. Selecting only 10 ITEMS FOR EACH TRIAL, repeat all of the same steps again. The number 10 represents how many individuals in the population survive to reproduce. Record your data in Table 2.

Phenotype	Trial 1	Trial 2	Trial 3
	Percent of Total	**Percent of Total**	**Percent of Total**
	Color 1: _____/10 = _____%	Color 1: _____/10 = _____%	Color 1: _____/10 = _____%
	Color 2: _____/10 = _____%	Color 2: _____/10 = _____%	Color 2: _____/10 = _____%
	Color 3: _____/10 = _____%	Color 3: _____/10 = _____%	Color 3: _____/10 = _____%

Interpretation

Compare the tables. Which shows greater genetic variation, or differences?

_____

Answer

Based on the data, what can you conclude about the effects of population size on genetic drift?

_____

_____

_____

Evidence

Organize the data for only one color.

Table 1: _____

Table 2: _____

Use the data to support your answer.

UNIT 2

Physical Science

Energy

Drive a car. Turn on a television. Watch a movie on your computer. Each of these tasks requires energy of some kind, and humans rely on a variety of energy sources to meet their demands for power.

In this chapter you will learn about:

Lesson 7.1: Energy
When a physicist uses the term *energy*, he or she means the ability to do work. In this lesson, find out about different forms of energy—such as potential, kinetic, and radiant—and learn how energy can change from one form to another.

Lesson 7.2: Waves
The movement of the vast ocean, the sound of a violin, and the quality of a microwave dinner may seem to have little in common, but they all depend on waves. Introduce yourself to wave theory and how it relates to the visible spectrum, high-energy waves, and sound.

Lesson 7.3: Electricity and Magnetism
Look around the room—how many electrical devices or appliances do you see? We depend on electricity for working, learning, entertainment, health, and nearly every aspect of daily life. Learn about electric currents, circuits, and how they relate to magnets in this lesson.

Lesson 7.4: Sources of Energy
Humans rely on different kinds of energy to do work. Some sources of energy are renewable, meaning if they are managed well, they continue to be available. Other kinds of energy are nonrenewable, meaning no matter how well they are managed, supplies are limited and cannot be re-created. Find out about a variety of renewable and nonrenewable sources of energy that humans depend on to do work.

Lesson 7.5: Endothermic and Exothermic Reactions
Some chemical reactions absorb heat, and the heat is stored in the chemical bonds of products. Other chemical reactions release heat to the environment. Examine the differences between endothermic and exothermic reactions.

Goal Setting

Nothing reminds us more of our dependence on electricity than a blackout! When the power goes out temporarily, we suddenly become very aware of all the conveniences we depend on that are powered by electricity.

How many electrical devices do you think you use in a day? First take a guess and write it down. Then make a list of as many electrical items as you can think of that you use in a typical day.

Was your estimate accurate? Would estimating be an important skill for a scientist to have? Think about situations when you might find estimating helpful in your everyday life as you read through this chapter.

Energy

Lesson Objectives

You will be able to
- Define energy
- Differentiate between kinetic and potential energy
- Recognize different types of energy and energy transformations

Skills

- **Reading Skill:** Determine the Central Idea of a Text
- **Reading Skill:** Determine Meaning

Vocabulary

contract
efficient
energy
expand
law of conservation of energy
transformation

KEY CONCEPT: Energy, the ability to do work, occurs in different forms that can be changed from one type to another.

When was the last time someone told you that you have potential? Potential is the possibility of becoming something special.

In science, the word potential *is usually paired with the word* energy. *You may have already read about different kinds of energy, such as the energy stored in the food you eat. The food has potiential energy, which changes to other kinds of energy after it is digested and transported to your body's cells.*

Energy

You may think of **energy** in many ways. However, scientists define energy as the ability to do work. There are many forms of energy. One of the most common is **kinetic energy**, which is the energy an object has when it is in motion. A car driving along a street and a hammer swinging down toward a nail have kinetic energy.

Another form of energy is **potential energy**, or the stored energy an object has because of its position. At the top of a hill, someone on a sled has no kinetic energy because he or she is not moving. The potential for movement is present, however, because of the sled's position at the top of the hill. Once the sled starts moving, the potential energy changes into kinetic energy.

Conservation of Energy

Like matter, energy does not appear and disappear. In the example of the sledder, the amounts of potential energy and kinetic energy should be exactly the same. This idea is known as the law of **conservation of energy**, which states that energy can be neither created nor destroyed.

While energy is always conserved, it often changes forms. This is called a **transformation**. For example, some of the sled's **mechanical energy**—the sum of potential and kinetic energy—is transferred out of the sled as heat. This may be described as heat loss, but energy is never really "lost," rather its form or place changes. A system is **efficient** if it achieves maximum productivity with only a small amount of heat lost during a transformation.

DETERMINE THE CENTRAL IDEA OF A TEXT

Authors write for a number of reasons, but whether they write to explain, inform, or even entertain, their work possesses a central idea. You can think of the central idea as a main idea, or the idea that the rest of the text supports.

You can use different strategies to determine a text's central idea.

1. Start by reading the title. Then scan the text to find headings and subheadings. Pause to ask yourself, *What is the main topic of this text?* If you can answer the question, you may already know the text's central idea.

2. An author usually presents the main idea of a section in the first paragraph, so skim the opening paragraphs in all of the sections.

3. Look for repetition. Check to see if the same idea appears throughout the text.

4. Form an idea of what you think the central idea is. Then check the visuals to see if they support your thinking. If they do, you may have identified the central idea successfully. If they don't, determine what central idea links the visuals.

Apply one or more of the strategies above to determine and write the central idea of this lesson.

_____

_____

_____

_____

_____

_____

Other Forms of Energy

The movement of electrons has a special name—**electrical energy**. Electricity is a form of electrical energy.

Energy stored in the bonds that hold atoms together is known as **chemical energy**. Chemical energy is another form of potential energy because it depends on the position or arrangement of atoms.

Splitting or changing the nucleus of an atom results in **nuclear energy**. A small amount of the atomic mass is converted to energy. The energy produced by stars is nuclear energy. Some of this nuclear energy is transformed into sunlight, or **radiant** energy.

The head, or title, of a
section usually reflects
the section's central
idea. And if you read
the first sentence or
two of the section's
opening paragraph, you
are likely to find that
the author states the
central idea plainly.
Visuals accompanying
the section usually
support the section by
representing the central
idea in a different way.

Read the section head
at the top of this page.
Then read the first
sentence of the first
paragraph and examine
the diagram. Use what
you read and see to
state the central idea
of this section.

Energy Changing Form

Every kind of energy can change form. For example, during nuclear reactions occurring in the stars, new elements are formed, and some of their mass is changed into energy. Some of the nuclear energy changes to radiant energy that travels through space. Plants use that radiant energy to produce food. The energy in the food flows through an ecosystem as animals eat the plants and then are eaten by other animals.

When a plant dies, the energy trapped within it eventually can be converted to coal, oil, or gas. These compounds are rich in chemical energy. At a power plant, the energy in coal may be converted to electrical energy, which in turn may be used to cook a healthful meal. The heat produces chemical changes in the food. If a person eats the food, the body converts the chemical energy in the food to mechanical energy for activities such as walking or running. Or some of the chemical energy may be used to make new body cells and tissues.

TYPES OF ENERGY

WRITE TO LEARN

Write one or two
paragraphs describing
examples of energy
conversions that occur in
a typical school day.
Choose a title and write
an opening sentence that
reflect your writing's
central idea.

THINK ABOUT SCIENCE

Directions: Complete the sentences below by filling in the blanks.

1. _____ energy is the energy of a moving object.

2. An object's position gives it _____ energy.

3. Energy stored in the bonds between atoms of molecules is _____ energy.

4. The movement of electrons produces _____ energy.

5. Splitting or changing an atom's _____ provides nuclear energy.

Heat

For many years, scientists struggled to understand heat. Some thought heat was a kind of fluid that flowed through the air. It took several years before scientists realized that heat is the transfer of energy from hot objects to colder ones.

When objects are hot, they increase slightly in size, or **expand**. This occurs because the particles in a hot object move around more, bumping each other and spreading apart. In cool objects, the movement of the particles slows, and the object **contracts**, or gets smaller.

Heat moves or is transferred by one of three processes: conduction, convection, and radiation.

Conduction

Conduction is heat transfer due to direct contact. A hot pan on a stove warms the food on top of it by conduction. Your hand feels warmer when it touches hot food because of conduction from the food to your hand.

Conduction occurs because the particles in an object are constantly moving and bumping into each other. Inside a hot pan, metal atoms are moving very fast. When these atoms hit molecules of food, those molecules begin moving faster, too.

Pots and pans are made of metals because metals are good conductors of heat. Pot handles are made of wood because wood is a poor conductor of heat. Materials that do not quickly conduct heat are called **insulators**.

Water is a poor conductor, as is plastic. Air is an especially poor conductor. Poor conductors do not feel cold when you touch them. This is because they do not absorb heat from your hand quickly. Cork and fiberglass both contain air, and both are poor conductors.

Convection

Heat can move through liquids and gases in a process called **convection**. When air is heated, its molecules move faster and farther apart. With its molecules farther apart, the hot air becomes less dense. It rises above the cooler air, which is denser and, therefore, sinks. The difference in air temperature causes the air to move. In the illustration on this page, notice that the hot air from the furnace rises into the room. The cool air sinks until it returns to the furnace and is heated again.

Reading Skill
Determine Meaning

When you read about energy, you may discover the term *thermal energy*. Thermal energy is related to heat, but the two terms have different meanings.

Atoms within matter are constantly vibrating. Their kinetic energy is measured as temperature. Higher temperatures indicate greater kinetic energy. If matter, such as steaming cocoa, has greater thermal energy than the air around it, its atoms slow as heat moves, or transfers, from the cocoa to the air. As heat enters the air, atoms in the air vibrate faster, resulting in greater thermal energy. The heat transfer results in changes in kinetic energy and of temperature. Explain how you can observe this change in the air termperature.

CONVECTION

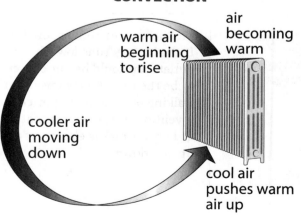

warm air beginning to rise

air becoming warm

cooler air moving down

cool air pushes warm air up

Radiation

Radiant energy is transferred by electromagnetic waves. You will learn more about electromagnetic waves in the next lesson. For now, understand that electromagnetic waves come from many sources, such as microwaves, infrared lamps, or sunlight. Because electromagnetic waves carry radiant energy, objects that absorb these waves receive energy.

Radiant energy can raise the temperature of an object when it is absorbed. Clothing and skin absorb radiant energy, which is why you feel warm in sunlight. Materials with dark, dull surfaces absorb a lot of radiation, which is why dark clothes often make you feel hot on sunny days. Materials with light, shiny surfaces reflect the radiation. Radiant energy passes through clear materials, such as air, glass, and water.

Vocabulary Review

Directions: Complete the sentences below using one of the following words or terms.

contract   efficient   energy   expand   law of conservation of energy   transformation

1. The _____ states that energy can be neither created nor destroyed.

2. Objects may _____ when they are heated.

3. A(n) _____ occurs when one form of energy changes into another.

4. When particles in an object slow down, the object will _____.

5. The less energy lost as heat, the more _____ a machine is.

6. The ability to do work or cause change is _____.

Skill Review

Directions: Look at the visual and read the passage below. Then respond to each question that follows.

Mobile phones and personal music players run on batteries. If the battery runs low when you are far from an electrical outlet, you could be out of luck. But that no longer needs to be the case. A new invention lets you use the energy of walking or running to power these portable devices. The invention is a generator that you attach below your hip or knee. It makes electricity every time your leg moves up and down.

This generator can produce enough power from a one-minute walk to power a cell phone for thirty minutes.

Skill Review (continued)

1. Which of these titles would give the most useful information about the passage to the reader?
   A. Walking Is Good for You
   B. Portable Power
   C. Human Energy Can Replace Batteries
   D. Cell Phones Use Batteries

2. What does the caption add to your understanding of the passage? _____

3. A man wants to charge his cell phone for 5 hours of use. How long must he walk with the generator on his leg? _____

Directions: Apply what you know about energy transformations to answer the question below.

4. What type of energy change does the generator in the passage perform?
   A. electrical to electrical
   B. potential to electrical
   C. electrical to kinetic
   D. kinetic to electrical

Skill Practice

Directions: Choose the best answer to each question.

1. A billiard ball rolls to a stop on a felt table. What best describes the energy change of this event?

   A. The ball lost its mechanical energy due to nuclear reactions.
   B. The ball lost its kinetic energy due to friction with the table.
   C. All of the kinetic energy of the ball changed into potential energy.
   D. All of the potential energy of the ball changed into kinetic energy.

2. Which of the following describes the change in energy when a log is burned?

   A. Mechanical energy is changed to chemical energy.
   B. Chemical energy is changed to heat energy and light energy.
   C. Mechanical energy is changed to radiant and chemical energy.
   D. Chemical energy is changed to heat energy and electrical energy.

3. Why are both iron and wood useful to make a frying pan?

   A. Iron is a good conductor, and wood is an even better conductor.
   B. Both iron and wood are good insulators.
   C. Both iron and wood are good conductors.
   D. Iron is a good conductor, and wood is a good insulator.

4. A heater for a fish tank works better along the bottom of the tank than at the top because

   A. water cannot move inside a tank.
   B. hot water sinks and cold water rises.
   C. hot water rises and cold water sinks.
   D. water in a tank moves only horizontally, not up and down.

Waves

KEY CONCEPT: Energy may travel in the form of a wave. The properties of a wave determine how much energy it has.

Have you ever seen a beach in person or in a photograph? On some days, the waves of water are smooth and gentle. They lap gently on the shore. But on other days, the waves are crested with white, foaming water. They break and crash along the shore. All waves carry energy, but different amounts of energy. The amount of energy in waves determines what it can be used for.

Wave Theory

Radiant energy is one type of energy that moves in waves. Radiant energy is also called electromagnetic radiation. The different forms of this radiation make up the **electromagnetic spectrum**, shown below. Only part of the spectrum is visible to humans.

ELECTROMAGNETIC SPECTRUM

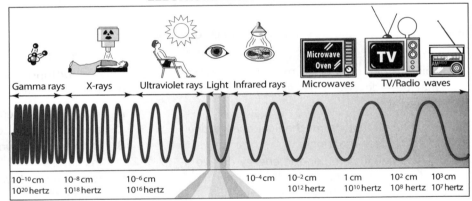

| Gamma rays | X-rays | Ultraviolet rays | Light | Infrared rays | Microwaves | TV/Radio waves |

| 10^{-10} cm | 10^{-8} cm | 10^{-6} cm | | 10^{-4} cm | 10^{-2} cm | 1 cm | 10^{2} cm | 10^{3} cm |
| 10^{20} hertz | 10^{18} hertz | 10^{16} hertz | | | 10^{12} hertz | 10^{10} hertz | 10^{8} hertz | 10^{7} hertz |

Parts of the electromagnetic spectrum have different properties. These properties depend on the **wavelength**, which is the distance between two waves. They also depend on the wave **frequency**, which is the number of waves that pass a point in a given amount of time. Waves with the greatest frequency carry the most energy.

All forms of electromagnetic radiation travel at the same speed through empty space. Their speed slows slightly in air or liquid.

Before reading a science text, it is helpful to scan the text for headings and visuals that may give you a sense of the text's main idea. Skimming the opening paragraphs in each section is also helpful.

Once you have identified a text's main idea, you can focus on specific details that tell about it. These details provide information you need to **support**, or strengthen, your conclusions. They are textual evidence.

Read the following paragraph. Find two details that support the main idea.

(1) Rainbows form when sunlight travels through tiny water droplets suspended in the air. (2) The drops act like prisms, refracting, or breaking, sunlight into different wavelengths, which appear as bands of color. (3) The next time a rainbow appears in the sky, look closely at the colors. (4) If you look from the outer edge to the inner edge, you will see the colors red, orange, yellow, green, blue, indigo, and violet. (5) The colors always appear in the same order.

Sentences 2, 4, and 5 present helpful details because they support the main idea that rainbows occur when sunlight moves through raindrops suspended in the air.

High-Energy Waves

The high-energy waves on the left side of the diagram can do the most harm to people. Gamma rays penetrate deeply. They have a lot of energy and damage living cells as they pass through. Gamma rays, however, can also be used to save lives by killing cancer cells. X-rays are frequently used by doctors to take a picture inside the body. Sunlight contains **ultraviolet (UV) rays**. Ultraviolet rays are powerful enough to tan skin and are used to kill bacteria.

At times, you are asked to answer a question or offer a conclusion related to something you have read. In either case, you provide evidence from the text to justify your response. When you justify, you offer details, or pieces of important information, that make your response clearly correct, or at least reasonable.

Citing textual evidence strengthens your responses. It makes your responses more sensible and even more powerful.

Consider the question *Is light white?* Write an answer to this question. Cite evidence from the lesson to support your answer.

Low-Energy Waves

The low-energy waves on the right half of the spectrum are very useful. Infrared waves can be used to cook food. Because all warm objects emit infrared rays, they are also used to map Earth from space and to locate people, such as hikers who become lost and cannot be seen in forests or rough terrain.

Microwaves are the shortest of the radio waves and are used in the transmission of radar signals. They also give us a fast way to cook. Because some microwaves have the same frequency as the water molecules contained in food, microwaves are used to cook moist food. As the water molecules in the food vibrate, the energy of the microwaves is converted into heat.

Radio waves bring us music and news, and they help airplanes and submarines navigate. The length of radio waves ranges from hundreds of feet down to fractions of an inch.

The Visible Spectrum

The visible spectrum is most important to us because we rely on light to see. Light can be produced when an atom gains energy. The atom must release this extra energy and return to its normal state. Often, the energy is released in the form of light. The bundle of energy released by the atom is called a **photon**.

For many years, people disagreed about the nature of light. Because it travels in a straight line, many were convinced that light was composed of tiny particles. Several experiments, however, proved that light was a wave. Finally, scientists found that light is a combination of particles and waves. When light is either emitted or absorbed, it behaves like a particle. When light is traveling long distances, it behaves like a wave.

Most of us have seen a sunbeam and think of light as white. A rainbow, however, offers proof that what we see as pure white light is actually a blend of many colors. When light passes through a **prism**, a triangular piece of glass, white light splits into a group of colors that we call the **visual spectrum**. The diagram of the prism shows the colors of light.

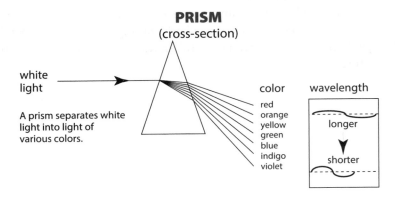

PRISM
(cross-section)

white light

A prism separates white light into light of various colors.

color
red
orange
yellow
green
blue
indigo
violet

wavelength

longer

shorter

Prisms break up the white light because each color of light has a slightly different wavelength. Each color bends at a different angle as it passes through the prism. Colors of shorter wavelengths bend more than colors of longer wavelengths.

How Light Travels

Light travels in waves that move in a straight line. As a result, light cannot go around objects. Instead, when light strikes an object, one of several things can happen.

Some objects, such as a thin pane of glass, allow light to pass through. These objects are **transparent**. Other objects reflect or absorb light. This causes a shadow, or dark area, on the side opposite the light. These objects are called **opaque**. A third type of object, such as wax paper or thick glass, scatters the light that passes through it. These objects are **translucent**.

Most surfaces **reflect** at least a little light. To reflect is to bounce away from. Because a mirror is smooth and shiny, it will reflect most of the light that strikes it. Light waves bounce off a mirror like a ball bouncing off a wall. Unlike other objects, a mirror does not scatter the light it reflects. The image that forms from a mirror is very clear and defined. The reflection from a rougher surface, such as a pool of water, appears distorted and poorly defined.

Light in Different Mediums

When light passes through transparent materials it continues to move in a straight line. However, light passes through different materials at different speeds. Light travels fastest through empty space, then slows down when it passes through air. It slows down even more in water and glass.

Air, water, and glass are each examples of a medium for light. When a ray of light travels from one medium to another, it bends, or **refracts**. You can see refraction by placing a pencil in a clear glass of water. Observe the pencil from an angle, and it appears to be broken at the water's surface. This is because the light slows down and bends at the point where it passes from air to water.

THINK ABOUT SCIENCE

Directions: Complete each sentence by filling in the blanks.

1. High-energy waves include _____, _____, and _____.

2. Low-energy waves include _____, _____, and _____.

3. Light shows characteristics of both particles and _____.

4. Because light cannot go around an object, it forms a _____ when an object blocks its path.

5. A prism acts to _____ light.

6. A mirror acts to _____ light.

Core Skill
Draw Conclusions

Conclusions are statements or explanations based on specific facts or details. They are logical interpretations of important information.

For example, as you read the text on this page, or after rereading the text, write a conclusion about why sonar is an effective method for mapping the ocean floor. Before writing your conclusion, review the text to identify useful details that support your thinking.

WRITE TO LEARN

Use print or online news sources to read an article of interest to you. Write a conclusion that summarizes the article. Include important details from the article that you used to justify your conclusion.

Sound

Light can travel through empty space. Sound, however, is a mechanical wave, meaning it needs a medium, such as a solid, liquid, or gas, to travel through. Sound also travels much more slowly than light.

You can hear sound waves and sometimes you can feel them, but you can't see them without an instrument called an **oscilloscope**. An oscilloscope produces a picture of a sound wave as a wavy line.

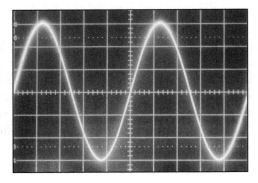

A sound wave is a transfer of energy as air particles vibrate, or move back and forth rapidly. Say a sound wave is moving horizontally from left to right through air. Particles of air vibrate as the wave passes. Each particle pushes on the particle next to it. Particles are pushed together, causing a high-pressure region called a **compression**. As the particles squeeze together, they create a low-pressure region behind them called a **rarefaction**. Air particles in the rarefaction spread out. A sound wave produces a series of compressions and rarefactions. This repeating pattern of high- and low-pressure regions is sometimes called a pressure wave.

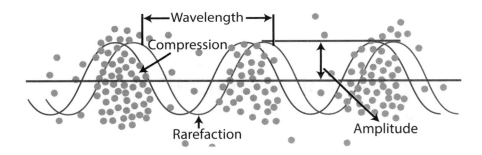

The high point of a wave is called the **crest**. The low point is called the **trough**. The distance from the resting point of the wave to the crest or to the trough is called the wave height, or the **amplitude**. The amplitude indicates the energy of the wave, which determines loudness. Louder sounds have greater amplitude.

A wavelength is the distance over which a disturbance in the medium occurs. In images like the one produced by the oscilloscope, a wavelength is the distance from one crest to the next, or from one trough to the next. One wavelength represents a wave cycle. The wave cycle forms a repeating pattern as the sound moves through air.

Frequency is a measure of how many wave cycles occur in a specific unit of time. Often, these measures are given in cycles per second, or Hertz.

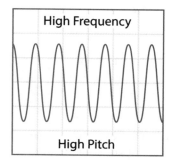

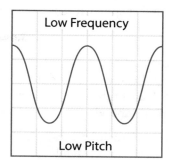

Frequency determines pitch. High-frequency waves produce high-pitch sounds. Low-frequency waves produce low-pitch sounds. The loudness or softness of a sound reflects the amount of energy the sound carries. A loudspeaker at a rock concert will create sound waves with a lot of energy, much more than the sound waves produced by a whisper.

Sound waves can be scattered, reflected, and absorbed in much the same manner as light waves. Sound waves also slow down or speed up when they travel through different materials. Sound waves travel slowest in a gas, where the molecules are more spread out. They travel faster in a liquid, and travel fastest through solids.

Just as light is reflected by a mirror, sound can be reflected by a hard surface. **Echoes** are reflected sound waves. The principle of reflected sound is applied in sonar (sound navigation ranging), which has been used to map the ocean floor. Bats and porpoises use sonar to find their way through the dark and to find food.

Vocabulary Review

Directions: Complete the sentences below using one of the following terms.

electromagnetic spectrum   frequency   prism   reflects   refracts   ultraviolet

1. The complete range of waves from gamma rays to radio waves makes up the
   _____.

2. Light _____ when it bounces off a surface.

3. Light _____ when it travels from air to glass.

4. A _____ splits white light into the colors of the visible spectrum.

5. Sunlight contains _____ rays that can burn the skin.

6. _____ is determined by the number of complete wave cycles per unit of time.

Directions: Read the following passage and respond to each question below.

> Some people think that skin cancer strikes rarely and randomly. Unfortunately, skin cancer strikes one in six Americans, especially those who spend long hours in sunlight. Research has shown that long exposure to sunlight is a significant risk factor for skin cancer. Other risk factors include suffering a bad burn in childhood.
>
> The dangerous part of sunlight is its ultraviolet rays, which are not the rays you can see. Sunscreen can protect your skin from harmful ultraviolet rays. By wearing sunscreen, anyone can still spend time in sunlight and stay healthy.

1. Choose the three factual details from the passage that explain why everyone should use sunscreen.
   _____ A. Some people think skin cancer strikes randomly and rarely.
   _____ B. Unfortunately, skin cancer strikes one in six Americans, especially those who spend long hours in sunlight.
   _____ C. Research has shown that long exposure to sunlight is a significant risk factor for skin cancer.
   _____ D. Other risk factors include suffering a bad burn in childhood.
   _____ E. Sunscreen can protect your skin from harmful ultraviolet rays.
   _____ F. By wearing sunscreen, people can still spend time in sunlight and stay healthy.

2. Which of these details would best support the main idea of the passage?
   A. Getting a suntan is fun and fashionable.
   B. Cancer can strike any exposed skin, including the arms, legs, and face.
   C. Some people dislike the smell of certain brands of sunscreen.
   D. Using tobacco products increases the risk of lung cancer.

Directions: Choose the best answer to each question.

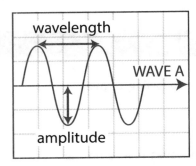

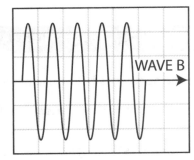

3. Wave A and Wave B are sound waves. What can you conclude about the pitch of these waves?
   A. Wave A has a higher pitch than Wave B.
   B. Wave B has a higher pitch than Wave A.
   C. The pitches of both sound waves are equal.
   D. Neither sound wave has a pitch.

4. If Wave A became louder, what could you conclude about how it would change?
   A. Its wavelength would increase.
   B. Its wavelength would decrease.
   C. Its frequency would increase.
   D. Its amplitude would increase.

Skill Practice

Directions: Choose the best answer to each question.

1. One way that sound waves differ from waves of light is that sound waves
   - A. have frequency and wavelength.
   - B. can travel through empty space.
   - C. need matter to travel through.
   - D. can bounce when they meet a smooth surface.

2. Which of these objects would be most useful to prove that white light is composed of light of different colors?
   - A. a prism
   - B. a mirror
   - C. a piece of wax paper
   - D. a window pane

3. Sound travels as longitudinal waves through matter. What property of sound changes when the amplitude of the sound wave increases?
   - A. higher pitch
   - B. lower pitch
   - C. greater volume
   - D. lower volume

4. Why do you often see a bolt of lightning several seconds before you hear the thunder it causes?
   - A. Light travels faster than sound.
   - B. Sound travels faster than light.
   - C. Sound waves echo many times in the air.
   - D. Sound travels as waves, and light travels as particles.

5. Explain the repeating patterns of compressions and rarefactions as a sound wave moves through air.

_____

_____

_____

_____

_____

_____

_____

_____

_____

Electricity and Magnetism

KEY CONCEPT: Electric current flowing through a circuit can be harnessed for its energy, and it produces magnetic effects.

Have you ever been on a ride at a water park? Riders climb aboard a small boat or sit inside an inner tube. Then the water carries them up and down hills and through turns, dips, and curves. The water and the boats or tubes travel in the same path all day long, over and over again.

An electric circuit is like a water-park ride. The moving parts, however, are much smaller than boats or tubes. The moving parts are electrons.

Electricity

The particles of an atom's nucleus are bound tightly together. They move very little. The electrons, however, are much freer to move from place to place and from atom to atom. The movement of electrons causes electricity. **Electricity** is the energy caused by the flow or separation of charged particles, such as electrons.

One way to get electrons to move from one atom to another is to rub two neutral objects together. In a clothes dryer, pieces of clothing rub against each other and against the drum of the dryer. As they rub, they lose and gain electrons. The result is that some bits of clothing repel each other and some attract each other. This is called **static electricity**.

The same thing happens when a comb is rubbed against hair. The comb removes some electrons from the hair, leaving the hair with more positive than negative charges. As the opposite charges on the comb and the hair build up, the two are attracted to each other, and the hair then stands up when the comb is brought near.

Electric Current

Electric charges can flow continuously in a circular path, called an **electric current**. Materials that carry electric currents are conductors. Materials that do not carry electric current are insulators.

You might think of resistance as a kind of frictional force. For example, when you walk, there is friction between your feet and the ground. It works in the opposite direction of your movement and slows your speed. **Resistance** opposes the forward flow of electrons, or an electric current, and is measured in ohms (Ω). The result is an increase in temperature. If the current passing through a wire is more than the wire can carry, heat builds up and can start a fire.

Compare an electric current to water streaming through a fountain. The strength of the flow of water depends on how hard the water is pushed up the fountain. In electricity, the push that gets the current flowing is measured in **volts** (V). The higher the volts, the stronger the push.

DETERMINE MEANING OF SYMBOLS

When you read about electricity, you may find diagrams of electrical circuits. The diagram may include special symbols. For example, examine the diagram below. Observe that a line, representing a wire, is attached to the positive terminal of a battery labeled 9V. The "V," which stands for volts, is a measure of the electric potential energy between the two ends of the battery.

Follow the arrows and you see 470 (Ω), or ohms. Ohms are a measure of resistance. Resistance is an opposition to the flow of electrons through the wire.

The letters LED stand for light-emitting diode. LED is a material that glows when voltage is applied, and it is used to make a variety of lights.

When you notice symbols in a scientific diagram, look for explanations in the diagram itself or in any captions that accompany the diagram. If explanations don't appear, look for corresponding symbols in the related text. If you still have a need for more information, remember that science symbols have universal meaning, and you can find explanations in print and online sources.

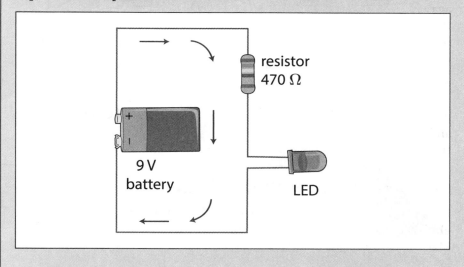

Circuits

An electric current flows only when it can follow a path from a starting point and back again. This path is called a **circuit**, and it must remain closed, meaning there can be no gaps or openings, for electrons to flow.

A typical circuit has at least three parts. The first part is a source of voltage, such as a battery. The battery provides a "push" for the electric current. It has two ends, marked by + and −. Inside the battery, electrons gather at the negative end and flow toward the positive end.

Wires are the second part of a circuit. They carry the current. The third part is the object that uses the electricity, such as a lamp or motor.

In a **series circuit**, current flows in one path only. If a break forms anywhere in the circuit, the circuit becomes open and electricity stops flowing. Some strings of holiday lights are connected in series. If one bulb burns out, the whole string of lights will go out as well.

In a **parallel circuit**, current flows in at least two different paths. If a break forms in one of the paths, current will still flow in the other path. Household electrical outlets are connected in parallel circuits.

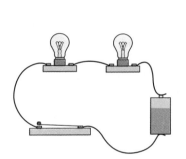

Series Circuit

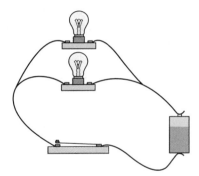

Parallel Circuit

THINK ABOUT SCIENCE

Directions: Match the words on the left with their description on the right.

_____ 1. resistance

_____ 2. volts

_____ 3. parallel circuit

_____ 4. series circuit

_____ 5. static electricity

A. a circuit with more than one path for the current

B. slows the flow of electricity

C. may result when objects rub together

D. the "push" of an electric current

E. a circuit with one path for current

Magnets

Magnets are objects that attract iron and a few other elements. They are surrounded by a region of attraction called a magnetic field, which decreases in strength with increased distance from the magnet. Many metallic materials, including certain ores of iron, are magnetic.

All magnets have poles, where the magnetic field is strongest. The poles are labeled north and south. Two unlike poles, such as a north and a south pole, attract each other. But two like poles **repel**, or push each other away.

If you have ever used a compass to find your direction, you have seen evidence of Earth's magnetic field. The needle of a compass is a thin magnet that swings freely around a post. From most places on Earth, one end of the needle points north. This is because Earth acts as if it has a magnet running through its center.

Scientists do not fully understand what makes Earth's magnetic field. Its likely source is the iron core in Earth's center. This core includes a liquid region, where the temperature and pressure are very high.

Curiously, Earth's magnetic field has changed many times throughout the planet's long history. Today, the north pole of the field is near—but not precisely at—Earth's geographic North Pole.

Electromagnetism

Electricity and magnetism are related. Each can be used to make the other. A coiled wire, for example, will act as a magnet if an electric current passes through it. Such a device is called an **electromagnet**. You use an electromagnet each time you use a hair dryer, doorbell, or telephone. Electric motors use electromagnets to change electric energy to mechanical energy.

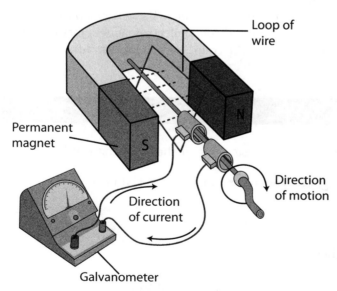

A galvanometer detects electric current.

Compound words are long words made up of two shorter words. You can often figure out the meaning of a compound word by looking at the meaning of the two smaller parts.

Reread this page and find three compound words. In one, the compound word is formed from the shortened form of two words. What are the two words that combine to form the compound word *maglev*?

Passing a magnet through the coils in the electromagnet shown in the diagram below creates an electric current—even without the battery. **Generators** convert mechanical energy into electrical energy using a magnet to create a steady flow of electrons.

Electric motors also use electromagnetism, but in the opposite way that generators work. Motors convert electrical energy into mechanical energy. The mechanical energy can be used to power a car.

In an electric motor, a coil of wire is positioned between two permanent magnets. The current from the battery magnetizes the coil, causing it to rotate between the magnets. The spinning coil provides the energy for the motor to do work.

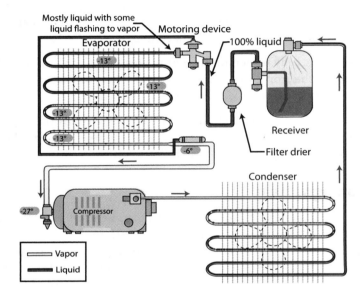

Another application of the electromagnetic principle is in the operation of maglev (magnetic levitation) trains. These trains provide a smooth and quiet ride. A current flows through electromagnets installed in the track and on the underside of the train. The magnetism lifts the train slightly above the track.

Vocabulary Review

Directions: Complete the sentences below using one of the following words:

circuit   electricity   electromagnet   generator   magnet   resistance

1. A doorbell is an example of a series _____.

2. _____ slows the flow of electrons.

3. A(n) _____ is a path along which electrons can travel.

4. A(n) _____ uses the motion of a magnet to produce electricity.

5. The energy caused by the flow of separation of charged particles is _____.

6. Any object that attracts or repels magnetic material is a _____.

Directions: Respond to each question below in the space provided.

1. The word part *electro* is the combining form of the word *electric*. Words that include *electro* relate to electricity. Using *electro* as a base, combine it with one of the following words to form a compound word that will match each definition below. Write the correct compound word.

   acoustics   mechanical   static   surgery   thermal

   _____ A. the use of electricity in surgery
   _____ B. generating heat from electricity
   _____ C. describes electric charges that do not move
   _____ D. conversion of sound into electricity or electricity into sound
   _____ E. a mechanical device run by electricity

Directions: Select the one best answer to the question below.

2. Two lightbulbs are connected in parallel into a circuit, as shown in the diagram on page 254. If one lightbulb is removed from its socket, what will happen to the other lightbulb?
   A. It will go out.
   B. It will stay lit.
   C. It will overheat, then go out.
   D. It will alternate between bright and dim.

Skill Practice

Directions: Choose the best answer to each question.

1. What best describes the structure of an electrical cord, which carries electricity to a household appliance?

   A. all conductor
   B. all insulator
   C. conductor inside, insulator outside
   D. insulator inside, conductor outside

2. What is an example of an effect of static electricity?

   A. Using an electromagnet to pick up metal paper clips.
   B. Turning on a flashlight's switch
   C. Rubbing a balloon on the carpet and sticking it on the wall.
   D. Taping paper to a refrigerator.

3. An electric motor is described as the opposite of a generator because it

   A. changes energy in the opposite way.
   B. rotates in the opposite direction.
   C. runs with an opposite amount of energy.
   D. is made of the opposite kinds of parts.

4. When a magnet is cut in half between its two poles, the two pieces that are formed each act like a magnet. What must have happened to each piece?

   A. Each piece lost either a north pole or south pole.
   B. A north pole or south pole formed along the cut edge.
   C. Electrical charges began flowing, creating an electromagnet.
   D. Electrical charges separated, creating static electricity.

Sources of Energy

Lesson Objectives

You will be able to
- Compare and contrast different sources of energy
- Distinguish between renewable and nonrenewable resources

Skills

- **Core Skill:** Analyze an Author's Purpose
- **Core Skill:** Analyze Text Structure

Vocabulary

biomass
crowdsourcing
energy density
magma
nonrenewable
nuclear fission
renewable
reservoir

KEY CONCEPT: Energy takes different forms, and each form can be used to do work.

You rely on energy to live. When you kick a soccer ball, draw a picture, or type on a keyboard, you use energy. When you shiver to stay warm or sweat to stay cool, you use energy. That energy comes from food that you eat. Your body breaks down the food into molecules. The energy your body needs to function is stored in the chemical bonds of those molecules.

Potential and Kinetic Energy

Kinetic energy is the energy of motion, and **potential energy** is stored energy. Imagine holding a chunk of coal in your hands. Potential energy is stored in the coals' chemical bonds. Or think about filling up your car with gasoline. Gasoline, which comes from crude oil extracted from Earth's surface, contains about 150 chemicals. The potential energy stored in those chemicals remains unusable until they are ignited. Then potential energy is converted to kinetic energy, which powers your car.

Potential forms of energy, such as coal and crude oil, are found in nature, but they only become useful when their potential energy is converted to kinetic energy. It is kinetic energy, or the energy in motion, that lets us do work.

Potential energy at the top of a roller coaster changes to kinetic energy as a gravitational force pulls the coaster train downward.

fotog/Getty Images

Renewable and Nonrenewable Sources of Energy

A resource is any living or nonliving material humans use to meet their needs or wants. Some resources are potentially **renewable**, meaning that if people manage them well, the resources can be replaced at a rapid rate. Biomass, wind, sunlight, water, and geothermal power are renewable resources.

Other resources are **nonrenewable**, meaning they exist in fixed amounts. Only lengthy geological, chemical, or physical processes can replace them. These processes take such long periods of time to occur that humans can exhaust, or use up, nonrenewable resources entirely. In other words, once these resources are used, they are gone for good. Coal, oil, natural gas, and nuclear power are examples of nonrenewable resources.

Biomass

Biomass is one source of stored renewable energy. It is carbon-based, organic matter derived from all living and once-living organisms. Crop stubble, for example, is an example of biomass. It is the plant material that is left behind after a crop has been harvested. Carbon is stored in the stubble's tissues. The carbon comes from the carbon dioxide, which plants absorb for use in photosynthesis, the process of converting the energy in sunlight into chemical energy, or food. When biomass is burned, this stored energy is released as heat. Burning biomass releases more than heat, however. It can also release thousands of chemical compounds, including carbon monoxide, carbon dioxide, water vapor, and small unburned particles of matter.

Some agricultural crops also are grown specifically with energy use in mind. For example, manufacturers add microbes to corn, soybeans, and other crops. The microbes break down the plant material into simpler forms they can use for energy. The process produces ethanol as a by-product. Ethanol, an alcohol, can be used alone or in a gasoline mixture to operate moving vehicles.

 THINK ABOUT SCIENCE

Some gas pumps have labels declaring that the substance you're pumping into your car is 10 percent ethanol. What might be an advantage of adding ethanol to gasoline?

_____

An author's purpose is the reason the author has written something. Generally, an author will write to persuade, describe, explain, or entertain. An author may also have more than one purpose for writing. Look for clues that will help you to determine what an author's purpose for writing is.

For example, three authors wrote a study examining the effects of the whooshing sound of a wind turbine on the health of people living near a wind farm. They concluded that the noise caused people to suffer "wind turbine syndrome," which resulted in panic attacks, nausea, and headaches. However, scientists rejected the study's conclusion for several reasons. First, the authors of the study had been outspoken opponents of wind power. Second, they were not medical experts. Third, their study included fewer than 20 people. Finally, they did not consider any prior global research on the same topic, which found no relationship between the noise from wind turbines and poor human health.

Given what you know about the way the study was conducted, what purpose do you think the authors may have had for seeking evidence of wind turbine syndrome?

Wind Power

Earth's surface is covered in both water and land. Neither is distributed equally, and each takes different forms. The different landforms and bodies of water absorb the Sun's heat differently. Consequently, the Sun heats Earth's surface unevenly, resulting in the motion of air, or more simply, wind. Like other moving objects, wind has kinetic energy, or energy of motion. The amount of kinetic energy produced depends on the object's mass and speed.

Wind causes the blades of a wind turbine to turn. The turbine's kinetic energy can be converted to mechanical energy to pump water or grind grain. The blades can also spin a shaft connected to a generator, which converts mechanical energy into electricity.

Wind turbines are often grouped together in wind farms. The kinetic energy from the turbines' spinning blades is converted to electricity at the site and then fed into the wires of a utility grid to reach customers.

As long as the Sun shines, winds will blow. Consequently, wind is a renewable energy resource.

Solar Power

The Sun provides a steady source of renewable energy. Sunlight strikes photovoltaic (PV) cells, or solar cells. A PV cell is made of semiconductor materials, the kinds of materials found in a computer. A cell can be as small as a single postage stamp, or it can be connected to other cells to form PV modules several feet long. PV modules can also be connected to form PV power-generating units to produce even more electricity.

When sunlight strikes a PV cell, it may reflect, be absorbed, or pass through the cell. It is the energy in absorbed light that is converted to electricity inside the cell. That energy knocks electrons loose from the atoms of the semiconductor materials inside the cell. The loose electrons flow, forming an electrical current.

Hydropower

People have relied on the energy of flowing water, or hydropower, for thousands of years. Long ago, farmers built wooden **turbines**. These spinning paddle wheels were attached to a shaft, which was attached to a large stone called a millstone. Another stone, called the bed, lay beneath the millstone.

Grain was spread on the bed. Then as moving water flowed over the paddles of the turbine, the shaft and the millstone turned, grinding the grain upon the bed.

Today, people use the energy stored in flowing water to generate electricity. To control the rate of flow, people construct dams to create water **reservoirs**—water-storage areas, like artificial lakes. Tunnels beneath the dam are gated. When the gates are lifted, water flows from the reservoir into the power plant, where it causes massive turbines to spin rapidly. The turbines, as a consequence, cause generators to move, producing electricity that is then transported through a utility grid.

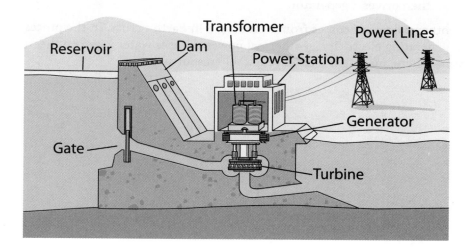

With the use of digital tools, modern scientists have access to more information than ever before. Such a wealth of information presents a new challenge: How is it possible to make sense of so much data?

Scientists can program computer software to analyze many millions of pieces of data quickly. Programs can organize data, graph data, and highlight data points to make it easier for scientists to interpret experimental results.

Today, some scientists, as well as a variety of other professionals, are depending on **crowdsourcing** to analyze this mass of data. Crowdsourcing is a process in which companies and other institutions outsource, or use independent human resources, to perform specific tasks. Scientists use crowdsourcing to recruit volunteers and others outside of their field of study to lend their computer skills, time, and creativity to help interpret scientific data.

What are examples of crowdsourcing that you use or are aware of?

Geothermal Power

There is tremendous heat stored in Earth's layers. People have tapped into this heat for centuries. Since ancient times, people have bathed in hot springs, which are pools of naturally heated water.

Magma is extremely hot, semiliquid rock beneath Earth's crust. In some places, magma bubbles upward, filling chambers or pockets in the crust. In some locations, the magma in these chambers heats groundwater. The heated water rises, forming hot springs. The water at the surface cools. As the water cools, there is less space between molecules. The water becomes denser and sinks. As it moves downward, warm, less dense water comes up to take its place. The water moves round and round in a **convection current**.

In the last century, people began using geothermal fluid—heated water and steam—to generate electricity. Today, there are three kinds of geothermal power plants.

1. *Dry steam power plants* pump steam from underground reservoirs through a turbine. The steam spins the turbine, which in turn, drives a generator that produces electricity.

2. *Flash steam power plants* pump hot water under high pressure into a tank with less pressure. When the pressure changes, the water flash evaporates, meaning it instantly boils and turns to vapor, or gas. The vapor drives a turbine, which in turn drives a generator.

3. *Binary cycle power plants* differ from the other kinds in that the geothermal fluid does not come in contact with a turbine. Instead, it is pumped with a secondary fluid that has a lower boiling point than water through a heat exchanger. *Binary* means "consisting of two." Heat from the geothermal fluid causes the secondary fluid to flash evaporate. Vapor from the secondary fluid drives a turbine, which then drives a generator.

Geothermal power comes from heat stored in Earth's crust. In the process of generating electricity, geothermal power releases water vapor to the atmosphere, making it a clean renewable resource.

Fossil Fuels

Like biomass, fossil fuels are carbon-based matter. Consequently, burning them releases the carbon stored in their chemical bonds into the atmosphere in the form of carbon dioxide. Unlike biomass, fossil fuels took several million years to form and are not renewable. The carbon stored in coal, oil, and natural gas was once stored in prehistoric plants and animals.

Coal, one type of fossil fuel, is buried at different depths beneath Earth's surface. Workers use giant machines to uncover and mine deposits. At the surface, the coal is processed to remove dirt, rock, and other unwanted materials. After processing, the coal is shipped by truck, train, or even pipe to its destination. The coal is then converted into other products, or burned directly.

Deposits of oil and natural gas, two other types of fossil fuels, also exist below Earth's surface. Companies drill through rock to reach the deposits and use pumps to bring the raw products to the surface. From there, natural gas is piped to facilities where it can be processed into fuels.

Raw, or crude oil is also piped or shipped to facilities, called refineries. At the refineries, oil is heated and converted to a number of products, including gasoline, jet fuel, and home heating fuel. Some of the converted product is shipped to power plants, where it is burned to generate electricity.

THINK ABOUT SCIENCE

Name one characteristic that makes biomass and fossil fuels similar. Name one characteristic that makes them different.

_____

_____

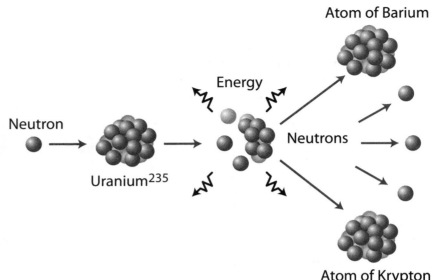

Informational texts include facts that inform readers about a topic by presenting main ideas and supporting details. Authors think carefully about the way they organize text. They apply specific organizational structures to help readers understand the content they are reading. The following are five common structures used to organize informational text.

Cause and Effect: Author explains how or why something happens.

Compare and Contrast: Author describes similarities and differences in information.

Problem and Solution: Author presents a problem and explains how it might be solved.

Order/Sequence: Author describes steps in a process or events in the order they happen.

Description/List: Author describes a topic by listing characteristics, features, and examples.

Review the text structures the author used to write this lesson. Identify the types of structures and how they help you understand the text more easily.

A nuclear power plant generates electricity much like other kinds of power plants do—it uses steam to drive turbines. However, it is the substance used to make that steam that sets nuclear power plants apart.

Nuclear power plants rely on the heat that occurs when atoms of uranium split in two, a process called **nuclear fission**. Uranium is a heavy metal that exists in a variety of forms, or isotopes, on Earth. However, not all isotopes are the same. Only isotopes of Uranium^{235} (U-235) are used to generate electricity.

U-235 splits or decays automatically into two more stable atoms. The process of decay produces radiation in the form of subatomic particles and energy. Although nuclear fission occurs naturally, the rate of decay is slow. So, scientists speed the process by inducing, or forcing, atoms of U-235 to split.

Before U-235 can be used as a fuel in a nuclear reactor, it must be enriched. In nature, U-235 represents about 0.7 percent of the uranium isotopes in any sample of uranium mined from Earth's crust. Scientists use one of several processes to increase the percentage of U-235 in a sample to about 5 percent. Once the fuel is fabricated, or created, it is delivered to a nuclear power reactor.

Rods of enriched uranium are assembled in a reactor vessel. There, they form the core of a nuclear power plant. This core generates heat.

In a pressurized light-water reactor, heat from the core is transferred to water flowing under pressure through a pipe to a steam generator, where steam is then directed to a turbine. In a boiling-water reactor, water moves up through the fuel core, absorbing heat from the nuclear reactions in the core. The resulting water and steam are then separated. The water is pumped back to the reactor vessel to pass through the core again. The steam moves through a steam line, which directs it to a turbine. Then the turbine drives a generator that generates electricity.

The Law of Conservation of Energy

In science a law describes an observable **phenomenon**—a fact or event that we experience. It takes extensive experimentation and collective knowledge to formulate a law. Although laws in specific fields of science are generally understood and accepted by scientists, they also remain open to challenge and revision.

To generate power, the kinetic energy of water and wind are converted to mechanical energy that drives turbines. Inside photovoltaic cells, the energy in sunlight is absorbed and transferred to currents of flowing electrons. In the reactor vessel of a nuclear reactor, heat given off by the decay of uranium atoms transfers to water, which becomes steam that drives turbines. The forms of energy within a system may change from one form to another or transfer from one substance to another, but the total amount of energy in the system remains the same. This is the Law of Conservation of Energy.

Choosing a Source of Energy

How do people decide which sources of energy to use? Economic availability is one consideration. In other words, are supplies plentiful enough to make them affordable? Another consideration is **energy density**, or how much energy is stored in a given quantity of a resource. Still a third consideration is the consequences of their use. That is, what are the disadvantages, or drawbacks, associated with using certain sources?

Almost all sources of energy have drawbacks. When biomass and fossil fuels, for example, are burned to produce heat, they release carbon dioxide into the air. Decaying uranium produces radioactive waste, such as used fuel and radioactive clothing, tools, and other objects, that pose health risks to living organisms, including people. The energy in sunlight is powerful and renewable, but the cost of the machines is high, and the efficiency of the machines used to harness that energy remains low.

People try to assign each drawback a numerical value, or financial cost. They contrast these costs with the amount of energy a source can produce.

As supplies and technologies change, so do the costs. Such changes may encourage political leaders and consumers to make different energy choices. For example, costs associated with using solar and wind energy to generate electricity have fallen as technologies have improved, making the costs of these energy sources comparable to the costs of using fossil fuels.

Vocabulary Review

Directions: Fill in the blanks with the term that best fits the sentence.

biomass   crowdsourcing   energy density   magma   nonrenewable nuclear fission   renewable   reservoir

1. Scientists rely on _____ as an inexpensive way to collect and interpret data.

2. Because there is a limited or finite supply of crude oil, it is a _____ resource.

3. In _____, an atom splits into two more stable atoms.

4. A source of heat of thermal hot springs is _____, molten rock from deep within Earth.

5. Corn and soybeans are sources of energy called _____.

6. The Sun is a _____ resource because it remains available for use for millions of years.

7. _____ is the amount of energy stored in a given quantity of a substance.

8. A _____ is an artificial lake created for hydropower production.

Skill Review

Directions: Read and complete each activity.

1. Explain the Law of Conservation of Energy to someone unfamiliar with the law.

_____

_____

2. What would you describe as the advantages and disadvantages of science communities using crowdsourcing. List at least three of each.

Advantages	Disadvantages
_____	_____
_____	_____
_____	_____

3. In terms of its energy needs, how is your body like a car?

_____

_____

Skill Review (continued)

4. Use a Venn diagram to compare and contrast renewable and nonrenewable resources.

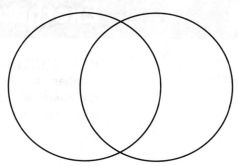

Skill Practice

Directions: Read each problem. Complete the activity.

1. A concept map is a graphic organizer that helps you connect important concepts and identify important details. Create a concept map like the one on the right. Write the words *Energy Sources* in the main circle in the center of your map. Use the satellite circles for important details about each source.

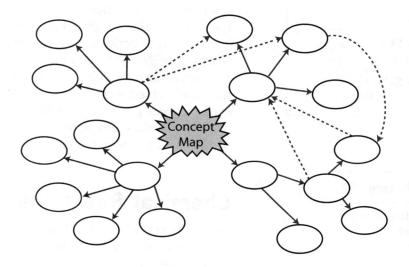

2. What suggestions could you give to community leaders who are looking for an alternative to burning coal or oil as a source of generating electricity?

_____

_____

_____

3. When is it practical to rely on wind energy to generate electricity?

_____

_____

_____

4. Recall that decaying uranium used in nuclear power plants produces radioactive waste, such as used fuel and radioactive clothing, tools, and other objects. Radioactivity poses health risks to living organisms, including people. What would you say to leaders who wanted to bury radioactive waste in your community?

_____

_____

Endothermic and Exothermic Reactions

KEY CONCEPT: Chemical reactions change one substance into another and change their potential energy. When energy is released, the chemical reaction is called exothermic. When energy is absorbed, it is called endothermic.

When you get a haircut you change the length of your hair, one of your hair's physical properties. Physical properties also include characteristics such as mass, volume, density, color, texture, and shape.

Like all matter, your hair has chemical properties, too. It has cells filled with pigments that give hair its color. But even after a haircut, chemical properties remain unchanged.

The chemical properties of matter describe how the matter reacts in the presence of other chemicals. For example, many science students build models of active volcanoes. They pour baking soda, or sodium bicarbonate ($NaCHO_3$), into the cone. Then, they add vinegar, CH_3COOH. The substances undergo two rapid chemical reactions. Carbonic acid forms but immediately changes into carbon dioxide and water. The carbon dioxide gas enters the air, and the liquid that is left is a solution of sodium acetate and water. In other words, the original matter is transformed into new matter.

Vocabulary

activation energy
catalyst
chemical reaction
compounds
endothermic
exothermic
potential energy
product
reactant

Chemical Reactions

All matter is made of chemical elements. A drop of water, a plant leaf, a skin cell, and a soap bubble all are examples of matter, and so they are also made of chemicals.

Some chemicals are pure, meaning they are made of one kind of atom. Other chemicals are **compounds**, meaning they are made of more than one kind of atom.

The smallest unit of a chemical that has all of the properties of that chemical is a **molecule**. A molecule of the chemical known as water is always made of two atoms of hydrogen bonded to one atom of oxygen. You may recognize its chemical abbreviation, H_2O.

A molecule of the chemical called table salt is always made of one atom of sodium bonded to one atom of chlorine. The abbreviation for a molecule of salt is NaCl.

Some chemicals react when they meet. Their atoms interact to make a new molecule of a different kind of matter with properties different from the properties of the original chemicals. This interaction and the production of a new chemical is called a **chemical reaction**.

A Familiar Example

Examine the following chemical reaction. It is probably familiar to you.

$$6CO_2 + 6H_2O \text{ in the presence of sunlight} \Rightarrow C_6H_{12}O_6 + 6O_2$$

Put into words, this statement says that in the presence of sunlight, 6 molecules of carbon dioxide interact with 6 molecules of water to produce new matter, one molecule of glucose and 6 molecules of diatomic (meaning "two atoms") oxygen.

Have you determined what this chemical reaction describes? It is the process of photosynthesis. Sunlight provides the energy necessary for carbon dioxide and oxygen to interact and produce sugar, or food for a plant. The process also produces oxygen, which leaves the plant as waste and enters the atmosphere.

Kinds of Energy

This is a good opportunity to recall the meanings of several words. The first is *work*. **Work** is when a force acts upon an object in the direction of its motion.

Two other words that are good to recall describe two kinds of energy, kinetic and potential. Remember that kinetic energy is the energy of motion. When you run, dance, and tie your shoes, the energy you have because of your movement is called kinetic energy. Kinetic energy exists at the molecular level, too, as atoms vibrate.

While kinetic energy is the energy of motion, potential energy is stored energy. It is the potential energy a substance has for doing work because of its position or its structure. Suppose, for example, that you hold a basketball above your head. Its position above the ground gives it **gravitational potential energy**. If you drop the ball, the force of gravity works upon the ball in a downward direction. The potential energy changes into kinetic energy.

Consider another example of potential energy. Imagine a rubber band. It has **elastic potential energy**. If you apply a force, such as pulling, the rubber band's potential energy is converted into kinetic energy.

Now consider a third kind of potential energy. The structure of a molecule gives it **chemical potential energy**. An attractive force exists between the atoms in a molecule, meaning they are attracted or drawn to each other. This force represents a kind of potential energy that changes to kinetic energy during a chemical reaction.

A **prediction** is an
educated guess based
on available evidence.
Scientists predict what
they expect to happen
during an investigation
based on what they
already know. You can
make predictions, too.
To make the most
accurate prediction
possible, it is helpful to
organize your evidence
in a chart that you can
extend to any size.

Prediction: *The chemical reaction is exothermic, resulting in an increase in temperature of the products.*
What I Know About Chemical Reactions

What kinds of evidence
regarding exothermic
chemical reactions could
you add to a chart
like this?

Exothermic Reactions

The prefix *exo–* means "outside," and the Greek base word *thermo* means "heat or hot."

The word *exothermic* means "outside heat," or releasing heat to the outside. When a force acts upon atoms and molecules, resulting in the release of heat, the process is called **exothermic**.

The chemicals that begin a chemical reaction are called the **reactants**. The chemicals that the chemical reaction produces are called the **products**. In an exothermic reaction, energy breaks the chemical bonds of reactants. The reactants rearrange themselves, forming new chemical bonds and releasing energy. An exothermic reaction feels warm to the touch. Some exothermic reactions release so much heat that it is dangerous to touch containers holding the products. Highly exothermic reactions can even explode.

Exactly how much heat is released during an exothermic reaction depends on the reactants, but all exothermic reactions release more heat than they absorb. You can use symbols to summarize an exothermic reaction.

Reactants   Products

$$A + B \longrightarrow C + D + heat$$

The burning of coal is an example of an exothermic reaction. Coal is a rock made of decayed plant material, or carbon. The purest coal, anthracite, is made almost entirely of carbon. Other kinds of coal are also made of carbon, but they contain other elements as well.

When coal is ignited, a chemical reaction occurs. Coal combines with oxygen to produce carbon dioxide. The reaction gives off heat, making it an exothermic process.

An exothermic reaction, such as burning coal, gives off heat.

THINK ABOUT SCIENCE

The National Aeronautics and Space Administration, or NASA, uses liquid hydrogen as rocket fuel. When combined with liquid oxygen and ignited, the fuel propels rockets into space. How do you know this reaction is exothermic?

_____

_____

Endothermic Reactions

Some chemical reactions are **endothermic**, meaning these reactions will not occur unless work is done. The prefix *endo–* means "within." Combined with *thermo*, meaning "heat," the word *endothermic* means "heat within." Endothermic reactions absorb heat from their surroundings.

Although endothermic reactions absorb heat at first, they cool as the reaction continues. In the end, the reactions absorb more heat than they release. Look at this summary of an exothermic reaction.

Reactants   Products

$$A + B + heat \longrightarrow C + D$$

Consider photosynthesis once more, a common example of an endothermic reaction. Photosynthesis doesn't occur without an input of energy, which comes from sunlight. The energy in the light breaks the chemical bonds of the reactants CO_2 and water. The bonds rearrange themselves to form a molecule of sugar. Heat is stored in the bonds of the new product.

Another example of an endothermic reaction involves ammonium chloride (NH_4Cl) and water (H_2O). Ammonium chloride dissolves readily in water, and it is used to make a variety of products, including firecrackers, fertilizers, cough medicine, some kinds of licorice, and shampoo. The chemical reaction, written below, absorbs heat but cools as the reaction progresses.

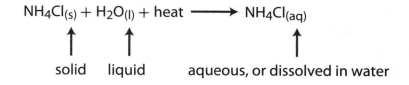

$$NH_4Cl_{(s)} + H_2O_{(l)} + heat \longrightarrow NH_4Cl_{(aq)}$$

solid   liquid   aqueous, or dissolved in water

Dissolved in cough medicine, ammonium hydroxide changes secretions in the lungs and makes them easier to cough up.

THINK ABOUT SCIENCE

Melting ice is not a chemical reaction. It represents a phase change, from a solid to a liquid. How is this phase change similar to an endothermic reaction?

_____

_____

Have you ever used technical directions to assemble a model or to conduct a chemistry experiment? Technical directions often rely on visual clues as much as they rely on text clues. Directions usually begin with labeled pictures of all of the materials required to complete the task.

Directions are usually chunked, or arranged in small blocks of text. They may be combined with diagrams that support or help explain the words. Blocks of text are numbered to help you complete each step in its proper order. There may be small text boxes, too, that provide helpful hints, reminders, or safety warnings.

Research instructions for creating a chemical reaction. For example, look for a project for making "hot ice," which results in an exothermic reaction. Follow the directions carefully in order to complete the process. Whether writers are explaining how to assemble a model or how to add a catalyst to reactants, they think carefully about every word and every diagram they choose. Why do you think it is important to follow directions precisely?

Potential Energy in Chemical Reactions

Reactants and products have different amounts of potential energy stored in the bonds that join their atoms. You can graph the change in potential energy that occurs in an endothermic reaction.

An endothermic reaction absorbs energy, so the products have more potential energy than the reactants do.

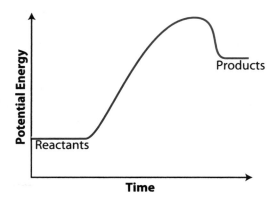

You can also graph the change in potential energy that occurs in an exothermic reaction.

An exothermic chemical reaction usually releases light, sound, or heat. So, the overall potential energy of the product is less than the potential energy of the reactants.

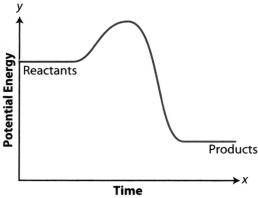

Notice the bump in each graph. These bumps represent a kind of barrier that must be overcome if a reaction is going to occur.

Recall that potential energy is stored in the bonds of atoms and molecules. Those bonds must be broken and made into new bonds for new products to form.

So what is required to break and possibly make new bonds? The answer is energy. It is called **activation energy**, and the amount of activation energy each kind of reaction requires is different.

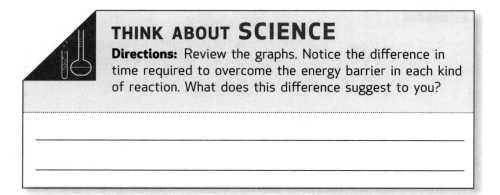

THINK ABOUT SCIENCE

Directions: Review the graphs. Notice the difference in time required to overcome the energy barrier in each kind of reaction. What does this difference suggest to you?

_____

_____

Catalysts

The activation energy required for some chemical reactions can be very high. A **catalyst** lowers that activation energy by providing an alternative pathway for the reaction to follow. Compare the process to a train ride.

Suppose a mountain range separates two points, A and B. Builders can lay train tracks up, over, and down each mountain between the points, or they can blast tunnels through the mountains and lay tracks inside. Trains traveling through the tunnels will reach point B far sooner than trains that cross over the mountains. The tunnels provide an alternative path, just as catalysts do.

In addition to speeding up a chemical reaction, a catalyst remains intact during the process, meaning it is not used up in the reaction. So, a small quantity of catalyst can be used to speed large quantities of reactants.

There are different ways a catalyst can speed up a reaction.

1. Positive and negative electrical charges attract one another. A catalyst can give a molecule a charge that will attract other reactants.

2. A catalyst can change the concentration of reactants, increasing the likelihood that reactants will collide.

3. A catalyst can change the shape of a reactant to make it easier for another reactant to react with it.

Although a catalyst decreases amounts of activation energy, it does not affect the potential energy stored in the products, as you can see in the following graph.

WRITE TO LEARN

Think about the kinds of chemical reactions that regularly occur in your kitchen. Give two examples of chemical reactions and explain why they represent either exothermic or endothermic reactions.

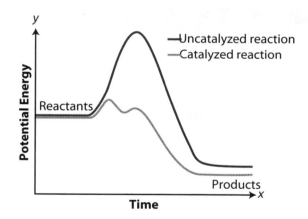

THINK ABOUT SCIENCE

Directions: Fill in the blanks to describe a catalyzed reaction.

Reactant 1 + Reactant 2 + _____ ⇒ _____ + Product 2

Directions: Use the terms to complete the sentences.

activation energy   product   chemical reactions   compounds   reactant   endothermic
exothermic   potential energy   catalyst

1. During _____, reactants change into products.

2. Chemicals made from more than one kind of atom form _____.

3. A roller coaster train at the top of a train track has _____.

4. A _____ modifies and increases the rate of a reaction without being consumed in the process.

5. For a chemical reaction to occur, _____ is required.

6. _____ chemical reactions release more heat than they absorb.

7. A _____ is a chemical that begins a chemical reaction.

8. _____ chemical reactions absorb more heat than they release.

9. A _____ is the chemical that the chemical reaction produces.

Skill Review

Directions: Examine the chemical sentences. Then complete the activities.

Reactants	Products	Reactants	Products
A + B ⟶	C + D + heat	A + B + heat ⟶	C + D

1. Explain how the processes are alike.

_____

2. Explain how the processes are different.

_____

Directions: Read the text and complete the activity.

3. In your car, a mixture of oxygen and gasoline is ignited, and a chemical reaction begins. What kind of reaction occurs? How do you know?

_____

4. Sodium (Na) is a poisonous element. So is chlorine (Cl). In a chemical reaction, the elements form NaCl, or table salt. How can you explain this change?

_____

Skill Review (continued)

5. Ammonia forms when the gases nitrogen (N_2) and hydrogen (H_2) combine under pressure. N_2 is very unreactive. What must manufacturers do to produce ammonia?

_____

6. Enzymes are proteins with specific jobs to do, and they are part of biochemical reactions. For example, an enzyme in your saliva begins breaking down food as you chew. How is an enzyme in a biochemical reaction like a catalyst in a chemical reaction?

_____

Skill Practice

Directions: Examine the chemical sentence. Then answer questions 1–3.

$$6CO_2 + 6H_2O + \text{energy in sunlight} \Rightarrow C_6H_{12}O_6 + 6O_2$$

1. What name is commonly given to this process?

_____

2. What kind of chemical reaction does the sentence represent? Explain your answer.

_____

_____

3. What happens to the energy that was absorbed by the reactants?

_____

_____

Directions: Use the chart to describe three kinds of potential energy.

4.

Gravitational Potential Energy	
Elastic Potential Energy	
Chemical Potential Energy	

5. Why do different industries use catalysts in the production of their products?

_____

_____

Review

Directions: Choose the best answer to each question.

1. What is one example of an object with high potential energy due to the position of its parts?

   A. a stretched bow with an arrow ready to be fired

   B. a bow after its been released, with the arrow in mid-air

   C. a bowling ball rolling down the alley toward pins

   D. an automobile battery in a car

2. Sound waves made in which way will have the greatest amplitude?

   A. pressing the key on a piano for a low note

   B. pressing the key on a piano for a high note

   C. whispering to a friend

   D. 40 friends shouting at once

3. Two magnets are placed side by side. If the magnets are moved together and then released, what will happen?

   A. They will move apart.

   B. The magnets will stick together.

   C. Neither magnet will move.

   D. The poles will exchange positions on one of the magnets.

4. How are biomass and fossil fuels alike?

   A. They are carbon-based.

   B. They fill reservoirs beneath the ground.

   C. They are renewable.

   D. They are nonrenewable.

5. What kind of energy is stored in the structure of a molecule?

   A. kinetic energy

   B. gravitational potential energy

   C. elastic potential energy

   D. chemical potential energy

Review

Directions: Choose the best answer to each question. Questions 6–8 refer to the following passage.

A refrigerator uses electricity to remove heat from air. The method combines the compression of gases and the process of evaporation.

When a gas is compressed, its temperature rises. When a gas is allowed to expand, or when a liquid evaporates, heat is absorbed by the gas.

The compressor in a refrigerator pumps a gas, usually Freon, through the entire system. The compressor squeezes the Freon and pumps it into the discharge line (1 on the diagram). As the gas is compressed, its temperature rises. This heated gas enters the condenser (2) on the back or bottom of the refrigerator, where heat is released into the air. As the compressed gas releases its heat, the gas cools and condenses into a liquid.

A narrow tube (3) carries the liquid freon from the condenser (2) to the evaporator tubing (4). This tubing is larger than the tube from the condenser, so as the liquid enters the evaporator tubing, the liquid begins to expand and become a gas.

In the process of passing through the evaporator tubing, the evaporating liquid absorbs the heat inside the refrigerator and cools the contents. The gas returns to the compressor through the suction line (5), to begin the cycle again.

6. What most likely is a property of Freon, the gas used in refrigerators?

   A. Freon is easy to compress.
   B. Freon reacts readily with metals.
   C. Freon breaks down quickly into other gases.
   D. Freon easily changes into a solid.

7. In a closed system, such as the tubes behind a refrigerator, liquid that is allowed to evaporate will

   A. be compressed.
   B. remain a liquid.
   C. absorb heat.
   D. release heat.

8. Based on the passage and diagram, which occurs in the condenser?

   A. Heat is absorbed by the gas.
   B. The gas is compressed.
   C. Heat is released by the compressed gas.
   D. The liquid evaporates into a gas.

Directions: Read the text and write a response.

9. Use combing your hair to explain the attraction of opposite electrical charges.

_____

_____

_____

_____

_____

10. Explain how ocean scientists use sound waves to map the ocean floor.

_____

_____

_____

_____

11. Explain the difference between an endothermic and exothermic reaction.

_____

_____

_____

_____

_____

12. Explain the source of heat that nuclear power plants convert to electricity.

_____

_____

_____

_____

_____

Review

Check Your Understanding

On the following chart, circle the number of any item you answered incorrectly. Next to each group of item numbers, you will see the pages you can review to learn how to answer the items correctly. Pay particular attention to reviewing those lessons in which you missed half or more of the questions.

Chapter 7 Review

Lesson	Item Number	Review Pages
Energy	1	238–243
Waves	2, 10	244–251
Electricity and Magnetism	3, 6, 7, 8, 9	252–257
Sources of Energy	4, 12	258–267
Endothermic and Exothermic Reactions	5, 11	268–275

Application of Science Practices

CHAPTER 7: ENERGY

Question

How is electrical power generated where you live?

Background Concepts

Spinning turbines are responsible for generating almost all of the electrical power used in today's homes and industry. Turbines spin electromechanical generators, which convert kinetic energy into electrical energy. The distinguishing factor among methods of electrical energy generation is the energy source used to spin the electromechanical generator.

For example, hydroelectric power plants use the kinetic energy of flowing or falling water to spin electromechanical generators to produce electrical energy. Most fossil fuel power plants burn fuel to create high-pressure steam, which spins the turbines of electromechanical generators in a way similar to the way hydroelectric power does.

Each method of electrical power generation has its own advantages and disadvantages related to economic cost, environmental health, and safety.

Investigation

Visit the website for your state's Public Utilities Commission. Explore the site to identify the sources of energy that power plants in your state use to generate electricity. Or use one of the contact links or phone numbers to ask for information.

List the energy source(s) in your state in the following table. Then use the website, other online sites, or print materials to identify the advantages and disadvantages of each energy source.

Energy Source	Economic Impact		Environmental Impact		Safety Impact	
	Advantages	Disadvantages	Advantages	Disadvantages	Advantages	Disadvantages

Application of Science Practices

Interpretation

Imagine that you are on your community council. Your community is considering constructing a new power plant. Based on the information you have collected, what recommendations would you make to business owners, environmental groups, and safety advocates? Write your recommendations to each group.

Business Owners: _____

_____

Environmental Groups: _____

_____

Safety Advocates: _____

_____

Answer

How is electrical power generated where you live?

_____

_____

In your opinion, which source of energy in your state has more advantages than disadvantages?

_____

_____

Evidence

Describe the evidence you have gathered to support your opinion.

_____

_____

Work, Motion, and Forces

Swing a bat and watch a baseball or softball soar. Propel yourself from home plate and race toward first base. Striking the ball was work. So was racing toward first base. Work happens when a force, such as a swinging bat or the contraction and relaxation of your body's muscles, is applied to move an object over a distance.

In this chapter you will learn about:

Lesson 8.1: Newton's Laws of Motion
A race car rounding a tight curve, speeding to the finish line, and then coming to a stop provides an exciting example of speed, velocity, and acceleration. The relationship between motion and forces is the subject of Newton's laws of motion covered in this lesson.

Lesson 8.2: Forces and Machines
What you think of as "work" or a "machine" might not be the same thing a scientist means by those terms. This lesson explains simple machines, compound machines, and the relationship between force and work.

Goal Setting

In a typical day, we use a variety of simple machines. For example, think about the tools found in your kitchen. You probably use knives, forks, and spoons every day. On occasion, you may also use kitchen scissors; a can opener; a pizza cutter; a cheese slicer; or a bottle opener.

Draw pictures of three simple machines in your kitchen. Then, after you have read the chapter, identify the simple machine or combination of simple machines each object represents. Explain how each machine makes work easier to do.

Newton's Laws of Motion

KEY CONCEPT: Forces must act upon an object to change its motion. Newton's three laws of motion explain the relationships between forces and moving objects.

"Turn left in 2.8 miles." This command could come from a passenger with a map. It could also come from a navigation system in your car. A satellite above Earth measures your car's position and velocity. A computer calculates the fastest route to your destination. About 300 years ago, Sir Isaac Newton explained the motion of objects. His work remains useful today.

Position and Motion

We often think of objects as either moving or not moving. The difference, however, is not always that simple. For an example, consider this riddle:

> A boy and girl are sitting on a blanket in the park. The girl stands up, walks 30 feet in a straight line, and then sits down on a bench. She rests there for several minutes, then stands up again and walks in the same straight line back to the blanket.
>
> "Did you see how I moved?" asks the girl.
>
> "Yes," says the boy. "I watched you move in a circle."
>
> How could the boy's observation be correct?

Give up? The answer is that the bench on which the girl sat was on a merry-go-round.

This riddle illustrates the importance of a frame of reference to describe position and **motion**, or a change in the position of an object. You can think of a **frame of reference** as the background against which motion occurs. In the riddle, the boy and other **stationary**, or nonmoving, objects in the park served as a frame of reference. The bench on the merry-go-round was not stationary compared to other objects, and this made the riddle puzzling.

We often think of a frame of reference as made of nonmoving objects. Remember that Earth—and everything upon it—is always moving in its orbit around the Sun. Nevertheless, Earth is a very useful frame of reference. Objects such as houses, streets, trees, and buildings are all stationary within Earth's frame of reference. Thus it is helpful to compare a position to them.

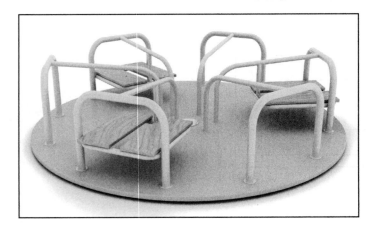

©iStockphoto/alxpin

DETERMINE THE CENTRAL IDEA OF A TEXT

Each text you read has a central, or main idea. A text's details, evidence, and visuals support the central idea in some way. They may provide explanations and examples, or they may present information in diagrams and other visuals.

Sometimes, an author states the central idea. The idea may appear in the first or last sentence of a text, for example. In other cases, the central idea is not identified so clearly. As the reader, you must consider the central idea of each paragraph and analyze the ideas to determine a central idea. You can also examine charts, tables, diagrams, illustrations, photographs and other visuals to determine which central idea they support.

Read the following text. Summarize its central idea.

A baseball travels at a speed of 30 meters per second, which is equivalent to 67 miles per hour. But the baseball's speed tells only half the story. In which direction is it moving? The baseball could be flying toward home plate or the outfield. Or it could be soaring straight up into the air or falling downward after a fly ball. Together, a moving object's speed and its direction determine its **velocity**.

What is the central idea of this text?

_____

_____

_____

_____

_____

21st Century Skill
Initiative and Self-Direction

Self-directed learners are interested in going beyond basic skills or conceptual understandings. Their curiosity drives them forward to learn more and to acquire real expertise in an area of personal interest.

Curiosity about baseball, for example, has led many scientists who study force and motion to gain a better understanding of this popular sport. These scientists use their knowledge to learn more about the physics of baseball. What factors influence the speed and trajectory of a baseball? Does the bat material—wood or aluminum—make a difference? What makes a knuckleball so hard to hit? High-speed equipment lets scientists observe and measure pitches and hits and learn much more about the science behind the game. Initiative, self-direction, and curiosity enrich these scientists' lives, making them more productive, and more interesting, too.

Describe an experience that left you wanting to know more. Explain what you did to reach your goal.

The passages on this page use common terms in a purely scientific way. Maybe the text has made you think differently about some of these common terms. Words such as *speed*, *distance*, and *accelerate* have specific meanings in a scientific context. When you read terms in informational texts such as this, it sometimes is a good idea to apply the terms to your own life and your own experience. For example, connecting the meaning of the scientific terms *accelerate*, *speed*, and *velocity* to the purpose of driving to a store or riding your bike can provide a deeper perspective to your understanding and help you to see the world in a different way.

Read again the passages on this page and think about how the information can be applied to details of a short trip you recently took. In a notebook, write a description of a trip using the scientific terms from the passages on this page.

Distance

Distance is a measurement of change in position. If you travel a distance of 800 kilometers (500 miles), you have followed a path that is 800 kilometers long. This path could be any shape, including straight lines, circles, and other shapes. You might travel 800 kilometers and end up exactly where you began!

Speed and Velocity

Speed is the change in the distance an object travels over time. Speed is always measured in a unit of distance per a unit of time. A car on the highway, for example, might be traveling at a speed of 60 miles per hour. A falling rock might hit the ground at a speed of 10 meters per second.

Velocity is defined as both speed and direction. The velocity of a car could be 60 miles per hour traveling east. The velocity of a falling rock might be 10 meters per second, straight down.

Acceleration

In common language, to accelerate is to go faster. Press the accelerator pedal on a car, for example, and the car's speed increases.

In physical science, however, an **acceleration** is any change in velocity. An acceleration can be an increase in speed, but it can be a decrease in speed as well. A car that is slowing down, perhaps because the driver is stepping on the brakes, is accelerating in the opposite direction of its motion. This is sometimes called a **deceleration**.

In addition, remember that velocity is both speed and direction. For this reason, a moving object that changes direction is also undergoing an acceleration. This means that a car turning left or right is undergoing an acceleration, even if its speed does not change. Other examples of acceleration include a roller coaster passing over a hill or through a loop, and a figure skater spinning or leaping.

THINK ABOUT **SCIENCE**

Directions: Read the sentence below and answer the questions that follow.

An airplane is flying over Minneapolis, Minnesota, at **500** miles per hour.

1. What does the sentence describe about the airplane's position, speed, and velocity?

   _____

2. How long will the airplane take to fly to New York City, which is approximately 1,000 miles away from Minneapolis? (Assume the airplane's speed does not change.)

   _____

Newton's Laws of Motion

British scientist Sir Isaac Newton (1642–1727) explained the relationship between motion and forces. He did so with three simple laws, which are described below. Although these laws are now hundreds of years old, they remain the model that scientists use.

Newton's First Law of Motion
The first law of motion states that an object at rest tends to remain at rest, while an object in motion continues to move in a straight line. This law is also called the law of **inertia**. Inertia is the tendency of an object to resist a change in its motion. When a person is riding in a car and the driver applies the brakes, the car stops but the person will move forward against the seat belt. This is because of the person's inertia.

Understanding the first part of this law is easy. We don't expect things to move unless something moves them. The second part of the law does not seem as reasonable. In our experience, objects do stop moving by themselves. This is because forces, such as friction and gravity, act upon them. A force is a push or a pull.

Newton's Second Law of Motion
The second law of motion states that a change in the motion of an object depends on the mass of that object and the amount of force applied to it. It can be written as:

$$\mathbf{F} = \mathbf{m} \times \mathbf{a}$$

$$\textbf{Force} = \textbf{mass} \times \textbf{acceleration}$$

One example of a force is the force of gravity, which on Earth acts to pull all objects downward. Friction is also a force. It acts to slow down motion when objects are in contact with each other.

To understand the relationship between mass and force, imagine a person struggling to push a piano. Then imagine the person using the same force to push a small book, which has a much smaller mass than the piano. While the piano might barely move, the book would move easily.

Newton originally described the second law in terms of a change in momentum rather than of mass and acceleration. **Momentum** is the product of velocity and mass. A bullet is small, but it has great momentum when it travels at high speeds. A heavy truck—even one that moves slowly—also has great momentum.

Newton's Third Law of Motion
Newton's third law of motion states that for every action, there is an equal and opposite reaction. The recoil, or kick, of a rifle illustrates this law. Gunpowder explodes in a rifle, forcing the bullet through the barrel. At the same time, the explosion pushes the rifle backward against the shooter's shoulder. In the same manner, a rocket is pushed up into the air when gas from the engine rushes out of the back. This equal and opposite reaction works just as well in space—even with nothing to push against.

Directions: Complete the sentences below using one of the following words:

acceleration   **distance**   **inertia**   **motion**   **speed**   **velocity**

1. Twenty meters from Point A to Point B describes _____.

2. A car's _____ may be 50 miles per hour heading north.

3. The tendency for an object to resist any change in its motion is _____.

4. Seventy miles per hour is the _____ at which a cheetah can run.

5. _____ is a change in an object's position.

6. A change in speed or direction is _____.

Skill Review

Directions: Respond to each statement below in the space provided.

1. Which of Newton's laws explains why people should use seat belts?

_____

2. Which of Newton's laws explains how rocket launches work?

_____

3. Which of Newton's laws explains why it is harder to lift a bowling ball than a Ping-Pong ball?

_____

Directions: Choose the one best answer to this question.

4. A hockey puck is shot down the ice. If no one touches the puck and it could travel for a very long distance, what would eventually happen to it?

   A. The puck would continue to travel with the same speed until it hit the net.
   B. The puck would stop because of the forces inside the puck.
   C. The force of gravity in the rink would slow the puck.
   D. The force of friction between the puck and the ice would slow the puck.

Skill Practice

Directions: Choose the best answer to each question.

1. Why is it more difficult to walk on ice than on dry pavement?

   A. Ice provides much more friction.
   B. Ice provides much less friction.
   C. You weigh less on ice.
   D. Ice melts when its temperature rises.

3. A rock with a mass of 20 kg falls off a ledge. It hits the ground at an acceleration of 40 $\frac{m}{s^2}$. If force equals mass times acceleration ($f = ma$) with what force does the rock hit the ground?

   A. 20 $\frac{kg \times m}{s^2}$
   B. 200 $\frac{kg \times m}{s^2}$
   C. 400 $\frac{kg \times m}{s^2}$
   D. 800 $\frac{kg \times m}{s^2}$

2. A truck driver steps on the brakes to allow an emergency vehicle to pass. After the vehicle passes, the driver resumes a speed of 30 miles per hour. Which statement describes the motion of the truck?

   The truck has experienced

   A. a change in acceleration.
   B. an increase in momentum.
   C. a decrease in inertia.
   D. a change in direction.

4. The driver continues driving 30 miles per hour as she drives onto an exit road that curves to the right. How has the truck's motion changed?

   The truck has experienced a change in

   A. speed.
   B. velocity.
   C. mass.
   D. volume.

Forces and Machines

According to the scientific definition, work involves a force that is applied to move an object over a distance. Even studying for tests does not qualify as work—unless you move your book across the desk!

Forces

A **force** is defined as a push or pull. Some forces act when two objects are in contact with each other. The force you exert when you lift or shove a heavy object is a contact force. Other forces can act without contact. Gravity is an example of a noncontact force.

When an object is at rest on Earth, it has at least two forces acting upon it. One is the downward force of gravity. The second is the upward force of the supporting object, such as the ground or floor. A book on a desk, for example, does not float in midair because of the force of gravity. It does not crash to the floor because of the upward force of the desk supporting it. Airplanes fly because the downward force of gravity is balanced by the upward force of the air pushing against the wings.

Objects may also have additional forces applied to them. When the forces are balanced, or in **equilibrium**, the object will not move. If the forces are not balanced, the object will move in the direction of the stronger force. In tug-of-war, if two teams are equally matched, the forces are balanced and the rope will not move. If one team adds a player, the rope will move in the direction of the greater force.

Force and Work

Most of us think we know what work is. In physics, however, the term **work** has a special definition. Work occurs when a force moves an object through a distance.

work = force × distance

The key is that both force and distance must be present for work to occur. Thus, a mover who struggles to budge a piano has not done any work upon it. Work occurs only when the piano actually moves.

Friction

Friction is the force that forms between a moving object and anything in contact with it. The contact slows motion and increases the work that must be performed. Friction also produces heat.

Different materials generate different amounts of friction. Motor oil, for example, is used in car engines because it generates very little friction. Without motor oil, the friction between the moving metal parts of the engine would be very high. The car would overheat.

DISTINGUISH BETWEEN FACTS AND SPECULATION

You have access to more scientific information today than ever before. You can read articles from scientific journals and popular science magazines. You can read encyclopedia entries and explore websites sponsored by science groups and science enthusiasts, or people who enjoy science but may or may not have science expertise. You can also access sites where people ask science questions and others post responses. Again, those responses may be provided by people with no formal science training or experience.

Given the breadth of science information that is available to you, it is important to **distinguish**, or understand the difference, between facts and speculation. Facts are true statements, or information that can be observed or proven to be true. Speculation is an assumption, or acceptance that something is true. Assumptions are not the same as facts, but they can appear in the printed and online materials you access. Consequently, it is important that you distinguish between facts and speculation as you read.

Read the following text.

Imagine that the moon had never formed. Earth would continue to exist, but not as we know it. Without the moon's influence, Earth would rotate more quickly, resulting in a day perhaps only one-third as long as we know it. The greater speed would result in more powerful winds that would scour Earth's surface. The planet would wobble on its axis, making temperatures less predictable, and with only the Sun to exert a gravitational force on Earth's oceans, the tides would be far less dramatic. As far as life goes, it would continue. But again, with different environmental forces, different kinds of life would have evolved over millions of years. The Earth we know would be far different from an Earth without a moon.

How do you know that this text is based on speculation and not fact?

_____

_____

_____

_____

_____

_____

Machines

Machines are designed to increase force, making work easier. Some machines are able to change a small movement into a larger one or change a small force into a larger one. Other machines are able to change the direction or position of a force and put it to bear where the force is needed most.

Machines can be simple or compound. **Simple machines** have few, if any, moving parts. Examples include levers, pulleys, wheels and axles, and inclined planes. **Compound machines** are made of two or more simple machines. An example is a handheld can opener, which consists of a wedge, a lever, and a wheel and axle.

Levers

One of the simplest machines is the lever. A **lever** is a rod or bar that rests on a pivot point called the **fulcrum**. One side of the lever is the effort arm, where a person applies an effort force. The other side of the lever is the resistance arm, where the output force is produced. There are three kinds of levers with different positions of the fulcrum, effort arm, and resistance arm. Some levers, such as a hammer and a wheelbarrow, increase the force that is applied to them. Other levers, such as a fishing rod, increase the distance that the force is applied.

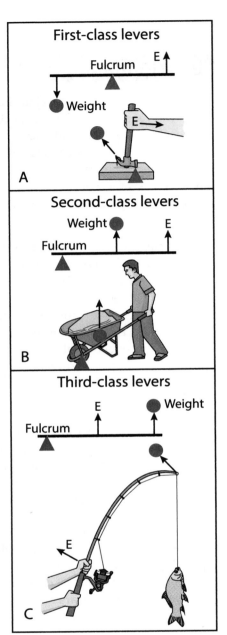

THINK ABOUT SCIENCE

Directions: Complete each statement below with the appropriate word.

1. Two or more simple machines make up a _____.

2. A _____ is a push or pull, or anything that can affect the motion of an object.

3. The force of _____ opposes the motion of an object.

4. If two forces acting on an object are equal, the object is in _____.

5. A _____ has few, if any, moving parts.

A first-class lever produces an output force that is greater than the effort force. To build a first-class lever, you need only a long, rigid bar and a small support, such as a small rock, to act as a fulcrum. Pushing or pulling the effort arm down is work. The person uses both force and distance. The following diagram shows what happens at the resistance arm of the lever. Notice that the distance is greatly reduced, but the force is increased. The work, however, is the same. A crowbar, shovel, and seesaw are all first-class levers.

LEVER

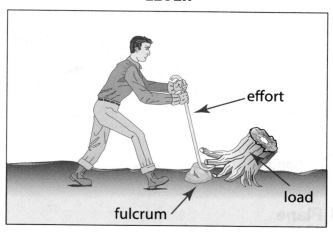

Pulleys

A **pulley** changes the direction of a force. By using a pulley, a person can lift an object by pulling down on it. One end of a rope is attached to the load, and force is applied at the other end to lift the load. Pulleys are used to lift flags on flagpoles and to lift some kinds of windows. Combinations of pulleys can also increase the amount of force.

PULLEY

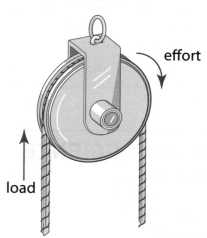

Core Skill
Distinguish Cause and Effect

A cause is why something happens. An effect is the result of the cause. When you read, you may have to distinguish between cause and effect to better understand what is being described or explained. In this lesson, you will read about many kinds of machines. Each has a cause-and-effect relationship between force and work. In a notebook, make a chart like the one below to distinguish cause and effect for each kind of machine.

Kind of Machine	Cause (Force)	Effect (Work)

WRITE TO LEARN

Learning to identify cause-and-effect relationships can help you better understand what you read. Think about a time you cooked or prepared a meal, or performed another common household task. In a notebook, list the actions you took (causes) and the results of your efforts, good or bad (effects). Then write two paragraphs about the event and organize your writing to clearly show the cause-and-effect relationship.

Wheel and Axle

A **wheel and axle** is a simple machine that is actually a special type of lever made up of two circular pieces. The larger circular piece, the wheel, is the effort arm. The smaller circular piece, the axle, is the resistance arm. Screwdrivers and the steering wheel of a car are wheel-and-axle machines. For the screwdriver shown below, the larger the handle (the wheel), the greater the increase in the force of the blade (the axle).

WHEEL AND AXLE

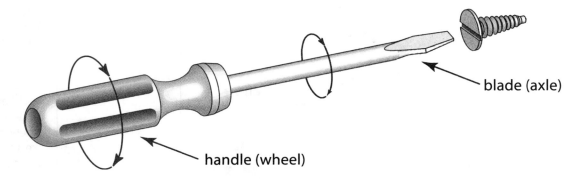

blade (axle)

handle (wheel)

Inclined Plane

An **inclined plane** is simply a flat surface that has one end higher than the other. Inclined planes multiply the effort force; however, the force must be applied over a longer distance.

A ramp is an example of an inclined plane. Pushing a heavy box up a ramp is easier than lifting it. Walking up a long gentle slope is easier than climbing up a steep slope. A wedge is also a type of inclined plane.

Compound Machines

Everyday tools, such as a pair of scissors, a shovel, a bicycle, and an automobile, consist of many different simple machines put together. The pair of scissors, for example, is two levers and two inclined planes. An axe is an inclined plane (the blade) and a lever (the handle).

An eggbeater is a compound machine composed of gears and wheels and axles. The handle turns a wheel, which itself turns a smaller wheel. Eggbeaters magnify force.

THINK ABOUT SCIENCE

Directions: Choose the term in the parentheses that best completes each sentence.

1. A wedge is a type of (lever, inclined plane).
2. The pivot point of a lever is called the (midpoint, fulcrum).
3. A doorknob is an example of a (wheel and axle, inclined plane).
4. A pulley changes the (direction, speed) of an applied force.

Gears

Bicycles and automobiles consist of many different types of machines. In these machines, the wheel-and-axle combination is turned by various sets of gears. **Gears** can change the direction of a force and can either increase the force or change its speed. Automobiles use gears to change the direction of a force (forward or reverse), to move forward slowly with much force (first and second gear), or to move rapidly with little force (third or fourth gear).

Engines

Engines are made of many different types of machines, but they add one critical element not seen in a simple machine. That element is a source of energy other than human muscle. Some of the earliest engines were windmills and waterwheels. Steam engines followed. James Watt, a Scottish inventor, was the first to popularize the steam engine. In an effort to sell his new, improved steam engine, he compared its power to the power of a horse. We still describe engines by using the term *horsepower*.

GEARS

Vocabulary Review

Directions: Match the words in the column on the left with their description on the right.

1. _____ compound machine

2. _____ equilibrium

3. _____ force

4. _____ distinguish

5. _____ simple machine

6. _____ friction

A. a lever, a pulley, inclined plane, or a wheel and axle

B. a push or a pull

C. when two forces acting on an object are equal

D. a force that opposes the motion of one object in contact with another

E. a combination of two or more simple machines

F. to understand as being different

Directions: Read the following passage and choose the best answer to each question.

> A screw is a simple machine. It consists of a long, thin ridge that wraps around a central bar. Because of its spiral shape, this ridge is similar to a circular ramp at a parking garage. By moving in a circle instead of just up and down, a screw makes work easier to do.

1. What is the difference between the ridge of a screw and a circular ramp at a parking garage?

   A. The ridge is flatter.
   B. The ridge is smaller.
   C. The ridge is steeper.
   D. The ridge is an inclined plane.

2. Why is a screw an example of a simple machine?

   A. A screw makes work easier to do.
   B. A screw can hold objects together.
   C. A screw is strong and tough.
   D. A screw can be turned easily.

Directions: Read the following passage and then choose the best answer to each question.

> Newton's second law of motion relates the force needed to lift an object to the mass of the object. The law explains why two people may be needed to lift a fully packed box of books. In contrast, a young person could easily lift another box, of equal volume, that is packed with pillows.

3. Which of the following is the cause for the box of pillows being easier to lift than the box of books?

   A. The boxes are different colors.
   B. The force of the boxes is different.
   C. The box of pillows has a smaller volume.
   D. The box of books has a greater mass.

4. Which cause-and-effect relationship could be used to illustrate Newton's second law of motion?

   A. A box of books remains at rest on a table.
   B. A skater moves backward when she pushes against a wall.
   C. A car stops, and the driver pushes forward against the seatbelt.
   D. A heavy book takes more effort to shove across a table than a lightweight book.

Skill Practice

Directions: Choose the best answer to each question.

1. Which of the following is a measurement of velocity?

   A. 5 kilometers per second
   B. 14 kilometers
   C. 25 kilometers per hour, heading north
   D. 10 meters per second

2. Pulling down on the rope of a pulley can lift an object attached to the other end of the rope. How does the pulley change the force applied to the rope?

   A. by increasing the strength of the force only
   B. by changing the direction of the force only
   C. by increasing the strength and changing the direction of the force
   D. by removing the force

3. What can gears change about a force that is applied to them?

   A. only the direction of the force
   B. only the strength of the force
   C. both the direction and strength of the force
   D. neither the direction nor strength of the force

4. In which of these situations is work being done?

   A. A woman pushes a mower across the lawn.
   B. A father holds a 30-pound child above his head.
   C. A student solves a math problem.
   D. A weightlifter pulls on a 300-pound barbell but fails to move it.

5. Where can an object move without friction?

   A. across a smooth floor
   B. through the air
   C. on ice
   D. in deep space

6. Which of these machines uses one or more inclined planes to make work easier to do?

   A. a round doorknob, which is a wheel that rotates to unlatch a door
   B. a seesaw, which is a board that moves up and down at both ends
   C. a conveyor belt, which is a flat surface that moves between pulleys
   D. a butter knife, which has sloped sides that can cut a stick of butter

7. A crowbar is a long, stiff bar that can be used to move heavy objects. What type of simple machine is it?

   A. lever
   B. wheel and axle
   C. pulley
   D. screw

Review

Directions: Choose the best answer to each question.

1. The newton is a unit that measures force. If you push against a brick wall with a force of 10 newtons, what does the brick wall do to you?

   A. pushes down on the ground with a force of 10 newtons
   B. pushes against you with a force that depends on the wall's weight
   C. pushes against you with a force of 10 newtons
   D. nothing (the wall exerts no force)

2. The forces are balanced on a length of rope when the rope is

   A. accelerating toward one team in a game of tug-of-war.
   B. tied around a heavy box that is not moving.
   C. used to lift a heavy box at rest on the ground.
   D. falling to the ground.

3. Two 50-pound barrels are at rest on the ground. Dave grabs one barrel and lifts it to a loading dock. Sonia rolls the other barrel up a ramp to the loading dock. Which of the workers, if any, did more work on the barrel against the force of gravity?

   A. Sonia did more work against the force of gravity.
   B. Dave did more work against the force of gravity.
   C. Both Dave and Sonia did the same amount of work.
   D. Neither Dave nor Sonia did work.

Directions: Answer the questions.

4. Explain the relationship among speed, distance, and velocity.

   _____

   _____

5. A driver steps on the brakes of her car. What makes this action an example of acceleration?

   _____

   _____

6. Imagine two drivers whose vehicles have suddenly stopped working. They must be pushed to the shoulder of a road for repairs. One driver is driving a compact passenger car. The other driver is pushing an oversized pick-up truck. Compare the force each driver applies to his or her vehicle in terms of Newton's Second Law of Motion.

   _____

   _____

7. Explain the relationship between Newton's Third Law of Motion and the launch of a rocket.

_____

_____

_____

8. Professional movers sometimes lift heavy objects to slide wheels beneath each corner. Explain the advantage of the wheels in terms of force. Use the term friction in your answer.

_____

_____

_____

_____

_____

9. Imagine pushing against a brick wall. Explain why your action is not considered work.

_____

_____

_____

10. Identify the simple machines found in a handheld can opener and explain how they work.

_____

_____

_____

Check Your Understanding

On the following chart, circle the number of any item you answered incorrectly. Next to each group of item numbers, you will see the pages you can review to learn how to answer the items correctly. Pay particular attention to reviewing those lessons in which you missed half or more of the questions.

Chapter 8 Review

Lesson	Item Number	Review Pages
Newton's Laws of Motion	1, 5, 6, 7, 9	284–289
Forces and Machines	2, 3, 4, 8, 10	290–297

CHAPTER 8: WORK, MOTION, AND FORCES

Question

How can you demonstrate Newton's Third Law of Motion?

Background Concepts

Newton's Third Law of Motion states that when a first body (B_1) exerts a force, F, on a second body (B_2), B_2 exerts a force that is equal to F, but in the opposite direction, on B_1. The equal and opposite force exerted by B_2 is sometimes called a **reaction force**, because B_2 is **reacting** to the force exerted on it by B_1.

For example, when you stand on the floor, your weight exerts a downward force on the floor. Think of yourself as B_1. In reaction to your weight, the floor, or B_2, exerts an upward reaction force on you. The two forces cancel each other out. What would happen if B_2 didn't push back? Your body would not be in equilibrium and you would fall through the floor!

Investigation

Materials required

scissors
compass or circle template
pen
balloon
flexible straw
4 pushpins
2 Styrofoam trays
tape

You are going to build a balloon-powered car to demonstrate Newton's Third Law of Motion. Gather the materials in the list. You can use Styrofoam trays that grocery stores use to package meat.

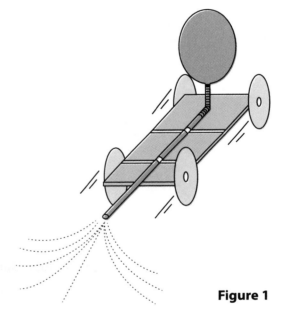

Figure 1

Refer to Figure 1 to complete the following steps:

1. Cut out a 3-inch x 7-inch rectangle from one tray. This will be the body of the car.

2. Use a compass to cut out four circles, each with a 1.5-inch radius, from the second tray. These will be the wheels of the car.

3. Bend the straw to make a right angle. Use tape to attach the straw to the body of the car.

4. Use pushpins to attach the wheels to the body of the car.

5. Place the car on a floor or table. Stop up the end of the long part of the straw with balled up paper held in place with tape to prevent air from escaping when you attach the balloon.

6. Blow up the balloon. Attach it to the short end of the straw. Wrap tape around the opening to seal the balloon to the straw.

7. Remove the stopper from the end of the straw when you're ready to "launch" your car.

8. Squeeze the balloon.

Application of Science Practices

Interpretation

What did the car do when you squeezed the balloon?

_____

_____

Answer

Describe the forces acting on the car.

_____

_____

Predict how different amounts of air in the balloon would affect the car's speed and distance traveled.

_____

_____

Evidence

Explain how Newton's Third Law of Motion is demonstrated by the car's movement.

_____

_____

Chemical Properties

Mix some flour, milk, butter, sugar, baking powder, and an egg in a bowl, and you probably won't find the soupy mixture too appetizing. But if you put it in the oven and heat it for a while, a delicious transformation will take place. The resulting cake is an example of chemistry at work!

Chemistry is the study of matter. Scientists define matter as anything that has mass and takes up space. Through their studies of matter, chemists have made many scientific breakthroughs that affect us all.

In this chapter you will learn about:

Lesson 9.1: Matter
You commonly see water in three different states of matter—ice, liquid, and vapor. This lesson investigates the states of matter, physical and chemical changes, and the relationship between energy and states of matter.

Lesson 9.2: The Atom
If you could take something—a lamp, a ball, a cardboard box—and take it apart piece by piece until you could no longer even see the pieces, you would still have something of that object in your hands. Look in the microscope and see the atoms, the basic building blocks of matter. This lesson investigates the structure and properties of atoms.

Lesson 9.3: Compounds and Molecules
Have you ever heard someone ask for a tall, cool glass of H_2O? That is the chemical formula for water. This lesson explains how individual atoms interact and bond to form compounds.

Lesson 9.4: Chemical Reactions and Solutions
Chemical reactions don't belong only in a laboratory—they occur in the kitchen and the laundry room, too. This lesson looks at different kinds of chemical solutions and the law of conservation of matter.

Lesson 9.5: The Chemistry of Life
You and every other living thing is made of carbon compounds. Find out more about these substances in this lesson about organic chemistry.

Lesson 9.6: Chemical Equations
Matter can neither be created nor destroyed. In this lesson, you'll learn how to balance chemical equations to account for every atom and molecule involved.

Goal Setting

Did you ever wonder how ingredients combine to make your favorite dish, or what makes your grandmother's secret recipes work? We use principles of chemistry in the kitchen every day.

Choose a simple recipe that combines ingredients and uses heat. In the chart below, write each step of the recipe in the first column. Then, after you read this chapter, write a sentence in the second column about the chemistry concept you read about in this chapter that explains what happens in each step.

Recipe Step	Chemistry Concept

Matter

KEY CONCEPT: Matter is anything that has mass and takes up space. Matter exists on Earth in one of four states—solid, liquid, gas, or plasma.

Think about different styles of dancing. During a slow dance, couples hold each other together and sway to the music. A lively square dance takes more energy to perform, as dancers swing and glide past each other. And a high-energy free-form dance takes even more energy, as dancers zoom around in different directions.

Particles of matter behave in different ways depending on how much energy they have. The energy level of their "dance" will determine if they are a solid, liquid, gas, or plasma.

The States of Matter

Scientists define **matter** as anything that has mass and volume. Matter takes up space. On Earth, most matter exists in one of four states—as a solid, liquid, gas, or plasma.

How the particles of matter are arranged influences whether a substance is a solid, liquid, gas, or plasma. In **solids**, the particles are packed tightly together. They have little room to move. As a result, solids, like a book or a pencil, have a definite shape and volume. In **liquids**, the particles are linked, but they are loosely linked. They can slide past one another. Liquids, such as water, have a definite volume, but no definite shape. In **gases**, the particles are free to move about in any direction. As a result, gases have no definite shape and no definite volume. Air is a gas.

Although **plasma** is not typically found on Earth, about 99 percent of matter in the universe is a plasma. The matter in stars is a plasma. On Earth, plasmas are found in fluorescent and neon lights, as well as in lightning and auroras.

You can think of a plasma as a gas that has become so hot its main properties change. Unlike a gas, a plasma can conduct electricity and be moved around by magnets. One way to create a plasma is to heat a gas to very high temperatures, approximating temperatures found in the Sun. Plasmas can also be created by introducing an electric current into a gas. This is how neon lights are produced.

Plasmas interest scientists because they could lead to important discoveries. For example, **fusion** can occur in the plasma state. Fusion is a type of nuclear reaction occurring when two or more atomic nuclei combine into a single nucleus. This reaction produces a great deal of energy, as demonstrated by the Sun. On Earth, though, one obstacle to harnessing the energy of nuclear fusion is containing the reaction. To overcome the challenges of the high temperatures and the plasma's reaction with the container walls, scientists use strong magnetic fields to hold the plasma in place.

COMPARE AND CONTRAST INFORMATION

Much of what you learn about science comes from reading textbooks, such as this one. However, you can also learn about science by doing experiments and making models. Experiments and models are **simulations**, or representations, of actual scientific phenomena.

Consider what you learn from reading about the states of matter. You learn how the particles of matter are arranged, you learn under which conditions matter changes from one state to another, and you learn the properties of each state of matter. You may also learn some historical context about experimentations with matter and read about scientists who did important work in discovering properties of matter.

Now consider what you might learn from a simulation about the states of matter. For example, you might build models of matter in the gaseous, liquid, and solid states. These models would simulate the different ways that the particles are arranged in each state. You could manipulate the models to simulate actual changes of state. This would provide a hands-on, very visual way to learn information about matter.

Read the following passage, and then determine what Maria learned first-hand through a simulation.

> Maria has been exploring the properties of liquids. She read in her textbook that liquids have particles that are loosely linked and can slide past one another. She also learned that they have a definite volume, but no definite shape. In the lab, she poured 250 mL of water into a beaker. Then she poured that water into other containers. She noticed that the water took the shape of whatever container she poured it into!

Through her experimentation, Maria learned that liquids take the shape of the container they are in. Science involves learning through doing—simulations, experiments, and models can all reinforce and provide added information to the text you read.

The **state of matter**—whether it is a solid, liquid, gas, or plasma—depends on temperature. Different types of matter change state at different temperatures. Water, for example, freezes at 0°C (32°F). At 100°C, water boils. Iron melts from a solid to a liquid at 1,535°C.

Review the image on
this page and the text
about what happens
when heat is added
to a substance. Now
imagine that you were
actually able to do
this experiment. Would
you be able to see the
particle action that is
described in the text
and the image? Why or
why not?

Compare and contrast
the information gained
from the image and the
text. Describe how the
particle model (a type
of simulation) shown
in the image provides
information that you
wouldn't obtain through
experimentation.

All matter is made up of particles that are in constant motion. The state of matter depends on temperature. When a substance is heated and the temperature rises, the particles gain more energy and move faster.

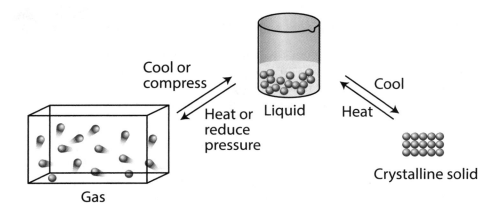

The particles of a solid are packed closely together. These particles move back and forth, but have a definite shape and volume.

When the particles in a solid gain more heat and their temperature rises, the particles eventually break the bonds that hold them together. This happens at a temperature called the melting point. When this happens, the matter changes state and becomes a liquid.

In the liquid state, particles slide past one another. A liquid has no definite shape, but takes on the shape of its container. A liquid does have a definite volume. If you pour a liquid from one container to another, its volume will remain the same.

If more heat is applied, the particles in the liquid gain more energy. Eventually they escape completely from the bonds holding them together. These particles form a gas. A gas has no definite shape or volume.

State of matter often depends on air pressure as well. The boiling point of water is 100°C at sea level. This is the temperature at which liquid water becomes a gas. In the mountains, air pressure is lower. Water boils at a slightly lower temperature on a mountain top than at sea level.

THINK ABOUT SCIENCE

Directions: Match the descriptions on the left with the correct state of matter on the right.

_____ **1.** vibrating particles that do not move from place to place.

_____ **2.** particles that slide past one another.

_____ **3.** particles that move freely.

A. gas

B. solid

C. liquid

Properties of Matter

The state of matter—solid, liquid, gas, or plasma—is just one of the **physical properties** of matter. Physical properties are those that can be observed without changing the identity of the matter. A stick of butter, for example, might be described as a solid, or by its shape, color, or smell. If you cut the butter, the shape changes, but it is still butter. Even if you melt the butter, it is still butter.

Some other physical properties of matter include hardness, density, and melting and boiling points. A **physical change** occurs when there is a change in a physical property such as state, size, or shape, but the substance remains the same. For example, if you melt butter, it becomes a liquid, but it's still butter. Its solid state can be restored by cooling it.

Chemical properties are those that describe how a substance reacts with another substance. For example, flammability and corrosiveness are chemical properties. Chemical changes result in new materials with new properties. A chemical property of wood is that it burns. After the reaction, however, there is no more wood. It has been changed to ash. Most chemical changes cannot be reversed. When wood is burned, the resulting ashes can't be turned into wood again.

Some physical changes are easy to confuse with chemical changes. Dissolving salt in water, for example, might seem like a chemical change. The salt no longer exists as tiny grains, but tasting the water will persuade you that the salt is still there. In fact, if the water evaporates, the salt will be left behind.

Elements

An **element** is a type of matter that cannot be broken down into a simpler substance. Each different type of matter is made of one or more elements. Some types of matter, such as oxygen, cannot be broken down into a simpler substance. Oxygen is an element. Water, on the other hand, can be broken down into two substances—oxygen and hydrogen—both of which are elements. Sugar is made of three elements—oxygen, hydrogen, and carbon.

Core Skill
Draw Conclusions

A **conclusion** is a general statement supported by reasoning and details To draw conclusions, readers apply their own background knowledge to details in the text to make reasonable judgments.

Review this list of changes and determine which are chemical changes and which are physical changes. Explain your conclusions.

- Frying an egg
- Breaking a pencil
- Adding food coloring to water
- Chopping wood
- Cutting a cake
- Baking a cake
- Iron rusting
- Milk spoiling

THINK ABOUT SCIENCE

Directions: Answer the questions below.

1. What is the difference between a chemical and a physical property of matter?

   _____

   _____

2. What is an element?

   _____

WRITE TO LEARN

Tell about something you saw on the news or read about. Then write a general statement, a conclusion, that you can draw about the event or incident based both on what you know and on details you learned from the information.

Early Ideas About the Elements

The idea of elements can be traced back to the ancient Greeks. Although the Greeks' definitions were not correct, they correctly recognized that matter is made of individual pieces.

The Greeks believed there were four elements—earth, air, water, and fire. "Earth" was associated with solids, "water" with liquids, and "air" with gases. These elements could combine in any number of ways to form common substances.

The Greek philosopher Aristotle introduced the ideas of properties. He assigned the qualities of warm, cool, wet, and dry to the four elements. He suggested that elements could be changed into one another if they shared the same qualities. That is, water could change into wet air, which could change back into water. He also suggested that two elements could transform into a third substance.

Many of the substances recognized today as elements were known in ancient times. However, they were not thought of as building blocks of matter. Substances such as gold, silver, iron, copper, and lead were recognized more for their usefulness in making tools and decorations. They were also used by alchemists, or early chemists. Alchemists were mainly concerned with changing common metals into gold. They always failed, for reasons scientists can easily explain today.

Properties of Elements

Scientists today have discovered more than 100 different elements. These elements are distinct. Mixing elements together does not change one element into another.

Although each element has distinct properties, certain groups of elements are similar. Some elements are hard and dull solids, some are colorless and odorless gases, and some are liquids. Metals conduct heat and electricity well, are bendable, and can be polished to a shine. Nonmetals are poor conductors and are brittle and dull. They usually exist as solids or gases at room temperature. Metalloids have properties of both metals and nonmetals.

The ten most common elements in the universe are hydrogen, helium, oxygen, carbon, neon, iron, nitrogen, silicon, magnesium, and sulfur.

PROPERTIES OF TEN MOST COMMON ELEMENTS ON EARTH				
Element	Physical Properties	Chemical Properties	Boiling Point	Melting Point
Hydrogen	odorless, colorless gas; lightest of all gases; mixture of three isotopes, H, H_2, H_3	most flammable of known substances; not very reactive at normal temperatures; highly reactive at high temperatures; found in all organic compounds	−252.762 °C	−259.2 °C
Helium	odorless, colorless, tasteless gas; second lightest element; less soluble in water than any other gas	doesn't form chemical compounds; nonflammable	−268.9 °C	−272.2 °C
Oxygen	odorless, colorless gas; most abundant element in Earth's crust; normally exists in diatomic form (O_2), but also occurs in triatomic form (O_3)	reactive; forms compounds called oxides with all other elements except helium, neon, argon, and krypton	−183 °C	−219 °C
Carbon	solid; dark gray to black nonmetal; exists freely as diamond and graphite	self-bonding; combines with hydrogen to make about one million organic compounds; also forms far fewer inorganic compounds	4,827 °C	3,652 °C
Neon	colorless, odorless, tasteless gas	chemically inactive; doesn't bond with other elements	−246 °C	−249 °C
Iron	silver-gray metal; has luster; malleable; ductile; exists in four crystalline forms	very active; combines readily with oxygen in moist air to form rust; reacts with very hot water and steam to produce hydrogen gas; dissolves in most acids; reacts with many elements	2,861 °C	1,536 °C
Nitrogen	odorless, colorless gas; normally found in diatomic form (N_2)	inactive gas at room temperature; does not combine with most elements; will combine with oxygen in the presence of lightning or a spark	−195.8 °C	−210 °C
Silicon	metalloid; metallic luster; brittle	never occurs as a free element; does not bond with oxygen or most other elements; more reactive at high temperatures	3,265 °C	1,410 °C
Magnesium	metal; silvery white; light	very chemically active; bonds with most nonmetals and almost all acids	1,091 °C	650 °C
Sulfur	pale yellow, solid, nonmetallic; odorless; tasteless; poor conductor of heat and electricity	insoluble in water; combines with hydrogen to make a poisonous gas that smells like rotten eggs; chemically reactive; combines with almost all elements	444.6°C	115.2 °C

Directions: Complete the sentences below using one of the following words:

chemical property   element   matter   physical property   state of matter

1. _____ has mass and takes up space.

2. A substance that cannot be broken down into anything simpler is a(n) _____.

3. Rose-colored is a(n) _____ of the rock quartz.

4. A substance's _____ may be solid, liquid, gas, or plasma.

5. A(n) _____ of oxygen is that it combines readily with hydrogen, iron, and other elements.

Directions: Use the Venn diagram to compare and contrast hydrogen and neon.

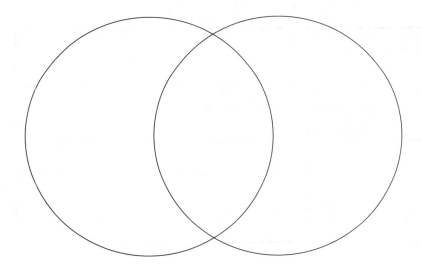

Directions: Choose the one best answer to each question.

1. Which of these ideas from ancient times remains valid today?

   A. Elements have distinct properties.
   B. The four elements are air, water, earth, and fire.
   C. Elements readily change from one to another.
   D. Iron and lead can be changed into gold.

2. When electricity passes through mystery substance X, it changes completely into hydrogen and oxygen. What can be concluded about mystery substance X?

   A. X is an element.   C. X is sugar.
   B. X is made of two elements. D. X is carbon dioxide.

Skill Practice

Directions: Choose the best answer to each question.

1. A marching band wants to model a solid. What would be the best instructions to give each musician?

   A. Walk slowly around the field, but always near another musician.
   B. Stand perfectly still in one spot.
   C. Stand in one spot as you sway back and forth.
   D. Run randomly around the field.

2. What is one way to investigate the chemical properties of an unidentified substance?

   A. Strike it with a hammer.
   B. Measure the heat released when it burns.
   C. Measure its mass and volume.
   D. Observe whether it floats in water.

3. What happens to the particles of a liquid when the temperature is raised to the boiling point?

   A. The particles slow down as they change into the gas state.
   B. Bonds form between the particles and the air.
   C. The particles escape from their attraction to one another.
   D. The particles form strong bonds with one another.

4. Where would the boiling point of water be lowest?

   A. on an ocean beach on a hot day
   B. on an ocean beach on a cold day
   C. at a ski resort in the mountains
   D. in deep space, where there is no air

5. Which is an example of a physical change?

   A. ice melting to water
   B. a lump of coal burning to become ash and gases
   C. flour, sugar, eggs, and other ingredients baking into a cake
   D. iron rusting when left out in the rain

6. Which is an example of a chemical change?

   A. an iron nail being magnetized
   B. sodium and chlorine combining to form salt
   C. a brass rod expanding when heated
   D. water turning to steam when heated

7. What can be concluded by studying the properties of the 10 most common elements in the universe?

   The elements

   A. are found in greater quantities on Earth than they are throughout the universe.
   B. are found in the same quantity on Earth as they are throughout the universe.
   C. have diverse properties and include metals and nonmetals.
   D. that are nonmetals have high melting and boiling points.

8. A scientist wants to find a way to harness nuclear fusion reactions. Studying which object or event in nature would most likely help her?

   A. a lightning strike
   B. a thunderstorm
   C. a waterfall
   D. a star

The Atom

KEY CONCEPT: Elements are made of tiny particles called atoms.

Every citizen of the United States has a Social Security number. Each number is unique, which means that no two people have the same identifying number. The numbers themselves, however, cannot tell you anything about the characteristics of the person.

The tiny particles that make up each of Earth's elements also have a number that identifies them. Like a Social Security number, each identifying number is assigned to only one kind of atom. However, unlike a Social Security number, once you learn how to interpret them, these numbers can give you clues about the characteristics of each atom.

Structure of the Atom

Elements are made of particles called **atoms**. An atom is the smallest unit of an element that still maintains the properties of that element. Atoms are extremely small. Early scientists developed **models**, or representations, of atoms based on experiments and observations. Niels Bohr developed a model that is still useful today. The model is not completely correct, but it explains the atom's basic structure.

THE BOHR MODEL

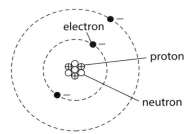

An atom is made of three kinds of particles: **protons**, **electrons**, and **neutrons**. Protons and neutrons are located within the **nucleus**, the central core of the atom. Protons have a positive electrical charge. Neutrons are **neutral**—they have no electrical charge. Electrons are negatively charged particles found outside the nucleus. The number of electrons is equal to the number of protons in an atom. This means that an atom has no overall charge.

Electrons travel in cloud-like regions around the nucleus in distinct orbitals. These orbitals form groups called shells around the nucleus. The shell closest to the nucleus can hold 2 electrons. Shells become larger and hold more electrons farther from the nucleus. The filling of these shells determines many of the atom's properties.

Atoms may gain or lose electrons. An atom then becomes an **ion**, carrying either a positive or a negative charge.

CITE TEXTUAL EVIDENCE

Details are words and phrases that support the main idea of a passage. In a science passage, **factual details** are textual evidence that you can use to support your analysis of the material. Before trying to locate factual details in a passage, first analyze the text to find the main idea. To understand what you read, you must grasp both the main idea and the textual evidence that supports the main idea.

When you have found the main idea in a passage, scanning can help you to locate the textual evidence that supports it. You scan a passage by reading it quickly but closely to find a specific fact or detail. As you read, pay special attention to titles, headings, and words in bold print. They may help you find information.

Read the following paragraph. Determine the main idea. Then, state the textual evidence that supports your analysis.

Atoms are very, very small. A drop of water from the tap is made of about 5×10^{21} atoms. That number is written as the numeral 5 followed by 21 zeros. Yet atoms are made of parts that are smaller still. The width of an atom's nucleus is about 10,000 times smaller than the width of the atom.

In this paragraph, the main idea is that atoms are small and that their parts are even smaller. The second and third sentences contain textual evidence to support your analysis of the main idea. The last sentence is a specific detail about the small size of a part of the atom.

Core Skill
Apply Scientific Models

The structure of an atom can be better understood if you make a model of it. Use simple items such as coins or small candies to represent protons, neutrons, and electrons. First, make your own model of the Bohr model shown here. As you learn more in this lesson, make models of atoms of different elements. Also make models of ions.

THE ATOM WITH ELECTRONS IN A CLOUD PATH

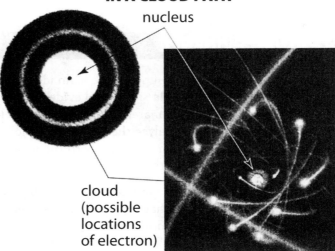

nucleus

cloud
(possible
locations
of electron)

Atomic Number

Atoms of an element always have the same number of protons. The **atomic number** is the number of protons in the nucleus. The element hydrogen, for example, always has 1 proton, and its atomic number is 1. Oxygen has 8 protons. Its atomic number is 8.

Atomic Mass

Most of the mass in an atom, or the **atomic mass**, comes from the protons and neutrons in the nucleus. Electrons have very little mass.

Atomic mass can be measured in grams or kilograms, but the values would be extremely small. Instead, scientists use units called atomic mass units (amu). The mass of a typical carbon atom, for example, is 12 amu.

The total number of protons and neutrons in the nucleus is the **mass number**. All atoms of the same element have the same number of protons, but they do not always have the same number of neutrons. Atoms with different numbers of neutrons are called **isotopes**. Isotopes are identified by their mass number.

THINK ABOUT **SCIENCE**

Directions: Choose the word or number that best completes each sentence.

1. An element that has 8 protons has (16, 8) electrons.

2. When an atom gains or loses an electron it becomes a(n) (neutron, ion).

3. The (atomic mass, atomic number) describes the number of protons in each atom.

4. (Electrons, Neutrons) move in the space around the nucleus of an atom.

Organizing the Elements

During the 1800s, scientists tried to classify the known elements. Russian chemist Dmitry Mendeleyev noticed a relationship between the atomic mass of elements and their properties. He listed the sixty elements known at the time in order of their atomic masses. Then he arranged the elements in a **table**, a graphic that organizes data in rows and columns. His table grouped elements with similar properties together. Mendeleyev's work resulted in the **periodic table of the elements**.

Using the Periodic Table

Today, the periodic table is arranged according to the elements' atomic number, not atomic mass. Atomic number is the key property that causes the pattern of properties in elements.

The first **period**, or row, of the table has two elements: hydrogen and helium. The periods that follow have even more elements.

The columns of the periodic table are called **groups**. The elements of a group often have similar properties. Group VIII is called the noble gases. These gases rarely, if ever, enter into chemical reactions.

The elements of Group VII are called the halogens. These elements react with metals to form brittle solids called salts.

Groups I and II are metals. The elements in the middle section of the periodic table are also metals. Most of the elements in the periodic table are metals. Nonmetals are located at or near the upper right corner of the table.

Elements called semimetals or metalloids have some of the properties of metals. These elements form a border between the regions for metals and nonmetals on the periodic table.

Why the Table Works

The elements within a group share properties because their electrons are organized in similar ways.

In the periodic table on the next page, look at the small numbers along the right side of each box. These numbers tell how the electrons are arranged in atomic shells. Within each group, the final number in each box is always the same. This is the number of outer electrons in an atom. It determines many of the atom's chemical properties.

21st Century Skill
Critical Thinking and Problem Solving

One of the most interesting things about Mendeleyev's original periodic table is that he left spaces for elements that hadn't been discovered yet. How did he know to do that? The evidence provided by his arrangement of the elements by atomic mass showed that there were "missing" elements. He explained that he left the spaces to be filled when those "missing" elements were discovered. Do some research and write an explanation of how those elements were discovered.

WRITE TO LEARN

A table presents information in rows and columns. One common kind of table is a bus (or train) schedule. In a notebook, write a paragraph that summarizes the information you find in a bus schedule. Then answer this question: *Which way of presenting a bus schedule is easier to understand: a paragraph or a table?*

THINK ABOUT SCIENCE

Directions: Write a short response to each question.

1. What observations helped Mendeleyev develop the periodic table?

2. Why can elements be found only in one place in the periodic table?

3. How are the elements within a group, or column, alike?

Group I   II

Period

Legend (Carbon example):
- atomic number: 6
- number of electrons in each energy shell: 2, 4
- name: Carbon
- symbol: C
- atomic mass (number of protons and neutrons): 12

Period	Group I	Group II							
1	①〔1〕 **H** Hydrogen 1								
2	③〔2,1〕 **Li** Lithium 7	④〔2,2〕 **Be** Beryllium 9							
3	⑪〔2,8,1〕 **Na** Sodium 23	⑫〔2,8,2〕 **Mg** Magnesium 24							
4	⑲〔2,8,8,1〕 **K** Potassium 39	⑳〔2,8,8,2〕 **Ca** Calcium 40	㉑〔2,8,9,2〕 **Sc** Scandium 45	㉒〔2,8,10,2〕 **Ti** Titanium 48	㉓〔2,8,11,2〕 **V** Vanadium 51	㉔〔2,8,13,1〕 **Cr** Chromium 52	㉕〔2,8,13,2〕 **Mn** Manganese 55	㉖〔2,8,14,2〕 **Fe** Iron 56	㉗〔2,8,15,2〕 **Co** Cobalt 59
5	㊲〔2,8,18,8,1〕 **Rb** Rubidium 85	㊳〔2,8,18,8,2〕 **Sr** Strontium 88	㊴〔2,8,18,9,2〕 **Y** Yttrium 89	㊵〔2,8,18,10,2〕 **Zr** Zirconium 91	㊶〔2,8,18,12,1〕 **Nb** Niobium 93	㊷〔2,8,18,13,1〕 **Mo** Molybdenum 96	㊸〔2,8,18,13,2〕 **Tc*** Technetium 98	㊹〔2,8,18,15,1〕 **Ru** Ruthenium 101	㊺〔2,8,18,16,1〕 **Rh** Rhodium 103
6	〔55〕〔2,8,18,18,8,1〕 **Cs** Cesium 133	〔56〕〔2,8,18,18,8,2〕 **Ba** Barium 137	〔57〕 to 〔71〕	〔72〕〔2,8,18,32,10,2〕 **Hf** Hafnium 178	〔73〕〔2,8,18,32,11,2〕 **Ta** Tantalum 181	〔74〕〔2,8,18,32,12,2〕 **W** Tungsten 184	〔75〕〔2,8,18,32,13,2〕 **Re** Rhenium 186	〔76〕〔2,8,18,32,14,2〕 **Os** Osmium 190	〔77〕〔2,8,18,32,15,2〕 **Ir** Iridium 192
7	〔87〕〔2,8,18,32,18,8,1〕 **Fr** Francium 223	〔88〕〔2,8,18,32,18,8,2〕 **Ra** Radium 226	〔89〕 to 〔103〕	〔104〕〔2,8,18,32,10,2〕 **Rf*** Rutherfordium 261	〔105〕〔2,8,18,32,18,11,2〕 **Db*** Dubnium 262	〔106〕〔2,8,18,32,12,2〕 **Sg*** Seaborgium 263	〔107〕〔2,8,18,32,13,2〕 **Bh*** Bohrium 262	〔108〕〔2,8,18,32,14,2〕 **Hs*** Hassium 265	〔109〕〔2,8,18,32,15,2〕 **Mt*** Meitnerium 266

Rare Earth Elements

Lanthanide Series	〔57〕〔2,8,18,18,9,2〕 **La** Lanthanum 139	〔58〕〔2,8,18,20,8,2〕 **Ce** Cerium 140	〔59〕〔2,8,18,21,8,2〕 **Pr** Praseodymium 141	〔60〕〔2,8,18,22,8,2〕 **Nd** Neodymium 144	〔61〕〔2,8,18,23,8,2〕 **Pm*** Promethium 145	〔62〕〔2,8,18,24,8,2〕 **Sm** Samarium 150
Actinide Series	〔89〕〔2,8,18,32,18,9,2〕 **Ac** Actinium 227	〔90〕〔2,8,18,32,18,10,2〕 **Th** Thorium 232	〔91〕〔2,8,18,32,20,9,2〕 **Pa** Protactinium 231	〔92〕〔2,8,18,32,18,21,9,2〕 **U** Uranium 238	〔93〕〔2,8,18,32,22,9,2〕 **Np*** Neptunium 237	〔94〕〔2,8,18,32,24,8,2〕 **Pu*** Plutonium 244

	III	IV	V	VI	VII	VIII
						(2) He Helium 4 — 2
	(5) B Boron 11 — 2,3	(6) C Carbon 12 — 2,4	(7) N Nitrogen 14 — 2,5	(8) O Oxygen 16 — 2,6	(9) F Fluorine 19 — 2,7	(10) Ne Neon 20 — 2,8
	(13) Al Aluminum 27 — 2,8,3	(14) Si Silicon 28 — 2,8,4	(15) P Phosphorus 31 — 2,8,5	(16) S Sulfur 32 — 2,8,6	(17) Cl Chlorine 35 — 2,8,7	(18) Ar Argon 40 — 2,8,8
(28) Ni Nickel 59 — 2,8,16,2	(29) Cu Copper 64 — 2,8,18,1	(30) Zn Zinc 65 — 2,8,18,2	(31) Ga Gallium 70 — 2,8,18,3	(32) Ge Germanium 73 — 2,8,18,4	(33) As Arsenic 75 — 2,8,18,5	(34) Se Selenium 79 — 2,8,18,6	(35) Br Bromine 80 — 2,8,18,7	(36) Kr Krypton 84 — 2,8,18,8
(46) Pd Palladium 106 — 2,8,18,18,0	(47) Ag Silver 108 — 2,8,18,18,1	(48) Cd Cadmium 112 — 2,8,18,18,2	(49) In Indium 115 — 2,8,18,3	(50) Sn Tin 119 — 2,8,18,18,4	(51) Sb Antimony 122 — 2,8,18,5	(52) Te Tellurium 128 — 2,8,18,18,6	(53) I Iodine 127 — 2,8,18,18	(54) Xe Xenon 131 — 2,8,18,18,8
(78) Pt Platinum 195 — 2,8,18,32,17,1	(79) Au Gold 197 — 2,8,18,32,18,1	(80) Hg Mercury 201 — 2,8,18,32,18,2	(81) Tl Thallium 204 — 2,8,18,32,18,3	(82) Pb Lead 207 — 2,8,18,32,18,4	(83) Bi Bismuth 209 — 2,8,18,32,18,5	(84) Po Polonium 209 — 2,8,18,32,18,6	(85) At Astatine 210 — 2,8,18,32,18,7	(86) Rn Radon 222 — 2,8,18,32,18,8
(110) Ds* Darmstadtium 281 — 2,8,18,32,*32,17,1	(111) Rg* Roentgenium 280 — 2,8,18,32,*32,18,1	(112) Cn* Copernicium 285 — 2,8,18,32,*32,18,2	(113) Uut* Ununtrium 284 — 2,8,18,32,*32,18,3	(114) Fl* Flerovium 289 — 2,8,18,32,*32,18,4	(115) Uup* Ununpentium 288 — 2,8,18,32,*32,18,5	(116) Lv* Livermorium 293 — 2,8,18,32,*32,18,6	(117) Uus* Ununseptium 294 — 2,8,18,32,*32,18,7	(118) Uuo* Ununoctium 294 — 2,8,18,32,*32,18,8

* = Manmade

(63) Eu Europium 152 — 2,8,18,25,8,2	(64) Gd Gadolinium 157 — 2,8,18,25,9,2	(65) Tb Terbium 159 — 2,8,18,27,8,2	(66) Dy Dysprosium 163 — 2,8,18,28,8,2	(67) Ho Holmium 165 — 2,8,18,29,8,2	(68) Er Erbium 167 — 2,8,18,30,8,2	(69) Tm Thulium 169 — 2,8,18,31,8,2	(70) Yb Ytterbium 173 — 2,8,18,32,8,2	(71) Lu Lutetium 175 — 2,8,18,32,9,2
(95) Am* Americium 243 — 2,8,18,32,25,8,2	(96) Cm* Curium 247 — 2,8,18,32,25,9,2	(97) Bk* Berkelium 247 — 2,8,18,32,26,9,2	(98) Cf* Californium 251 — 2,8,18,32,28,8,2	(99) Es* Einsteinium 252 — 2,8,18,32,29,8,2	(100) Fm* Fermium 257 — 2,8,18,32,30,8,2	(101) Md* Mendelevium 258 — 2,8,18,32,31,8,2	(102) No* Nobelium 259 — 2,8,18,32,32,8,2	(103) Lr* Lawrencium 260 — 2,8,18,32,32,9,2

Vocabulary Review

Directions: Match the description on the right to the correct term on the left. Write the letter on the line provided.

1. _____ atom
2. _____ proton
3. _____ neutral
4. _____ neutron
5. _____ electron
6. _____ model
7. _____ table

A. negatively charged particle

B. smallest unit of an element

C. neither positive nor negative

D. particle with no electric charge

E. positively charged particle

F. a drawing or plan that represents a real object

G. a structure for organizing data

Skill Review

Directions: Find and underline the factual details in the passage below.

(1) The work of several scientists led to the modern model of the atom. (2) As scientists often do, each scientist built on the work of others. (3) In 1803, John Dalton suggested that atoms were solid. (4) Further experiments showed that this was incorrect. (5) In 1898, J. J. Thompson showed that atoms contained electrons. (6) Further work improved on this model. (7) Experiments by Ernest Rutherford showed that atoms were made up of a dense, positively charged nucleus surrounded by negative electrons.

Directions: Analyze the section of the periodic table below. Then choose the best answer to each question.

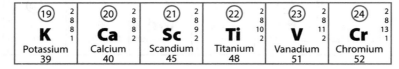

1. How many protons are in an atom of calcium (Ca)?

   A. 2
   B. 10
   C. 20
   D. 40

2. What is the atomic mass of chromium (Cr)?

   A. 8
   B. 13
   C. 24
   D. 52

3. How are the elements arranged from potassium (K) to chromium (Cr)?

   A. by the first letter of their name
   B. by increasing atomic number
   C. by similar physical properties
   D. by similar chemical properties

Skill Practice

Directions: Choose the best answer to each question.

1. Imagine that an atom were enlarged to the size of a peach. What would best represent the size and position of the nucleus?

   A. the outer edge of the pit
   B. a dot in the center of the pit
   C. the skin of the peach
   D. the yellow flesh of the peach

4. To find elements with similar properties to iodine, where should you look in the periodic table?

   A. anywhere in the table
   B. in a diagonal line from iodine
   C. in the same row as iodine
   D. in the same column as iodine

2. Nitrogen-14 and nitrogen-15 are two isotopes of nitrogen. What property makes them different?

   A. number of electrons
   B. number of protons
   C. number of neutrons
   D. atomic charge

5. Which value will be true for all carbon atoms?

   A. an atomic mass of 12
   B. a nuclueus that has 6 protons
   C. a nucleus that has 6 neutrons
   D. a nucleus that has 7 neutrons

3. In his first periodic table, what did Mendeleyev do to keep elements with similar properties in the same group?

   A. He included blanks for the isotopes of elements.
   B. He included blanks for elements yet to be discovered.
   C. He arranged the elements in order of atomic number.
   D. He invented values for atomic mass and atomic number.

Compounds and Molecules

KEY CONCEPT: Individual atoms form compounds by making chemical bonds with one another. To do this, each atom gains, loses, or shares electrons.

In the language you hear every day, a bond is an attraction between two things or two people. The glue on the back of a sticky note forms a weak bond when you place it on a note board or on another piece of paper. Other kinds of glue form much stronger bonds.

Atoms also stick together, and they do so with bonds of varying strengths. Weak bonds are easily broken, while strong bonds require much more energy to break.

Compounds and Bonding

Most elements combine with two or more other elements to form **compounds**. A compound is matter made up of two or more elements that are chemically combined.

Atoms of different elements have different numbers of electrons in their outer shells. Atoms that have full outer shells are very stable. They rarely, if ever, form compounds. Chemists describe them as **inert**. Atoms with only partially full outer shells are unstable. They become stable by forming **chemical bonds**. A compound forms when bonds unite atoms of two or more different elements.

Bonding

Bonding, the joining of atoms, can occur in many ways. **Ionic bonds** form when atoms give up or gain the electrons in their outer shell. The formation of sodium chloride, or table salt, shows this. An atom of sodium (Na) has one electron in its outer shell. An atom of chlorine (Cl) has seven electrons in its outer shell.

A sodium atom will readily lose that one electron and become a positively charged ion. The chlorine atom grabs the available electron from the sodium to fill its outer shell. The additional electron makes it a negative ion. The negative chlorine ion bonds with the positive sodium ion to form sodium chloride (NaCl).

SODIUM CHLORIDE

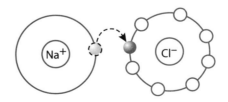

The Na electron has transferred over to the Cl atom.

DETERMINE MEANING

In chemistry, each element is represented by a **symbol**. A symbol is a thing that represents or stands for something else. Chemistry symbols can be one, two, or three letters. Scientists all over the world use the same symbols, which are shown in the periodic table. An element's symbol may be directly related to its name—for example, the symbol for helium is He. Other symbols are not as closely tied to their English names. For example, the symbol for lead is Pb. This symbol is derived from the Latin name for lead, *plumbum*. Element names are derived from many sources, including other languages, locations, and the names of notable scientists.

Elements combine to form compounds. Likewise, symbols combine to make formulas. A formula is the shorthand way to write the name of a compound. For example, sodium (symbol: Na) combines with chlorine (symbol: Cl) to form sodium chloride. The formula for sodium chloride, NaCl, is a combination of the elements' symbols.

Read the following passage. Determine the meaning of the symbols as they are used in the context of this passage.

> Many car owners are concerned about iron oxide, a crumbly, ugly substance that weakens steel. Also known as rust, iron oxide forms when iron and oxygen react with one another. This reaction takes place much faster when water and salt are present. A protective coating on a metal surface will slow down the formation of iron oxide. The chemical formula for iron oxide is Fe_2O_3.

Fe is the symbol for iron. *O* is the symbol for oxygen. When they combine they form iron oxide.

Core Skill
Understand Text

Examining cause and effect in a passage will help you to better understand the text. This can be especially true in science textbooks, because there are many processes and cycles that occur in the natural world. A **cause** is what makes something happen. An **effect** is the result of the cause.

As you read about types of chemical bonds on this page and the next, identify cause-and-effect relationships. Look for words such as *since* and *because* to locate possible cause-and-effect relationships.

To be *accountable* means that you are responsible for your actions. There are several things that scientists need to do to be accountable for their investigations and results. For example, scientists must be aware of common sources of error in their experimentation. Errors can be introduced in several ways, including personal bias, random errors in measurement, and systemic errors—errors inherent to the entire experiment. Explain how checking for errors can make scientists more accountable to themselves, their colleagues, and the scientific community.

A compound, typically, is quite different from the elements that form it. On its own, sodium is a silvery metal that reacts violently with water. Chlorine is a poisonous gas. Sodium chloride, however, is a white, nonpoisonous solid that dissolves in water.

Most ionic compounds are solids at room temperature because the bond between their particles is very strong. As solids, they do not conduct electricity. As liquids or when dissolved in water, however, they are good electrical conductors. Most ionic compounds have high melting points.

The other bonding process that forms a compound is called **covalent bonding**. In this process, electrons are not transferred from one atom to another. Instead, they are shared. Their electron shells overlap and both of their outer shells are filled.

Covalent bonds can be as strong, if not stronger, than ionic bonds. Atoms joined by covalent bonds act as a unit that is called a **molecule**. A molecule may form from as few as two atoms or from thousands, millions, or even billions of atoms.

Many nonmetals exist in their element form as molecules. Among them are oxygen (O_2) and nitrogen (N_2), both of which are gases in Earth's atmosphere. Most molecules, however, are part of compounds, meaning they are made of two or more different elements. Molecular compounds may be solids, liquids, or gases at room temperature. Nearly all are made from nonmetals.

Water is an important molecular compound. As you can see in the diagram, a pair of electrons is shared between the oxygen atom and each of the two hydrogen atoms. The atoms are more stable together than they were alone.

SEPARATED HYDROGEN AND OXYGEN ATOMS

A WATER MOLECULE

● = electrons shared by hydrogen and oxygen atoms

Chemical Formulas

Chemists use **formulas** to represent the structure of molecules. These formulas are written in a particular way. For instance, H_2O is the formula for water, a molecular compound. The small number 2, a subscript, tells us there are two atoms of hydrogen (H) in the molecule. Because there is no subscript after oxygen, we know the molecule contains only one atom of oxygen (O). The formula for the ionic compound sodium chloride (NaCl) indicates that there is one ion of sodium (Na^+) for each chloride ion (Cl^-) in the compound.

In some chemical equations, there may be a number in front of a formula. A formula written $4H_2O$ describes four molecules of water.

One molecule of water (H_2O) has two atoms of hydrogen and one atom of oxygen. Four molecules of water ($4H_2O$) have eight atoms of hydrogen ($4 \times 2 = 8$) and four atoms of oxygen ($4 \times 1 = 4$).

The formula and structure of a molecule of water is fairly simple. For more complex molecules, a diagram of the structure is helpful to indicate what kinds of atoms are present and how they are connected. For example, propane is a fuel used for heating and in torches or grills. The formula for propane, C_3H_8, shows that a propane molecule has three atoms of carbon and eight atoms of hydrogen. The formula gives no information of the arrangement of the atoms. The molecule's structure is shown below.

A MOLECULE OF PROPANE

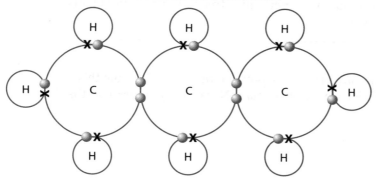

● = electrons from carbon
X = electrons from hydrogen

THINK ABOUT SCIENCE

Directions: Complete each statement below with the appropriate word.

1. In a _____, electrons are shared.

2. All _____ are made up of two or more elements that are chemically bonded.

3. A _____ gives information about what elements are in a compound and how those elements are arranged.

Directions: Match the words in the column on the left with their description on the right.

1. _____ molecule

2. _____ compound

3. _____ bonding

4. _____ formula

A. forms from bonds among two or more elements

B. shows the elements of a compound

C. two or more atoms joined by covalent bonds

D. the process of combining two or more atoms

Directions: Read the following passage and respond to the statements below.

You can think of atoms as seekers. They seek to become more stable by gaining, losing, or sharing electrons, which is why they form chemical bonds. But sometimes a bond breaks. This changes a stable atom or molecule into two unstable units called free radicals. Free radicals seek to regain stability by "stealing" an electron from another atom or molecule. This, in turn, changes the "victim" into a free radical.

In the human body, uncontrolled free radicals can damage cells and tissues. Compounds called antioxidants donate electrons and, therefore, neutralize free radicals. Antioxidants are found in many fruits and vegetables.

Directions: The passage describes a chain of cause-and-effect relationships. To show this chain, fill in the blanks below with the proper words or phrases. Note that you may use the same word or phrase more than once.

1. Being unstable causes atoms to form _____.

2. A broken bond causes _____.

3. A _____ causes more free radicals by "stealing" electrons.

4. In the human body, uncontrolled free radicals cause _____.

5. _____ cause free radicals to become neutral molecules.

Directions: Read the following passage. Then choose the best answer to each question.

A compound and its component elements have very different physical and chemical properties. For example, consider the two elements that make up sodium chloride (NaCl). Sodium is a silvery-white metal. Sodium is so volatile that usually it is kept in oil. If it comes in contact with water, it reacts explosively! Chlorine, in contrast, is a green-colored gas. It also is very reactive.

When sodium and chlorine combine chemically, they form sodium chloride (NaCl), a compound better known as table salt. This compound shares very few properties with its component elements.

Skill Review (continued)

6. What does the word *volatile* mean in the context of the passage?

   A. unpredictable   **C.** explosive

   B. stable   **D.** spontaneous

7. Which of the following examples could also be used to illustrate the main idea of the passage?

   A. helium and argon are elements that do not combine.

   B. nitrogen atoms combine to form molecular nitrogen

   C. aluminum foil is torn into tiny pieces

   D. oxygen and hydrogen combine to form water

8. Which of these sentences would have been most useful for the writer to add to the end of the second paragraph?

   A. Potassium chloride (KCl) is also a type of salt.

   B. Roman soldiers were paid in salt.

   C. Too much salt can raise blood pressure.

   D. Salt is a hard, brittle solid that dissolves in water.

Skill Practice

Directions: Choose the best answer to each question. Questions 1 and 2 refer to the following diagram.

CARBON TETRACHLORIDE

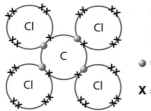

   ● = electrons from carbon

   X = electrons from chlorine

1. The diagram shows the structure of carbon tetrachloride, an industrial cleaner. What is the chemical formula for carbon tetrachloride?

   A. C_4Cl **C.** $4CCl$

   B. $1C_4Cl$ **D.** CCl_4

2. What type of bonds make up a molecule of carbon tetrachloride?

   A. hydrogen bonding

   B. metallic bonding

   C. covalent bonding

   D. ionic bonding

3. A scientist is studying a compound that has the molecular formula $C_6H_{12}O_6$. What additional information would most help her identify the compound?

   A. the number of each kind of atom in the molecule

   B. a diagram of the molecular structure

   C. the number of ionic bonds in the molecule

   D. the number of nitrogen atoms in the molecule

4. The element argon is the third most common gas in Earth's atmosphere, yet it never forms compounds with other elements. What property of an argon atom best explains this fact?

   A. It becomes stable by gaining only one electron.

   B. It has 18 protons in its nucleus.

   C. It has a full outer shell of electrons.

   D. It becomes stable by losing only one electron.

Chemical Reactions and Solutions

KEY CONCEPT: Matter can neither be created nor destroyed. When elements or compounds enter chemical reactions, the numbers of their atoms always remains the same.

Do you keep a change jar at home? Have you ever exchanged change for paper currency? When you take change to the bank and exchange it for paper currency, the amount of money remains the same. You may walk in the bank with 20 quarters, 30 dimes, 25 nickels, and 75 pennies and walk out with a ten dollar bill. You have neither gained nor lost money.

When atoms of different elements join together to make a compound, their totals also remain the same. The resulting product may look different, but the same numbers and kinds of atoms are still there.

Chemical Reactions

When elements combine to form compounds, a chemical reaction takes place. During the reaction, the bonds between atoms break apart and new bonds form. This reaction is expressed in an **equation**, a statement that says two things are the same or equal. Here is the equation for one example of such a reaction:

$$2H_2 + O_2 \longrightarrow 2H_2O$$

The equation shows that two molecules of hydrogen combine with one molecule of oxygen to yield two molecules of water. Each side of the equation shows four hydrogen atoms and two oxygen atoms. Equations must be **balanced** so the same number of atoms of each element are on each side of the equation.

The equation below describes photosynthesis, the process plants use to make food. During this reaction, carbon dioxide and water combine to form glucose.

PHOTOSYNTHESIS

$$6CO_2 + 6H_2O \xrightarrow{\text{light}} C_6H_{12}O_6 + 6O_2$$

carbon   water   glucose   oxygen
dioxide

The numbers in front of the molecules show the ratios that the reaction requires. These numbers are necessary to balance the atoms in the equation.

The Law of Conservation of Matter

Chemical equations have the identical number of atoms of each element on both sides of the equation. This is the **law of conservation of matter** that states that matter is neither created nor destroyed during chemical reactions.

COMPARE AND CONTRAST INFORMATION

In addition to a textbook, there are many ways to learn about all kinds of subjects, including science. For instance, you could watch a video, use computer or hand-held models, read a website, or even play a game to learn about chemistry!

In fact, you may end up using more than one type of media to learn a subject. Comparing and contrasting information from many sources can allow you to gain a broader understanding of the topic. You'll have a better chance of seeing the material from many points of view.

But what if you review multiple sources of information and they contradict each other? Actually, this provides another opportunity for you to strengthen your skills in comparing and contrasting information. If you find that different sources contradict each other when you are learning about a subject, you need to determine which is a better source to trust.

Read the next page. Compare and contrast the information with this blog passage from the internet. Which source do you think is more reliable?

> Chemical reactions take place when elements are put together. I think the reaction is usually written as an equation. Photosynthesis is a pretty cool chemical reaction. It's the process that plants use to make water.

In this case, your textbook is a more reliable source of information. Blogs are not considered to be good sources for learning concepts, as they often include the blogger's opinion. Also, you can't always be sure of a blogger's background. When using internet material, try to stick with government or educational websites.

Core Skill
Analyze Structure

Just as a chemical compound is the combination of two (or more) elements, a compound word is formed by combining two words. You can often understand the meaning of a compound word by looking at the meaning of the two smaller words. On this page, *photosynthesis* is a compound word. The words *wastebasket*, *dishwasher*, and *footprint* are all compound words.

In a notebook, make a list of compound words you find in this chapter. Write the smaller words that form each compound word you find. Then write the meaning of each compound word.

While chemistry, like all science disciplines, is a very "hands-on" subject, computer models are actually used quite a bit by chemists. Three-dimensional computer models can be used to model reactions to predict what will happen when substances are combined. Such models can help chemists in many ways: they can help scientists know what to expect when they do the actual experiment; they can help determine if hypotheses are correct; and they can save scientists money by using fewer material resources in the lab. What limitations do models have in representing chemical reactions?

Consider the reaction in which nitrogen (N_2) combines with hydrogen (H_2) to form ammonia (NH_3):

$$N_2 + 3H_2 \longrightarrow 2NH_3$$

As the equation shows, two atoms of nitrogen (N_2) plus six atoms of hydrogen ($3 \times H_2$) yield two molecules of ammonia. If you count the total number of atoms on each side, you will find that they are equal. The equation agrees with the law of conservation of matter.

Solutions

When salt is placed in water, it seems to disappear. The law of conservation of matter, however, tells us this is not possible. Tasting the water will convince you that the law is correct. The salt is still there. Single molecules of salt have simply been broken off by the water. The resulting salty water is called a **solution**. A solution is a mixture in which one substance is completely and uniformly dissolved in another. Sand in water does not form a solution because the sand does not dissolve. After a short time, the sand simply settles to the bottom.

Solutions have practical applications. Soaps and detergents, for example, rely on solutions with water. One end of soap molecules dissolves in water and one end can dissolve grease molecules. When dispersed in water, the soap molecules surround the grease particles, allowing them to be washed away.

In a solution, the **solute** is the substance that dissolves. The **solvent** is the substance (often a liquid) in which the solute is dissolved. Gases, liquids, or solids can all be solutes or solvents. For example, in a cup of sweet tea, sugar is a solute, as are materials from the tea leaves. Water is the solvent.

THINK ABOUT SCIENCE

Directions: Write a short response to each question.

1. Describe the law of conservation of matter in your own words.

    _____

    _____

2. Explain the chemical reaction involved in photosynthesis using the law of conservation of matter.

    _____

    _____

When substances held together by ionic bonds are dissolved, they usually break apart into ions. An ion is an atom or group of atoms that have gained either a positive or a negative charge. When salt (NaCl) dissolves, it forms positive ions (Na^+) and negative ions (Cl^-). The electron does not return to its original atom.

A substance that forms ions when dissolved in water is called an **electrolyte**. This name comes from the fact that a solution of ions and water can conduct electricity.

A compound that releases hydrogen ions into a solution is called an **acid**. For example, when hydrochloric acid (HCl) dissolves in water, it breaks down into a positive ion (H^+) and a negative ion (Cl^-). A hydrogen ion is one proton. Unlike the sodium ion (Na^+) produced when salt breaks down, the hydrogen ion (H^+) is very reactive. You can imagine it seeking out the electron that it lost. Strong acids are very harsh, dangerous compounds.

Bases are compounds that release hydroxide ions (OH^-) into a solution. A basic solution can also be described as **alkaline**. Like acids, strong bases such as lye or bleach can be hazardous.

When an acid and a base are mixed together, they often form water and a compound called a **salt**. In everyday language, *salt* describes table salt, a specific compound that is sodium chloride (NaCl). In chemistry, sodium chloride is just one example of a salt.

Nonliquid Solutions

People sometimes think of solutions only as solids dissolved in liquids, but solutions involving other states of matter also exist.

Metal solutions are made by melting different metals together. The result is called an alloy. The alloy brass is a solution of copper and zinc. Copper is attractive and easy to shape. Adding zinc makes the resulting brass stronger and tougher than copper would be alone.

The air we breathe is a solution of oxygen, nitrogen, and other gases. Other solutions involve liquids only. Ammonia and water can easily mix together to form a solution.

THINK ABOUT SCIENCE

Directions: Match the descriptions on the left with the correct state of matter on the right.

_____ 1. solute

_____ 2. acid

_____ 3. solvent

_____ 4. base

_____ 5. salt

A. substance that releases hydroxide ions in a solution

B. the substance that dissolves in a solution

C. formed by combining an acid and a base

D. compound that releases hydrogen ions in solution

E. the substance into which something is dissolved

pH Scale

Acids and bases are not all the same strength. Strong acids, such as hydrochloric acid and sulfuric acid, would cause you great pain if they touched your skin. But citric acid, which is much milder, causes the tart taste of many fruits. Likewise, some bases can cause serious burns, while milder bases are found in household products such as soap and baking soda.

The **pH scale** is a way to measure the strength of an acid or a base. It measures the relative amounts of hydrogen ions and hydroxide ions in a solution. The neutral value of the scale is 7.0. This is the pH of a sample of pure water. Values of pH less than 7 are acidic. The lower the pH, the stronger the acid. Likewise, values greater than 7 are basic. The higher the pH, the stronger the base.

THE pH SCALE

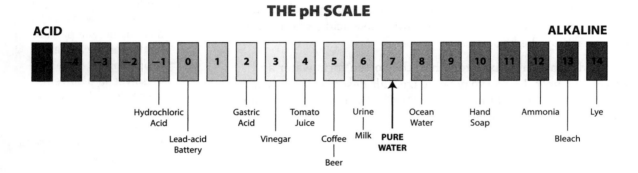

Vocabulary Review

Directions: Complete each sentence using one of the terms below:

acid   balanced   base   law of conservation of matter   solution

1. A compound that releases hydroxide ions in a solution is a _____.

2. The _____ states that matter cannot be created or destroyed.

3. The air we breathe, consisting of oxygen and other gases dissolved in nitrogen, is a _____.

4. To obey the law of conservation of matter, a chemical equation must be _____.

5. The sour taste of a lemon comes from a(n) _____.

Directions: Read the following passage. Locate three compound words. Write each word, identify both parts of the word, and write a definition.

> Chemistry students, preparing for a lifetime career, find that not everything they need to know is in a textbook. One of the skills needed by any scientist is teamwork. Communication among scientists can result in timesaving information.

1. _____

2. _____

3. _____

Directions: Answer the following questions by applying ideas from this lesson.

4. A log with a mass of 50 kilograms is burning in a fireplace. After it burns, the ashes have a mass of 2 kilograms. What idea from this lesson accounts for the missing 48 grams?

5. A glass of orange juice has a pH of 4.4. More water is added to the juice and the pH changes. What idea from the lesson explains the change?

Skill Practice

Directions: Choose the best answer to each question.

1. The law of conservation of matter states that matter is neither created nor destroyed during chemical reactions.

   Which of these statements is an example of this law?

   A. Matter changes states after a chemical reaction.
   B. Matter exists in the same state before and after a chemical reaction.
   C. Old atoms are destroyed and new atoms are created in a chemical reaction.
   D. The same number of atoms are present before and after a chemical reaction.

2. Why is gasoline a solution?

   A. it mixes well with water
   B. it mixes poorly with water
   C. it is a mixture of many substances
   D. it burns in oxygen releasing energy

3. When substance X is mixed throughly in water, a solution forms that conducts electricity very well. Which is substance X most likely to be?

   A. oxygen
   B. sodium chloride
   C. glucose
   D. sand

4. A chemist measures the pH of a solution to be 7.8. Based on the pH scale, what would be the best choice for bringing the solution closer to a neutral pH? (You may refer to the illustration on page 330.)

   A. adding battery acid
   B. adding lemon juice
   C. adding milk
   D. adding lye

The Chemistry of Life

KEY CONCEPT: Carbon forms strong bonds and is the basis for the chemistry of living things.

Do you know anyone who knits? Give a skilled knitter a few balls of yarn. Within a few days, you could get back a sweater, a scarf, a blanket, or a host of other objects. With yarn as a material, all sorts of patterns, designs, and shapes are possible.

Carbon atoms can join together to form long strands, just like a strand of yarn. These chains form rings and bond with other atoms to create more than two million different kinds of molecules, many quite large. Life depends on many of the molecules that carbon makes.

The Chemistry of Life

Molecules that organisms make and use are called **biomolecules**. Nearly all of these molecules contain carbon. The study of carbon and its compounds is called **organic chemistry**, or the chemistry of life. However, it concerns nearly all carbon compounds, not just those of living things. A carbon atom has four electrons available to form bonds. Carbon atoms form in long, stable chains.

Carbon chains may also include other elements, such as oxygen or nitrogen. Chains can also form rings. Carbon atoms are part of many large, complex molecules.

There are several types of organic compounds.

Hydrocarbons

BENZENE   **METHANE**   **ACETYLENE**

Hydrocarbons are made of carbon and hydrogen atoms. The structures shown above are examples of small hydrocarbons. Benzene (C_6H_6) is used in industry. Methane (CH_4) is also called natural gas, and is a fuel. Acetylene (C_2H_2) is also a fuel.

Fossil fuels (coal, petroleum, and natural gas) are hydrocarbons. These formed from the compressed remains of ancient plants and animals.

Petroleum is a mixture of hydrocarbons. They are separated by a process called **distillation**. The distillation of petroleum produces the components of gasoline and other fuels, as well as compounds used to make perfumes, medicines, plastics, and other products.

ANALYZE AUTHOR'S PURPOSE

Everything you read is written for a reason. This reason is called the **author's purpose**. The basic reasons for writing are to inform, to entertain, or to persuade. Identifying the author's purpose will help you better understand the meaning of a piece of writing.

An important part of the scientific method is to share the results of an experiment. Scientists must share information about their procedures and results to the scientific community. Through this sharing, other scientists can review, analyze, and question the scientist about the experiment. More importantly, other scientists can use the information to try to **replicate**, or repeat, the experiment to see if the same results occur. Replication is an extremely important part of determining that the scientific results are valid—that they are true and accurate.

To identify the author's purpose, ask yourself: *What is the author writing about? Is the author giving information? Is the author writing to entertain? Is the author trying to persuade me to think or act a certain way?*

Read the following paragraph and identify the author's purpose.

My experimental analysis reveals that the three plants that received the most water daily did not grow as well as the three plants that received a moderate amount of water. The three plants that received a minimal amount of water actually have grown taller than the plants receiving the most water.

The author's purpose is to inform. The author is providing the results of a scientific experiment. Note that the author does not include any opinions and does not include any persuasive writing.

Reading Skill
Understand Text

On the previous page, you learned that hydrocarbons are separated by the process of distillation. Distillation is a method that is often used in laboratories to physically separate the components of a substance. Do internet or library research to learn about the process of distillation in chemistry, for hydrocarbons and other substances. Have you ever heard of distilled water? Use your research to hypothesize how water is distilled.

WRITE TO LEARN

It is important to recognize bias when reading. If information is not balanced, the reader may not get all the information that is necessary to make an informed decision.

In a notebook, write a paragraph describing a recommended diet with the bias of someone who never eats meat, or who does not like vegetables. Then write a second paragraph, presenting a more balanced description of a healthy diet.

Organic Polymers

Recall that carbon atoms can form long chains. Often, these chains are made from units that repeat and repeat. A compound made from repeated units is called a **polymer**. Organic polymers include the compounds that make up plastics. The structure of the polymer determines whether a plastic is thin and stretchy, like food wrap, or hard and tough, like billiard balls. Other organic polymers are made and used by living things. They include the three classes of biomolecules described below.

Carbohydrates

Compounds called **carbohydrates** are made of carbon, oxygen, and hydrogen. They include simple sugars, such as glucose ($C_6H_{12}O_6$), which have molecules of only six carbon atoms. Complex carbohydrates are polymers of simple sugars. Some are thousands or even millions of units long!

All living things, including humans, use glucose and other carbohydrates as a source of energy. Many substances in food, including starches, are complex carbohydrates that the body breaks apart into glucose. Other complex carbohydrates include cellulose, which is part of wood and other tough plant parts. Only certain animals can break apart cellulose.

Proteins

Proteins are also polymers. The unit of a protein is called an amino acid. There are twenty amino acids that living things typically use to make a protein. Each contains carbon, oxygen, hydrogen, and nitrogen. Some amino acids contain sulfur, too.

Living things use proteins in many ways, including for growth and repair, and to help break apart glucose for its energy. Foods rich in proteins include meats, beans, fish, and eggs.

Lipids

Like carbohydrates, **lipids** are made from atoms of carbon, hydrogen, and oxygen. But the arrangement and numbers of these atoms is different in lipids. Lipids include fats, oils, waxes, and steroids. One type of lipid is the key ingredient in the cell membrane, which forms the outer boundary of a cell.

Many people try to limit fat and oils in their diet. However, the body needs certain kinds of lipids that it cannot make on its own.

THINK ABOUT SCIENCE

Directions: Fill in the blanks with the appropriate word.

1. A compound made from repeating units is called a(n) _____.

2. Unlike carbohydrates, proteins include the elements _____ and _____.

3. Fats and oils are examples of _____.

Complex carbohydrates include starches such as bread, pasta, and rice. They are polymers made from units of small, simple sugars. When you eat a complex carbohydrate, your digestive tract breaks it down into simple sugars. The simple sugars are small enough to be absorbed by the bloodstream.

Digesting starches takes longer than digesting simple sugars such as the sucrose found in many "junk foods"—soft drinks, candy, and some breakfast cereals. When you eat these foods, sugar very quickly enters your blood. You should be very careful about the amount of junk food you eat.

Core Skill
Analyze Author's Purpose

Most scientific writing is meant solely to inform. However, if enough scientific evidence is available to show that a particular outcome is certain, you may find persuasive writing in science materials. Read the two paragraphs on this page. Look for clues that let you know the author's purpose. What does the word *should* in the last sentence tell you about the author's purpose in the second paragraph? What does this tell you about scientific knowledge regarding the effect of junk food on health?

Vocabulary Review

Directions: Complete each sentence using one of the terms below:

biomolecule   hydrocarbons   organic chemistry   structure   replicate

1. To _____ means to repeat or reproduce in the same way or by the same method.

2. Sometimes called the "chemistry of life," _____ is the study of carbon and its compounds.

3. An organic polymer produced and used by living organisms is a(n) _____.

4. Organic compounds made up of only carbon and hydrogen atoms are _____.

5. A polymer may be thin and stretchy or hard and tough, depending on its _____.

Directions: Read the following passage and choose the correct answer to the question that follows.

Landfills consume many acres of valuable land. Once filled, the garbage deposited there may last for centuries. Burning the garbage is not the answer, however. This blackens the air with pollution from burning plastics.

Recycling and reusing plastics is a key to saving space in landfills and reducing pollution. Waste plastics can be used to make new materials, conserving fossil fuels for other important uses.

---

1. What is the author's purpose in writing this article?

   A. Inform people of the dangers of plastics to their health.
   B. Entertain people with interesting facts about landfills.
   C. Persuade people not to use plastics.
   D. Persuade people to recycle and reuse plastics.

Directions: Read the following passage and choose the best answer to each question below.

Weight loss can happen quickly and easily by eating a diet low in carbohydrates. Carbohydrates, such as bread and pasta, can raise the level of sugar in the blood, preventing fat breakdown. A diet low in carbohydrates lowers blood sugar, allowing the body to break down stored food for energy. Following a low-carb diet is easy when you switch to proteins or fats.

---

2. What makes this article biased toward a low-carbohydrate diet?

   A. The passage presents facts only in favor of this diet.
   B. The passage provides information about the risks and benefits of this diet.
   C. The passage explains why a low-carbohydrate diet works.
   D. The passage describes how to follow a low-carbohydrate diet.

3. How could the author make the article more balanced?

   A. Make it seem like weight loss is not very important.
   B. Give more details about a low-carbohydrate diet.
   C. Provide more information about the risks and benefits of the diet.
   D. Explain what a carbohydrate is.

Skill Practice

Directions: Choose the best answer to each question.

1. Which of these compounds is a carbohydrate?

   A. galactose ($C_6H_{12}O_6$)
   B. methane (CH_4)
   C. carbon dioxide (CO_2)
   D. ammonia (NH_3)

3. What is one way that the polymers in plastics are like the polymers used by living things?

   A. Both are made of carbon atoms that are randomly organized.
   B. Both are made of repeated units that contain oxygen and hydrogen.
   C. Both are made of repeated units that contain carbon.
   D. Both are relatively small, simple molecules.

2. What is one way that hydrocarbons differ from biomolecules, such as carboydrates and proteins?

   A. Hydrocarbons contain carbon only.
   B. Hydrocarbons contain carbon and hydrogen only.
   C. Hydrocarbons contain nitrogen, sulfur, and carbon.
   D. Hydrocarbons contain oxygen, hydrogen, and carbon.

4. Which of these sets of foods is the best source of sulfur?

   A. foods rich in proteins
   B. foods rich in carbohydrates
   C. foods rich in sugar
   D. foods rich in fats and oils

Chemical Equations

KEY CONCEPT: Chemical reactions can be expressed symbolically in the form of chemical equations. Familiarity with chemical equations makes it easier to understand and predict some reactions.

In a bowl, you mix flour and baking powder. In another bowl, you beat butter and sugar together. You add one egg and then another. Two teaspoons of vanilla extract are added to the batter before you add the flour mixture and some milk. You stir completely, pour the mixture into a greased pan, and bake it. In the oven, a series of chemical reactions occur. The result is a cake.

The ingredients in the cake mix went through chemical changes to create a new product. You have practiced successful kitchen chemistry.

Chemical Equations

Open any math book, and you're likely to see mathematical equations. Like the examples below, mathematical equations use numbers, letters, and special notation.

$x = 2y$
$12 \times (5 + 2^2) = 5^2 + x$

Now look at these equations.

$2H_2(g) + O_2(g) \qquad 2H_2O(l)$
$CH_4(g) + 2O_2(g) \qquad CO_2(g) + 2H_2O(l)$

Like mathematical equations, chemical equations also use numbers, letters, and special notation. Chemical equations summarize chemical reactions.

Chemical reactions require reactants, or starting substances. They also require some amount of activation energy, or energy that reactants must have to begin changing. Once the changes begin, the result is a new product. The product has chemical and physical properties different from the substances that made it.

Chemical equations are useful tools. You can use them to describe what you need to know about a reaction in a small amount of space. Knowing how to read, write, and balance chemical equations helps you describe the changes that occur during chemical reactions.

Understanding the Symbols

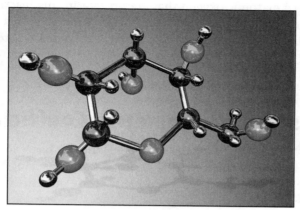

This model represents the bonded atoms within a molecule of glucose.

When you learned to write mathematical equations, you first learned the meanings of different symbols. It is helpful to understand the symbols in chemical equations, too.

Recall the discussion of the symbol O_2. The capital letter "O" is an abbreviation for the chemical oxygen. The small number 2, called a subscript because it is written "under" the letter, represents the number of atoms of oxygen. This element, O_2, is called diatomic oxygen. *Di* is Greek, meaning "two," and *diatomic* means "consisting of two atoms." The diatomic atom O_2 is essential for life. It takes up about 20 percent of Earth's atmosphere, and it is found in compounds like water. Plants and animals take it in, and it must be present for most things to **combust**, or burn.

Now look at another symbol: $C_6H_{12}O_6$

The capital letters are abbreviations for the chemicals carbon, hydrogen, and oxygen. In this molecule, 6 atoms of carbon, 12 atoms of hydrogen, and 6 atoms of oxygen are chemically bonded. The result is one molecule of glucose, a sugar found in plants. It is also found in animal tissue and in the blood, where it provides energy to a body's cells.

When you write a chemical equation, a symbol separates the reactants from the product. Those symbols include →, ↔, and ⇔. These symbols will make more sense later, but for now, know that a chemical reaction always includes one of these symbols. Look at these examples.

$$CH_4(g) + 2O_2(g) \rightarrow CO_2(g) + 2H_2O(l)$$

Sometimes, you will see the letters *g, l, s,* and *aq* in an equation. The letter *g* tells you that a substance is in the form of a gas, and the letter *l* identifies a liquid. The letter *s,* which doesn't appear in this example, identifies a solid. Another symbol not seen is this example is *aq.* These letters tell you that a substance is aqueous, meaning contained or dissolved in water.

The letters CH_4 form the symbol for a molecule of a chemical compound called methane, a gas that is a component of fossil fuels, including natural gas, coal, and oil. Animals such as cows, sheep, and goats release methane during digestion. Organisms that break down organic wastes buried deep in landfills where oxygen doesn't reach also release methane.

The large number 2 before the abbreviation $O_2(g)$ tells you the number of molecules of diatomic oxygen that are present in this reaction. This number, which also appears before the abbreviation $H_2O(l)$ in the products, is called a **stoichiometric coefficient**. When no stoichiometric coefficient is written, it is implied, or understood, that the number is 1.

What Is a Stoichiometric Coefficient?

The Greek word *stoikheion* means "element," and the base word *metry* is used to make words that describe "a process of measuring." So, a stoichiometric coefficient is a measurement of the quantities of reactants and products.

Now return to the examination of the chemical equation:
$$CH_4(g) + 2O_2(g) \rightarrow CO_2(g) + 2H_2O(l)$$

The letters CO_2 are an abbreviation for the chemical compound carbon dioxide. It is an odorless gas that animals give off during respiration. It is also produced when organic matter decays. It is also present in bubbling, or carbonated, beverages.

H_2O is the abbreviation for a molecule of water. Most life on Earth depends on this odorless and tasteless liquid for survival.

The arrow in the chemical equation tells you which chemicals are present at the beginning of the reaction and which are present at the end. When you read the equation, let the arrow represent the word *yield*, which means "produce." So, reading from left to right, you read, "One molecule of methane gas reacts with 2 molecules of diatomic oxygen to yield one molecule of carbon dioxide gas and 2 molecules of water."

Most chemical reactions involve far more than one or two molecules. However, no matter how many molecules are involved, the proportion between them will remain balanced.

The Law of Conservation of Mass

The Law of Conservation of Mass states that a chemical reaction can rearrange atoms, but it cannot change the overall number of each atom. In other words, if a reaction begins with fifteen hydrogen atoms, it will also end with fifteen hydrogen atoms. In the product, those hydrogen atoms may appear in different molecules than in which they started, but they will still be there. So, the number of each atom on the left side of an equation will always equal the number of the same kind of atoms on the right side of the equation.

When you don't know the stoichiometric coefficients for each chemical in an equation, you can apply the Law of Conservation of Mass. Suppose you know nitrogen (N_2) and hydrogen (H_2) react to yield ammonia (NH_3), but you do not know the proportion for each. You can use a system of trial and error to identify the correct proportions.

As you work, remember that you can change the coefficient but never the subscript.

$$N_2(g) + H_2(g) \rightarrow NH_3(g)$$

Count the total number of N atoms. There are 2 atoms in N_2, and 1 atom in NH_3. If you write the coefficient 2 in front of NH_3, the number of nitrogen atoms is balanced.

$$N_2(g) + H_2(g) \rightarrow 2NH_3(g)$$

Now count the total number of H atoms. There are 2 atoms in H_2, and with the new coefficient, there are 2×3, or 6 atoms in $2NH_3$. If you use the coefficient 3 for H_2, the number of hydrogen atoms is balanced.

$$N_2(g) + 3H_2(g) \rightarrow 2NH_3(g)$$

Always check your work when you balance an equation. Count the number of nitrogen and hydrogen atoms again. If the same number of each kind of atom appears in the product as in the reactants, your solution works, and the equation is balanced.

THINK ABOUT SCIENCE

Directions: The following equation describes the process of respiration. Use coefficients to balance the equation.

$$C_6H_{12}O_6 + O_2 \rightarrow CO_2 + H_2O$$

Core Skill
Interpret Information in Text and Graphical Form

When you read a text, pay close attention to diagrams, pictures, flowcharts, and graphs that accompany the text. These visual, or graphical, elements can serve several purposes. They can use images to restate the text, or they can give an example of the text. They can also extend the information given in a text. Read the following. Then examine the diagram.

The gas NO, called nitric oxide or nitrogen monoxide, forms naturally, but most forms from the burning of fossil fuels. The gas, which appears in car exhaust, enters the atmosphere, where it reacts with oxygen to form NO_2, or nitrogen dioxide. NO_2, a brown gas, can damage plants and reduce their growth. As an ingredient of smog, NO_2 can make breathing difficult and make organisms more vulnerable to disease.

What is the purpose of the diagram that accompanies this text?

Different Types of Chemical Reactions

You can divide simple chemical reactions into four broad categories. They are synthesis reaction, decomposition reaction, single replacement reaction, and double replacement reaction.

Synthesis Reaction

In a **synthesis reaction**, two reactants combine to form a single product. A simple equation for a synthesis reaction is:

$A + B \rightarrow AB$

Consider these examples of synthesis reactions.

$N_2 + H_2 \rightarrow NH_3$

One molecule of diatomic nitrogen and one molecule of diatomic hydrogen combine to make one molecule of ammonia. Ammonia is a colorless gas with a sharp smell. It is used to make fertilizers and is dissolved in liquids to make industrial products, such as cleaning solutions.

$C + O_2 \rightarrow CO_2$

In this synthesis reaction, one atom of carbon combines with one molecule of diatomic oxygen to make one molecule of carbon dioxide.

CO_2 is found naturally in Earth's atmosphere, but it has become more abundant due to a number of human activities, including the burning of fossil fuels in the presence of oxygen, or combustion. Earth's surface absorbs some of the energy in sunlight and radiates lower-energy waves of light back into space. Because CO_2 absorbs this energy before it can escape into space, the heat is trapped much like heat in a greenhouse. So, CO_2 is called a greenhouse gas, which contributes to a widespread warming of Earth's atmosphere.

Decomposition Reaction

A **decomposition reaction** is the opposite of a synthesis reaction. When something decomposes, it breaks apart. In a decomposition reaction, a reactant is broken down into two or more products. Some decomposition reactions require a great deal of energy before they can occur. A general equation representing a decomposition reaction is:

$AB \rightarrow A + B$

When water decomposes, or breaks apart to make oxygen and hydrogen gas, the reaction is called "water splitting." Plants engage in water splitting when they use the energy in sunlight for photosynthesis. During photosynthesis, carbon dioxide and water decompose to form sugar and oxygen.

$6CO_2 + 6H_2O \rightarrow C_6H_{12}O_6 + 6O_2$

When an electric current passes through water between the positive and negative poles of a battery, the water splits into oxygen and hydrogen. The energy in the hydrogen gas can be used to run an engine or launch a rocket into space.

$2H_2O \rightarrow O_2 + 2H_2$

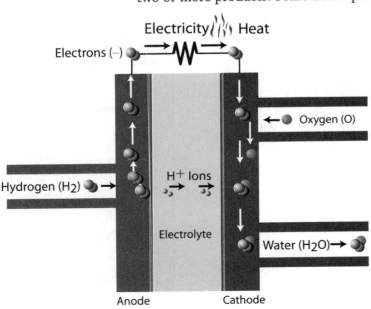

Electricity Heat

Electrons (−)

Oxygen (O)

Hydrogen (H₂)

H⁺ Ions

Electrolyte

Water (H₂O)

Anode   Cathode

Emissions from hydrogen-powered vehicles are in the form of water vapor.

Single Replacement Reaction

In a **single replacement reaction**, one reactant atom displaces, or takes the place of, an atom in a new product compound. The general equation for a single replacement reaction is:

$$A + CB \rightarrow C + AB$$

When an atom is likely to react with another atom, it is called reactive. There is a range of reactivity, with some elements being more reactive than others. Reactivity measures how easily an atom will form new bonds with other atoms. If one type of metal is more reactive than another, it is likely that it will replace a less reactive metal whose chemical bonds are not as strong.

In a single replacement reaction, a reactive atom usually takes the place of a less reactive atom. The result is a new chemical compound and the release of an element. Most single replacement reactions involve metals that are added to a solution.

In a single replacement reaction, iron (Fe) combines with hydrochloric acid (HCl) in solution to produce iron dichloride and hydrogen gas. The atom of iron displaces the hydrogen.

$$Fe(s) + 2HCl(aq) \rightarrow FeCl_2(aq) + H_2(g)$$

In the laboratory, iron dichloride is used to make other iron compounds. Outside the laboratory, the compound is used in wastewater treatment plants to remove harmful substances and odors from drinking water.

Another example of a single displacement reaction produces copper. Iron filings are added to a solution of copper sulfate, which is blue. When the reaction occurs, the iron displaces the copper, and the solution turns green. Copper falls out of the solution.

$$Fe + CuSO_4 \rightarrow FeSO_4 + Cu$$

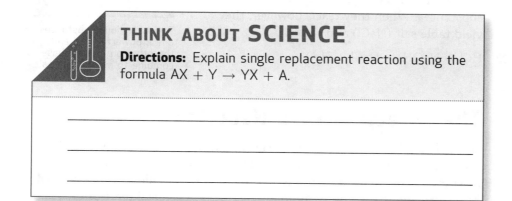

THINK ABOUT **SCIENCE**

Directions: Explain single replacement reaction using the formula $AX + Y \rightarrow YX + A$.

_____

_____

_____

21st Century Skills
Information and Communication Literacy

Members of the scientific community express and interpret their ideas and communicate the results of their investigations. Before they communicate their findings to the world at large, they carefully describe their methods and results to a variety of expert scientists. In this process, called a peer review, fellow scientists evaluate the quality of the work through a variety of methods and procedures. Scientists often rely on databases, or organized collections of data, to share data and conclusions.

Following the professional peer review, scientists publish the results of their work in major scientific journals, which are sold in print or online for the public to read. Some scientists also use social media networks to distribute their scientific information. Distributing information via the internet and social media brings wider attention and possibly more interest to the scientific study. Some scientists also establish websites where interested readers can learn about the work and donate money to support continued research.

What do you think of scientists self-publishing their research on websites and through social media? List some of your reactions.

Good examples of writing have a central idea that is supported in the text. Often, an author states the central idea clearly in the introductory paragraph. This statement, which contains the central idea, is called the thesis statement. The **thesis statement** presents the writing's main argument or idea. Read the following introductory paragraph.

"Beginning in the 1850s, scientists relied on weather stations scattered around the globe for collecting temperature data. Weather stations grew in number and in sophistication as technologies changed. Today, there are thousands of weather stations on the ground and on ocean buoys across the world's waters. Analysis of temperature records gathered from these sites over more than a century show not only a general warming of Earth's atmosphere but also particularly rapid warming in recent decades."

Which sentence in the paragraph serves as a thesis statement?

Double Replacement Reaction

In a **double replacement reaction**, atoms in two reactants displace atoms to make two new product compounds. The general equation for a double replacement reaction is:

$$AY + XB \rightarrow AB + XY$$

You may recall reading about acids, bases, and electrically charged atoms called ions. When an acid is added to a solution, it releases hydrogen ions ($H+$). When a base is added to a solution, it releases hydroxide ions ($OH-$). Both hydrogen and hydroxide ions are very reactive, making them dangerous substances.

When an acid and a base combine, a double replacement reaction occurs. This particular double replacement reaction is called **neutralization** because the reactant ions lose their charges, forming the neutral products salt and water.

Consider this example. Although your stomach produces gastric juices that contain hydrochloric acid (HCl), the acid is diluted, or watered down. It assists in the digestive process and helps destroy harmful disease-carrying organisms that enter the digestive system.

Cl is a solution of hydrogen chloride in water used in a variety of manufacturing industries. It is a poisonous acid that can irritate eyes and corrode, or eat away, skin tissue as well as membranes in the digestive and respiratory systems.

Sodium hydroxide (NaOH) can have many of the same results. It can also burn skin and cause hair to fall out. NaOH is a very strong base used to manufacture chemicals and cleaning agents, and it is also used to change petroleum into a variety of products.

Both HCl and NaOH are dangerous substances. When they react, however, they yield table salt (NaCl) and water.

$$HCl + NaOH \rightarrow NaCl + H_2O$$

Two hazardous substances, one a powerful acid and the other a powerful base, combine to make table salt.

The Direction of a Reaction

Throughout the lesson, you have read chemical reactions containing this symbol: $\rightarrow$. This symbol indicates that the reaction changes a set of reactants into one or more products. This process is called a **net forward reaction**, and it is common in many real-world chemical reactions.

For example, combustion is a net forward reaction. The equation for the combustion of octane (C_8H_{18}), one of the main ingredients of gasoline, is

$$2C_8H_{18} + 25O_2 \rightarrow 16CO_2 + 18H_2O$$

Net forward reactions happen in only one direction. That is, reactants make products, but the process cannot be reversed. You cannot combine carbon dioxide and water, for example, to re-create octane and oxygen.

A **reversible reaction**, however, travels in both directions. In other words, products can be combined to re-create reactants.

Riou/Photographer's Choice RF/Getty Images

Recall this reaction from earlier in the lesson when you were learning to balance equations.

$$N_2 + 3H_2 \rightarrow 2NH_3$$

This reaction of nitrogen (N_2) and hydrogen (H_2) to produce ammonia (NH_3) is actually reversible. So, the proper way to write the equation is with the symbol $\leftrightarrow$.

$$N_2 + 3H_2 \leftrightarrow 2NH_3$$

When nitrogen and hydrogen begin to change into ammonia, some of the ammonia begins to decompose into nitrogen and hydrogen. Eventually, the rate at which ammonia is being created and the rate at which ammonia is decomposing will be equal. This means that for every new ammonia molecule that is made, another is unmade. This point is called **chemical equilibrium**. The term *equilibrium* means "in a state of balance." When equilibrium is reached, you write the equation with the symbol $\Leftrightarrow$.

$$N_2 + 3H_2 \Leftrightarrow 2NH_3$$

Changing the pressure, temperature, volume, or concentration of the chemicals in a reversible reaction will change the equilibrium point. Still, the reaction will ultimately find a new equilibrium.

SYMBOLS USED IN CHEMICAL EQUATIONS

Symbol	Explanation
$\rightarrow$	Indicates a result of a reaction; yields
$+$	Separates two or more formulas
$\leftrightarrow$	Indicates a reversible reaction
$\Leftrightarrow$	Indicates chemical equilibrium
(s)	A reactant or product in the solid state
(l)	A reactant or product in the liquid state
(aq)	A reactant or product in an aqueous solution (dissolved in water)
(g)	A reactant or product in the gas state

WRITE TO LEARN

Choose your favorite cupcake recipe or find one in a printed or online cookbook. Write an introductory paragraph to describe the cupcake and its ingredients. Then write the reactants and the product as a chemical equation. You do not need to use real chemical symbols for each ingredient. Instead, use words and also symbols, such as $+$ and $\rightarrow$.

Describe the changes you observe as you combine ingredients. Then, in a final paragraph, explain what makes the cupcake-making process a net-forward reaction.

Vocabulary Review

Directions: Match each vocabulary word with its correct definition.

1. _____ chemical equilibrium

2. _____ synthesis reaction

3. _____ combust

4. _____ decomposition reaction

5. _____ double replacement reaction

6. _____ net forward reaction

7. _____ reversible reaction

8. _____ single replacement reaction

9. _____ stoichiometric coefficient

A. when the rate of reactants becoming products matches the rate of products becoming reactants

B. a reaction in which a complex reactant yields two or more simpler products

C. a reaction in which two or more simple reactants yield a more complex product

D. a reaction that cannot easily be reversed

E. a reaction in which one type of atom replaces a similar type of atom

F. a reaction in which the products can be transformed back into the reactants that made them

G. to combine fuel and oxygen in a reaction that releases heat

H. a number that describes the relative amount of each type of molecule

I. a reaction in which atoms are exchanged between two molecules or compounds

Directions: Read and complete the activities.

1. The reactants Fe and Cl yield $FeCl_3$. Write the stoichiometric coefficients that balance the chemical equation.

   _____ Fe(s) + _____ Cl(g) → _____ $FeCl_3$(s)

Directions: Read the following chemical equation. Then answer the questions.

Cl_2(g) + 2KBr(aq) → 2KCl(a) + Br_2(l)

2. What makes this an example of a displacement reaction?

   _____

   _____

3. How do you know that chlorine (Cl_2) is more reactive than bromine (Br_2)?

   _____

4. What do you know about acids and bases that explain why their neutralization will yield water?

   _____

   _____

Directions: Read and complete the activities.

When gray, powdered zinc metal reacts with grayish-purple iodine, the reaction is powerful. The heat of the reaction is so great that some of the iodine not used in the reaction changes from a solid to a gaseous state, producing a purple vapor.

Zn(s) + I_2(s) → ZnI_2(s)

5. What kind of chemical reaction is this? Justify your answer.

   _____

   _____

   _____

6. Use the Law of Conservation of Mass to explain the reaction.

   _____

   _____

   _____

   _____

Skill Practice

Directions: Read and complete the activities.

1. The reactants HCl and Mg yield $MgCl_2$ and H_2. Write the stoichiometric coefficients that balance the chemical equation.

   _____ HCl + _____ Mg → _____ $MgCl_2$ + _____ H_2

2. The reactants C_6H_5COOH and O_2 yield CO_2 and H_2O. Write the stoichiometric coefficients that balance the chemical equation.

   _____ C_6H_5COOH + _____ O_2 → _____ CO_2 + _____ H_{20}

3. The reactants C_4H_{10} (a fossil fuel) and O_2 react to yield CO_2 and H_2O.

   $2 C_4H_{10}(l) + 13 O_2(g) \rightarrow 8 CO_2(g) + 10 H_2O(l)$

   What are the stoichiometric coefficients in this reaction? _____

   What do the letters l and g tell you? _____

   What does the arrow in the equation mean? _____

   What tells you that this is a combustion reaction? _____

4. Use the letters A, B, C, E, and F to write a model for each of the following reactions.

   Combination: _____   Single replacement: _____

   Decomposition: _____   Double replacement: _____

Directions: Read the text. Then answer the questions.

Automobile air bags contain three chemicals, sodium azide (NaN_3), potassium nitrate (KNO_3), and silicon dioxide (SiO_2). A car's sensor tells the bag to inflate. The following reactions occur:

   a. $NaN_3(s) \rightarrow Na(s) + N_2(g)$
   b. $Na(s) + KNO_3(s) \rightarrow K_2O(s) + Na_2O + N_2(g)$
   c. $K_2O(s) + Na_2O(s) + SiO_2(s) \rightarrow$ small bits of safe glass

---

5. What causes the bag to inflate? _____

6. What is the final product? _____

7. What kinds of reactions occurred? Justify your answer.

   _____

   _____

Directions: Choose the best answer to each question.

1. Scientists describe matter as anything that has

   A. solids and liquids
   B. mass and volume
   C. mass and shape
   D. volume and temperature

2. What is the difference between a physical property and a chemical property?

   A. Physical properties can be observed without changing the identity of the object.
   B. Chemical properties can be observed without changing the identity of the object.
   C. Chemical properties are internal, and physical properties are external.
   D. There is no difference between physical properties and chemical properties.

3. The atomic _____ is the number of protons in the nucleus.

   A. mass
   B. weight
   C. number
   D. group

4. The substance formed when two or more elements are chemically combined is known as a

   A. chemical bond
   B. chemical symbol
   C. chemical formula
   D. chemical compound

Review

5. In which type of bond do atoms give up or gain electrons in their outer shell?

   A. covalent
   B. ionic
   C. both covalent and ionic
   D. neither covalent nor ionic

8. Which of the following is an organic polymer that all living things use for energy?

   A. carbon atom
   B. protein
   C. lipid
   D. carbohydrate

6. Which of these correctly states the law of conservation of matter?

   A. Matter can be created, but not destroyed, in chemical reactions.
   B. Matter can be both created and destroyed in chemical reactions.
   C. Matter is neither created nor destroyed in chemical reactions.
   D. Matter can be destroyed, but not created, in chemical reactions.

9. _____ summarize chemical reactions.

   A. Chemical formulas
   B. Chemical equations
   C. Chemical elements
   D. Chemical compounds

7. The study of carbon and its compounds is called

   A. organic chemistry
   B. inorganic chemistry
   C. organic distillation
   D. hydrocarbonation

Directions: Choose the best answer to each question.

THE BOHR MODEL

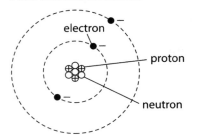

Most events in chemistry can be explained by the Bohr model of the atom. According to this model, an atom is made of three types of tiny particles: positively charged protons, negatively charged electrons, and neutrons that lack charge. The protons and neutrons are packed tightly together in a central core, called the nucleus. Electrons move in the space around the nucleus.

Yet as scientists continue studying atoms, they find that atoms are much more complex than this model suggests. For example, they have discovered hundreds of other atomic particles. The particles include muons, tauons, bosons, gluons, at least six kinds of quarks, and the list goes on. Some of these particles combine to form protons and neutrons. Most have roles yet to be fully understood.

The motion of electrons is also very complex. From the Bohr model, you might conclude that electrons move in circle-shaped, well-defined pathways. In fact, an electron's motion is random and varied, more like a moth around a lamp than a planet or a race car in their circular paths. Electrons also move very, very rapidly. They move so rapidly that it is impossible to measure both their speed and position at any one time.

10. What best describes the Bohr model of the atom?

   A. accurate and useful
   B. useful but too simple
   C. useful but too complex
   D. inaccurate and not useful

11. Which question is most likely the subject of current scientific research?

   A. Are atoms made of only three particles?
   B. Where are neutrons found in an atom?
   C. What charge does a proton carry?
   D. What is the role of muons in an atom?

12. In what way is an electron like a moth around a lamp?

   A. relative size and shape
   B. path of motion
   C. speed of motion
   D. purpose of motion

Review

Directions: Answer each question as directed.

13. Categorize each of these properties as physical or chemical by placing a P (physical) or C (chemical) in the space.

    state _____

    flammability _____

    color _____

    corrosiveness _____

    density _____

    boiling point _____

    chemical reactivity _____

    shape _____

14. Use the periodic table in this chapter to determine the identity of the following elements:

    atomic number of 55 _____

    atomic mass of 10.8 _____

    chemical symbol Na _____

    chemical symbol O _____

    atomic mass of 237.0 _____

    atomic number of 22 _____

15. You pour flavored drink crystals into a pitcher of water and stir to make fruit punch. Identify the solute, the solvent, and the solution in this scenario.

    solute _____

    solvent _____

    solution _____

16. Distinguish which of these illustrations shows ionic bonding and which shows covalent bonding.

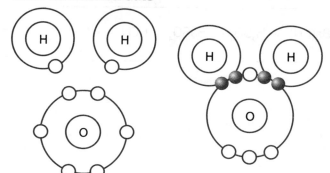

SEPARATED HYDROGEN AND OXYGEN ATOMS   **A WATER MOLECULE**

⬤ = electrons shared by hydrogen and oxygen atoms

_____

SODIUM CHLORIDE

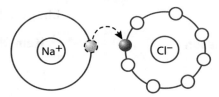

The Na electron has transferred over to the Cl atom.

_____

17. Which of these structures represents an organic compound? Explain your reasoning.

_____

_____

_____

_____

18. Balance each of these chemical equations.

$Fe + O_2 \rightarrow Fe_2O_3$

_____

$Al + CuSO_4 \rightarrow Al_2(SO_4)_3 + Cu$

_____

$Ba(CN)_2 + H_2SO_4 \rightarrow BaSO_4 + HCN$

_____

19. Categorize each of the following reactions as a synthesis reaction, a decomposition reaction, a single replacement reaction, or a double replacement reaction.

$Fe + O_2 \rightarrow Fe_2O_3$

_____

$Al + CuSO_4 \rightarrow Al_2(SO_4)_3 + Cu$

_____

$Ba(CN)_2 + H_2SO_4 \rightarrow BaSO_4 + HCN$

_____

Check Your Understanding

On the following chart, circle the number of any item you answered incorrectly. Next to each group of item numbers, you will see the pages you can review to learn how to answer the items correctly. Pay particular attention to reviewing those lessons in which you missed half or more of the questions.

Chapter 9 Review

Lesson	Item Number	Review Pages
Matter	1, 2, 13	304–311
The Atom	3, 10, 11, 12, 14	312–319
Compounds and Molecules	4, 5, 15	320–325
Chemical Reactions and Solutions	6, 16	326–331
The Chemistry of Life	7, 8, 17	332–337
Chemical Equations	9, 18, 19	338–347

Application of Science Practices

CHAPTER 9: CHEMICAL PROPERTIES

Question

How can a chemistry experiment help you solve a chemical reaction equation?

Background Concepts

The Law of Conservation of Mass states that a chemical reaction can rearrange atoms, but it cannot change the overall number of each atom. So, for example, if a reaction begins with five hydrogen atoms, it will also end with five hydrogen atoms. Those hydrogen atoms may appear in different molecules than in which they started, but they will still be there. So, the number of each atom on the left side of an equation will always equal the number of the same kind of atoms on the right side of the equation.

In this experiment, you will combine vinegar (acetic acid) with baking soda (sodium bicarbonate). The vinegar is an acid, and the sodium bicarbonate is a base.

Investigation

Materials required

baking soda   measuring cup
vinegar   measuring spoons
water   1 empty plastic water or soda bottle (16 oz)
funnel   several regular-sized balloons
toothpick or straw

Refer to the figure on the right to complete the following steps:

1. Use the funnel to pour $\frac{1}{2}$ cup of vinegar and $\frac{1}{4}$ cup of water into the plastic bottle.

2. Rinse and dry the funnel thoroughly.

3. Place the funnel into the neck of a balloon. Pour about 2 teaspoons of baking soda into the balloon. You may need to use a toothpick or straw to push the baking soda through the neck of the funnel into the balloon.

4. Attach the balloon securely to the top of the bottle. Keep the balloon flopped to one side of the bottle to avoid spilling any of the baking soda.

5. Turn the balloon upright so that the baking soda falls into the bottle. This is Trial 1. Record your observations in the table below.

6. Repeat this experiment using half as much vinegar. This is Trial 2. Record your observations in the table below.

7. Repeat this experiment using half as much baking soda. This is Trial 3. Record your observations in the table below.

Trial 1	Trial 2	Trial 3

Interpretation

1. Based on your observations, what kind of product did the chemical reaction yield?

_____

_____

2. Why did the chemical reaction stop in each trial?

_____

_____

Answer

How can a chemistry experiment help you solve a chemical reaction equation? Use your observations to balance the incomplete chemical equation for the reaction.

$NaHCO_3$ + CH_2COOH → _____ + _____ + Na^+ + CH_3COO^-
baking   vinegar   ?   ?   sodium and
soda   (acetic acid)   acetate ions

Evidence

What physical state are the missing products in the equation?

_____

_____

UNIT 3

Earth and Space Science

Earth and Living Things

Nonliving elements that we see and feel every day—the air, soils, water—sometimes may seem less important than living things. However, these nonliving substances play a vital role in the health and survival of plants and animals. Nutrients are endlessly reused as they travel between the living and nonliving worlds in biogeochemical cycles. These cycles work together to ensure that living things receive the chemical substances they need to survive.

In this chapter you will learn about:

Lesson 10.1: Cycles of Matter
Where does the water that falls as rain come from? What happens to the carbon dioxide in the air you exhale? Materials that are essential to life on Earth—carbon, oxygen, water, nitrogen, and phosphorus—move in continuous cycles throughout Earth's surface and its atmosphere. In this lesson, you will investigate these cycles and their impact on living organisms.

Lesson 10.2: Fossil Fuels
Each of us relies upon fossil fuels such as oil, natural gas, and coal to power our daily activities. Along with energy, however, the burning of fossil fuels yields byproducts that can sometimes harm the environment. Learn more about the sources of fossil fuels and the consequences of their use in this chapter.

Goal Setting

Why is it important to study the Earth and living things?

Here are two reasons:

• To understand the biogeochemical cycles that sustain life on Earth

• To recognize the balance between the importance of fossil fuels and the need to protect the environment

To help you set goals for learning more about cycles of matter as you read this chapter, illustrate and label the cycles of matter. Identify the living and nonliving components for each cycle. Look for these concepts as you read. Be sure to create separate diagrams for each of these biogeochemical cycles:

- Carbon
- Oxygen
- Water
- Nitrogen
- Phosphorus

Cycles of Matter

Lesson Objectives

You will be able to
- Define a biogeochemical cycle
- Identify five kinds of biogeochemical cycles

Skills

- **Core Skill:** Follow a Multistep Procedure
- **Core Skill:** Draw Conclusions

Vocabulary

algae
biogeochemical cycle
detritivore
nitrogen fixers
nutrient
producers
weathering

KEY CONCEPT: All living things depend on specific nutrients, such as carbon, oxygen, hydrogen, nitrogen, and phosphorus. Although matter is neither created nor destroyed, it may change form. Nutrients are matter. They move in cycles through living organisms, rocks, soils, water, and chemical compounds. So, these cycles are called biogeochemical cycles.

The bottom of the box is warm. You open the lid and the scent of pizza reaches your nose. Your family has ordered your favorite kind of pizza. Within the hour, the pizza is gone, and only the box is left.

It's likely that the pizza box is made of layers of recycled paper or a combination of recycled paper and fresh plant material, such as wood or cotton. Cotton is a fluffy fiber that grows around a cotton plant's seeds. Like all things that have mass and take up space, cotton is matter. The Law of Conservation of Mass, which is also identified as the Law of Conservation of Matter, states that matter continues to exist. It may change form, but it is never lost nor destroyed. You can no longer identify the cotton in your pizza box, but it's there, in a different form.

What Are Nutrients?

Earth is home to millions of species of living things. Those living things interact with and depend upon nonliving things, like air, water, and soil. Nonliving matter is made of chemical substances, some of which are nutrients for living things. A **nutrient** is a substance that a living thing must have if it is to live, grow, and reproduce.

Living things need some nutrients in large amounts, including carbon, oxygen, hydrogen, nitrogen, and phosphorus. They need other nutrients, too, but in much smaller quantities. These nutrients include iron, copper, chlorine, and iodine.

Biogeochemical Cycles

Recall that the prefix *bio* means "life," and the prefix *geo* means "Earth." The word *chemical* refers to substances like carbon and oxygen, each with specific properties, structures, and atomic compositions. Together, the word and word parts form the word *biogeochemical*, or "life, Earth, chemical." A **biogeochemical cycle** is a nutrient cycle. Energy from the sun and gravity drive the cycle. Together, they keep the nutrients that are essential for life constantly cycled through living things and back to nonliving form.

Decomposers and Biogeochemical Cycles

The word *detritus* is a Latin word meaning "wearing away." In science, detritus refers to worn-away rocks, bits of plant and animal matter, animal waste, and dead organisms.

Detritivores eat detritus. By feeding on waste and dead organisms, they prevent these things from accumulating. Equally important, they make the matter of once-living things available to other living things, including decomposers. **Decomposers**, which are mostly bacteria or fungi, break down the complex molecules in decaying matter into simpler molecules. They absorb some of the nutrients and return others to the soil, where they become available to plants.

Other organisms like worms and insects eat the decomposers. So, nutrients move in a biological circle, from one organism to another until death, when the cycle begins again.

DECOMPOSERS RECYCLE NUTRIENTS

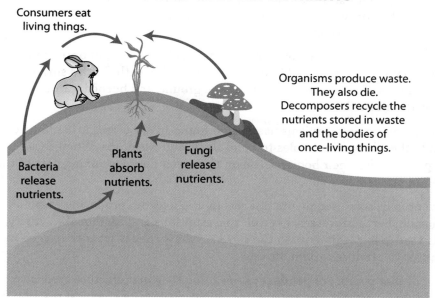

Consumers eat living things.

Organisms produce waste. They also die. Decomposers recycle the nutrients stored in waste and the bodies of once-living things.

Bacteria release nutrients.

Plants absorb nutrients.

Fungi release nutrients.

THINK ABOUT SCIENCE

What would happen if there were no decomposers?

_____

_____

Science investigations often require accurate measurements. Consider the measurements that are part of the procedure for analyzing the decomposition rate of rubber latex balloons.

Before beginning, it is necessary to collect some materials: a grow light, a pie pan, garden soil, peat moss, rubber latex balloons cut in strips, and 20 grams of activators. The activators are active bacteria that can be purchased at a garden shop. These bacteria decompose once-living material.

1. Mix the soil, peat moss, and activators in the pie pan.
2. Wet the soil evenly.
3. Blow up, deflate, and then cut rubber latex balloons into strips.
4. Bury the pieces in the soil. Measure to make the distance between pieces equal.
5. The light should be 8 inches above the pan.
6. Add the same amount of water to the pan each day to keep the soil moist.
7. Check the rate of decomposition each day for five days. Note the amount of decay.

What might happen if you did not measure the height of the grow light accurately?

The Carbon Cycle

THE CARBON CYCLE

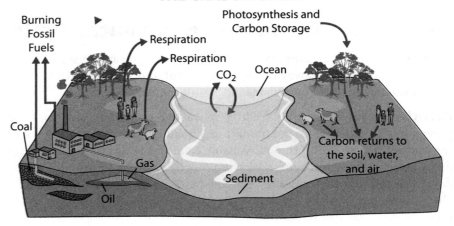

The **carbon cycle** depends on the carbon present in carbon dioxide, a gas stored in Earth's atmosphere, land, and in bodies of water. About 93 percent of carbon dioxide is stored in oceans, but it moves quickly between air and land. Carbon dioxide occupies only about 0.038 percent of air.

All living things store carbon in their cells. This carbon is released as carbon dioxide when decomposers consume once-living things or the waste they produce in life.

Because plants and animals formed fossil fuels, the carbon once stored in their bodies is now stored in oil, gas, and coal beneath Earth's surface. When these fuels are removed from the ground and burned, carbon dioxide enters the air.

In oceans, some organisms use carbon found in dissolved carbon dioxide and other carbon molecules to build shells and skeletons. When these organisms die, their bodies settle on the ocean floor, becoming part of the sediment, or fine soil.

Eventually, some sediments dissolve in ocean water, producing carbon dioxide. Earth processes, like volcanic eruptions and earthquake activity, also can bring sediments to the surface, where they react with oxygen in the air to produce carbon dioxide.

Recall that plants are **producers**, meaning they manufacture their own food. In plant cells, the energy in sunlight activates a chemical reaction, in which carbon dioxide gas reacts with water to produce sugar and oxygen.

Plants rely on the sugar they manufacture as a source of energy for growth, development, and reproduction. The bonds between the atoms in each sugar molecule store potential energy. When plants respire, they break apart sugar molecules, releasing the energy stored in the chemical bonds. Sugar reacts with oxygen to produce water and carbon dioxide. The producer uses the water and releases carbon dioxide to the atmosphere. This cycle of use and release keeps carbon dioxide available.

Producers, of course, depend on carbon dioxide for food production. Carbon dioxide is important in another way, too. It helps maintain Earth's heat balance.

THINK ABOUT SCIENCE

A wooden fence will begin to weather over time. What happens to the wood as it weathers and breaks apart?

_____

_____

Vocabulary Review

Directions: Write the word or term that matches the description.

**algae   biogeochemical cycle   detritivore   nitrogen fixers
nutrient   producers   weathering**

1. _____ an organism that eats rotting plant and animal matter

2. _____ a substance that a living thing requires for growth

3. _____ wind and water breaking rocks into smaller pieces

4. _____ organisms that make their own food

5. _____ the movement of nutrients like carbon, oxygen, and phosphorus

6. _____ plantlike organisms that life depends on for oxygen

7. _____ organisms that change nitrogen into forms plants can use

Skill Review

Directions: Read and complete the activity.

1. Explain what drives biogeochemical cycles.

_____

_____

_____

2. How do detritivores and decomposers contribute to biogeochemical cycles?

_____

_____

_____

Skill Review (continued)

3. Explain how plants contribute to the carbon cycle.

_____

_____

_____

4. Explain why algae are critical to the oxygen cycle.

_____

_____

_____

5. About 70 percent of Earth is covered in water. Explain why the water cycle is critical to meeting the water needs of living things.

_____

_____

_____

Skill Practice

Directions: Read and complete the activity.

1. Explain how biogeochemical cycles are related to the Law of Conservation of Matter.

_____

_____

_____

2. Some people build compost bins in their gardens. They mix soil and plant waste in the bin and add organisms such as earthworms. Earthworms eat bits of organic matter in the soil. Eventually, the soil becomes rich with organic compounds that make it a natural fertilizer that gardeners can add to their gardens. How do earthworms contribute to decomposition?

_____

_____

_____

3. Explain how carbon dioxide levels affect the atmosphere.

_____

_____

_____

Skill Practice (continued)

4. Compare different kinds of nitrogen fixers.

NITROGEN FIXERS

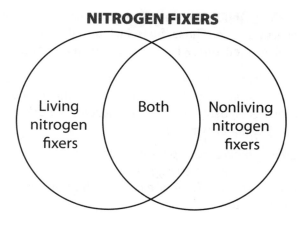

Living nitrogen fixers | Both | Nonliving nitrogen fixers

5. Explain the relationship between weathering and the phosphorus cycle.

_____

_____

_____

6. Use the chart to identify how living things use chemical nutrients.

Carbon	Nitrogen	Phosphorus

Fossil Fuels

KEY CONCEPT: People depend on fossil fuels to meet their energy needs. Those fuels include oil, natural gas, and coal. Each fuel formed through a process of decay over millions of years.

Your family takes a trip. You fill the car with gasoline, drive a short distance, stop for lunch, and then continue driving. Eventually, you reach your destination—an amusement park where thousands of electric lights welcome you.

Energy makes work possible. There is energy stored in the gasoline in your car. There is energy locked in the food you eat for lunch. And there is energy in the form of electricity that powers the lights and rides at the amusement park.

What Are Fossil Fuels?

Recall that the Latin word *fossilis* means "found by being dug up." Fossils are the remains of once-living organisms that are "dug up" from beneath Earth's surface. Some fossil material is used to make fuels to operate cars, trucks, and tractors, and to generate electricity.

Fossil fuels are made of different **hydrocarbons**, or molecules containing hydrogen and carbon. They occur in different lengths and shapes. Some are single chains, while others form branched chains or rings. CH4, or methane, is the smallest hydrocarbon, and it exists as a gas. As hydrocarbons add more carbon atoms, they get longer. Longer hydrocarbons exist in different states. They form liquids, and if they grow longer, they become tar-like solids.

When humans extract crude oil, natural gas, and coal from beneath Earth's surface, they are removing fossil fuels. Those fuels are made into products and used as sources of energy.

Crude Oil

Crude oil is also called **petroleum**. The Latin prefix *petr–* means "rock or stone," and the word *oleum* meaning "oil." So, petroleum can be defined as "rock oil." This thick, gooey rock oil is made mostly of hydrocarbons. It also contains small amounts of chemical compounds containing oxygen, nitrogen, and sulfur. This sticky substance is buried beneath Earth's crust, where it often collects in holes and cracks in rock.

Reaching buried oil requires drilling holes into the ground to make wells. Gravity causes the oil to flow into the wells, where it can then be pumped to the surface. Even after pumping is complete, some oil remains in a well. To capture this remaining oil, water is pumped into nearby wells. This water flows into the empty well, and the oil floats to the top. This oil is then pumped to the surface.

At the surface, most oil travels through pipes to a **refinery**. A refinery is a plant where the oil is **distilled**, or heated at very high temperatures until it boils. Because the different hydrocarbons in the oil have different boiling points, they change from liquid to vapor at different rates. Vapors move into different containers, where they rise and eventually cool. Eventually, the vapors condense to form liquids, which are collected and stored. These separate hydrocarbons are sold to manufacturers to create new products, such as asphalt, grease, wax, diesel oil, heating oil, jet fuel, and gasoline.

The Source of Crude Oil

Crude oil comes from organic, or once-living organisms. Marine plants and animals probably produced most of the oil humans have extracted or that remains buried in Earth. These marine organisms are called **plankton**. They are single-celled organisms that float on water, including salt water, freshwater, and mixtures of both. Plankton have been abundant on Earth for hundreds of millions of years.

For all that time, when the plankton died, they settled into the muddy sediments at the bottom of bodies of water. Bacteria in the sediments decomposed the plankton, producing a mixture of hydrocarbon compounds. This was the first stage in the formation of oil. Next came burial.

People and oil companies often describe dinosaurs as the source of petroleum, but it's not true. Plankton, which lived for hundreds of millions of years before dinosaurs appeared, produced most of Earth's petroleum supplies.

As sediments piled upon sediments, pressure increased. As pressure increases, so does temperature. Heat altered the hydrocarbons in the rock, producing oil.

Corey Ford/Stocktrek Images/Getty Images

THINK ABOUT SCIENCE

Plants continue to die and decay, so the process of oil formation continues. If the process continues, why do people need to worry about running out of oil?

_____

_____

21st Century Skill
Information, Communication, and Technology Literacy

Scientists have learned to identify rocks where oil is likely to be stored. To find these rocks, scientists use sound waves, listening devices, and computers.

First, scientists lay lines of listening devices on the ground. Then on land, large vibrator trucks send sound waves beneath the ground. Under the sea, submerged air guns send bursts of sound energy into the seabed.

Sound waves bounce off different layers of rock at different rates. The listening devices measure how long it takes for the sound waves to return to the surface. Scientists add the data collected by the listening devices into computers, which process the data. The result is a three-dimensional model of the ground beneath Earth's surface.

Consider the fact that crude oil comes from once-living marine organisms called plankton. Why, then, does so much oil exploration occur on land? Investigate the places on Earth that have a high concentration of oil and find out the reasons why.

Natural Gas

As more sediment layers accumulated, temperatures continued to rise. Some oil changed to gas, which explains why oil and natural gas are often found together beneath the ground.

When people began extracting oil from the ground, they thought natural gas was a waste product. They burned the gas, producing enormous flames that astronauts could see from the space shuttle. Today, most countries recognize how much energy is stored in natural gas and no longer burn it. In the United States, for example, natural gas represents one-fifth of all energy sources.

Natural gas is made mostly of methane, a simple hydrocarbon that forms a colorless, odorless gas. Natural gas is highly **flammable**, however. It ignites and burns easily. To prevent explosions in homes and other places where natural gas is a source of energy, gas manufacturers add a chemical to make it smell. The chemical smells like rotting eggs to make it easy to detect and repair natural gas leaks.

Before people understood the potential energy in natural gas, they burned it as a waste product.

Coal

According to the US Department of Energy, slightly more than one-fourth of Earth's supply of coal is buried in the United States. Coal miners work in 26 states, digging up this fossil fuel. Most of the coal they bring to the surface is transported to power stations that use it to generate electricity. Coal provides about one-fourth of US energy, and its abundance makes it the world's leading fuel.

How Coal Formed

Imagine Earth 300 to 400 million years ago. It was covered with thick vegetation, such as trees and ferns. When these plants died, they fell to the ground, where they began to decay.

In some places, ocean water flowed onto coastal land, covering the sediments on the ground. The ocean water added new chemicals to the sediments, like sulfur. In other places, freshwater covered the ground. Large swamps of decaying plant matter formed.

In the same process that produced oil and natural gas, sediment layers settled on older sediment layers beneath the water. The sediments squeezed together. Increased squeezing created increased pressure, which led to higher temperatures. The temperatures "cooked" the compounds, producing coal.

The kinds of plants that grew, died, and decayed changed over time. So did the time it took for sediment layers to form and cook. Consequently, different kinds of coal formed.

THINK ABOUT SCIENCE

Describe the relationship between pressure and temperature in the formation of fossil fuels.

_____

_____

Four Kinds of Coal

Today, miners dig four different kinds of coal from beneath the ground. They are lignite (LIG-nite), subbituminous (sub-bih-TYOU-mih-nus), bituminous (bih-TYOU-mih-nus), and anthracite (AN-thruh-site).

There are four kinds of coal, each taking a different amount of time to form.

Lignite is a soft, brownish black substance that formed in shallow water. Most lignite is younger than other forms of coal, beginning to develop about 250 million years ago. It is not fully decayed, making some bits of plant matter visible. As lignite burns, it produces the least heat of any other kind of coal.

Subbituminous coal is darker than lignite, giving it the name black lignite. However, subbituminous coal is more fully decayed than lignite and produces slightly more heat. Scientists estimate that of all known reserves of coal, about one-half are made of either lignite or subbituminous coal.

Bituminous coal is also called soft coal. It is dark brown or black and has visible bands. When burned, this coal produces more heat than lignite and subbituminous coal. It is also the most plentiful kind of coal and easy to transport. Consequently, it is widely used to generate electricity.

Anthracite is the hardest of the four kinds of coal, and so it is also called hard coal. Hard coal is black or steel gray, shiny, and clean to the touch. At one time, people burned anthracite to heat their homes. Today, however, people use less-expensive alternatives, such as oil or natural gas. The supply of hard coal is too limited and too costly to mine to make it a practical choice for home heating.

The Consequences of Burning Fossil Fuels

The United States is dependent on oil, natural gas, and coal to meet energy demands. But burning fossil fuels has harmful environmental consequences.

Global Warming

When fossil fuels burn, the carbon dioxide, or CO_2, that the plants used in photosynthesis is released to the atmosphere, where it traps heat. Scientists predict that if levels of the gas continue to rise, the planet will experience increased temperatures that will likely cause glaciers and ice caps to melt. That, in turn, will raise water levels and flood coastal areas. Warmer temperatures are also more likely to generate extreme weather, such as storms and droughts.

Core Skill
Make Predictions

Predictions are educated guesses about what will happen in a text based on clues the writer provides and on the reader's own knowledge. Coal-Direct Chemical Looping (CDCL) may help energy providers burn coal without releasing harmful chemicals, or pollutants, into the air. CDCL grinds coal into a powder, mixes in tiny beads of iron oxide, and heats the mixture. A chemical reaction produces carbon dioxide and water vapor. The water vapor is used to power steam turbines to generate electricity. Slightly less than 100 percent of the carbon dioxide is captured, preventing it from escaping into the atmosphere. The iron beads, once exposed to air, convert back to their original form, making them reusable. All that remains is a small amount of very fine coal particles, or ash.

If CDCL is used more widely at coal-fired power plants in the United States and around the world, what can you predict about carbon dioxide levels in the atmosphere?

©iStockphoto.com/aeduard

Readers can compare
and contrast media
sources to organize
ideas and to gain a
more complete view of
a topic. Comparing and
contrasting helps you
to identify consistent
information presented
by the sources and to
discover and investigate
details that seem in
conflict. Information
is available through
print and digital media.
Digital resources include
e-books, e-journals,
e-newspapers, blogs,
websites, videos, still
images, and other
related materials.

Consider the topic
of acid rain. A single
internet search will
produce numerous
articles and documents
on the topic. Narrow
your search to videos,
and you'll find many
choices. Skim the search
results to find the video
clips created and posted
by reliable sources, such
as NASA, the National
Atmospheric and
Space Administration.
Take notes on the
information presented.
Then compare what
you learned to what
you read in this lesson.
What information was
the same? What was
different? Write a
paragraph explaining
what you know about
acid rain from both
sources and what you
would like to investigate
further.

Air Pollution

Burning fossil fuels releases other substances, too, including carbon monoxide (CO). CO gas can cause headaches and stress the heart.

Nitrogen compounds react with oxygen in the presence of sunlight to form ozone (O_3). **Ozone** is a gas that forms naturally in the upper layer of Earth's atmosphere. There, it prevents harmful ultraviolet radiation from the Sun from reaching Earth's surface. But when high levels of ozone exist in the level of the atmosphere closest to Earth, it can be harmful. The chemical reacts powerfully with other chemicals, creating **toxins**, or poisons. These toxins can make it difficult to breathe, and they can lead to lung disease. They stunt the growth of crops and trees and weaken plants, making them more vulnerable to disease.

Ozone and other compounds can also combine with particulates to produce clouds of poisonous gases, called **smog**, a combination of smoke and fog. **Particulates** can include tiny bits of unburned fossil fuel, fumes from vehicles and industry, dust, and plant pollen. These particles can be inhaled, taken deeply into the lungs, where they can lead to breathing disorders and disease.

Smog endangers people's safety.

Acid Rain

Sulfur and nitrogen compounds react with water vapor in the air. The reaction produces sulfuric and nitric acids. These acids dissolve in rain and snow and fall as precipitation called **acid rain**. Scientists think acid rain weakens trees and other plants instead of killing them directly. Acids damage their leaves and enter the soil, where plant roots absorb them. Acid rain may cause essential plant nutrients to wash away, leaving trees with fewer nutrients. Acids also accumulate in rivers, ponds, and lakes, where they make the water too acidic to support life.

THINK ABOUT **SCIENCE**

Explain how ozone can be both good and bad for life on Earth.

_____

_____

Don Bayley/E+/Getty Images

Land and Water Pollution

Mining coal and pumping oil and gas affect land and water quality. So do transporting and burning the fuels. Miners, for example, dig up large areas of land to reach beds of coal beneath the surface. The land around the mine is stripped bare of vegetation. Water flowing through the mine deposits acids in the upper layers of soil. Materials buried with the coal are left as solid waste on the ground.

Ships can leak oil into water. The oil usually floats, affecting organisms that live and feed on or near the surface. Sticky oil attaches to the fur of seals and the feathers of birds, destroying the animals' ability to keep dry and warm. Oil coverings can prevent mothers from identifying their young. Animals that accidentally swallow oil or eat oil-contaminated food can suffer from illness, including internal bleeding, severe breathing problems, and organ damage. The oil disturbs animals' reproductive systems, makes eggshells thinner, and harms eggs and larvae. Animals' damaged immune systems make them vulnerable to pests.

Ocean currents carry tar balls onto shore from oil spills many hundreds of miles away. Tar balls are spheres of crude oil that have been weathered by wind and waves.

Alternative Fuels

If you examined a drop of pond water beneath a microscope, you would probably see a variety of plankton, including algae. More than 100,000 species of algae exist on the planet, and these algae may provide an alternative to fossil fuels.

Recall that algae are photosynthetic. Inside their cells, a chemical reaction initiated by the sun's energy produces the food they need for growth and development. What makes algae different from plants is how they store that food. Instead of sugar, the food is stored as oil, which can be the raw material for the fuels that keep cars, trucks, and other vehicles moving. Because living organisms are the source of that fuel, it is called a **biofuel**.

The production of this biofuel begins at an algae farm, where algae grow in large ponds. A new crop of algae grows every few weeks. The algae are harvested, and chemicals or sound waves are used to break down their cells. Oil is extracted from these cells and shipped to a biorefinery. In the future, it may be shipped to regular oil refineries for processing. Algae farms may one day be built next to power plants that currently use fossil fuels to generate electricity. Burning fossil fuels puts carbon dioxide into the air. Researchers hope the algae will use the carbon dioxide in photosynthesis.

dtimiraos/E+/Getty Images

Imagine you are preparing a report for an oil-producing corporation. You were asked to list and explain the advantages and disadvantages that could result from switching from fossil fuels to biofuels. Create an outline in which you list your main points. Then write the main details for each main point.

Vocabulary Review

Directions: Write the word that matches the description.

acid rain   biofuel   hydrocarbons   ozone   particulates   petroleum
plankton   smog   toxins

1. _____ another name for oil

2. _____ poisons

3. _____ organisms that died and decayed to form fossil fuels

4. _____ fine solid materials produced by burning fossil fuels

5. _____ fuel made from living things

6. _____ clouds of poisonous compounds

7. _____ precipitation containing sulfur and nitrogen compounds

8. _____ a compound that reflects ultraviolet radiation from the sun

9. _____ the main chemical compound found in fossil fuels

Skill Review

Directions: Read and complete the activities.

1. Explain what classifies methane as a hydrocarbon.

   _____

   _____

2. Describe the process a manufacturer uses to create a product such as diesel oil from petroleum.

   _____

   _____

3. Explain the relationship between plankton and fossil fuels.

   _____

   _____

4. Predict what people should do if they detect the smell of rotten eggs in buildings supplied with natural gas.

   _____

   _____

5. Describe the consequences of excessive levels of ozone in the lower layer of Earth's atmosphere.

   _____

   _____

Skill Practice

Directions: Read the text. Then complete the activity.

1. Explain how hydrocarbon length is related to petroleum's physical state.

   _____

   _____

   _____

2. Compare the process of oil and natural gas formation.

Oil	Natural Gas

3. Explain the role of water in the formation of coal.

   _____

   _____

   _____

4. Given the environmental costs of burning coal, write an argument for its continued use.

   _____

   _____

   _____

5. Describe the chemical components of smog.

   _____

   _____

   _____

Directions: Choose the best answer to each question.

1. Which of the following kinds of organisms acts as a decomposer?

   A. plants
   B. algae
   C. bacteria
   D. animals

2. Water changes to water vapor during

   A. infiltration.
   B. condensation.
   C. infiltration.
   D. evaporation.

3. Which of the following consequences might occur if algae vanish?

   A. Oxygen will decrease.
   B. Carbon dioxide will decrease.
   C. Plants will grow more slowly.
   D. Earth will grow cooler.

4. Which of the following describes the role of decomposers in the nitrogen cycle?

   A. They release nitrogen from rocks.
   B. They capture nitrogen from the air.
   C. They release nitrogen from dead organisms.
   D. They take in nitrogen during photosynthesis.

5. Trees rely on bacteria in the soil to meet their need for what element?

   A. oxygen
   B. nitrogen
   C. carbon
   D. hydrogen

6. Scientists think that coal-burning power plants may contribute to rising temperatures on Earth's surface. This is because the burning of coal

   A. releases heavy metals.
   B. adds carbon dioxide to the atmosphere.
   C. adds heat to the atmosphere.
   D. removes water from the atmosphere.

7. Which of the following biogeochemical cycles does not include movement through air?

   A. water cycle
   B. carbon cycle
   C. nitrogen cycle
   D. phosphorus cycle

8. The release of sulfur and nitrogen compounds can cause

   A. acid rain.
   B. increased temperature.
   C. ozone accumulation.
   D. severe drought.

Directions: Answer the following questions.

9. A corn plant absorbs a carbon dioxide molecule during photosynthesis. Describe the next steps that a carbon atom in this molecule could take as it moves through the carbon cycle.

_____

_____

_____

10. Describe the benefits and drawbacks of using coal as a fuel.

_____

_____

_____

11. Explain how oil is distilled at a refinery to create different hydrocarbons.

_____

_____

_____

_____

12. What is smog? What conditions are needed for smog to form?

_____

_____

_____

13. Compare the role played by detritivores with that played by decomposers. How are their roles similar? How are they different?

_____

_____

_____

_____

Check Your Understanding

On the following chart, circle the number of any item you answered incorrectly. Next to each group of item numbers, you will see the pages you can review to learn how to answer the items correctly. Pay particular attention to reviewing those lessons in which you missed half or more of the questions.

Chapter 10 Review

Lesson	Item Number	Review Pages
Cycles of Matter	1, 2, 3, 4, 5, 7, 9, 13	358–367
Fossil Fuels	6, 8, 10, 11, 12	368–375

CHAPTER 10: EARTH AND LIVING THINGS

Question

What threats to life and property do natural disasters pose, and how do we avoid or lessen their effects?

Background Concepts

Hurricanes, tornadoes, volcanoes, wildfires, earthquakes, tsunamis, and floods are natural disasters that can result in significant loss of life and property damage, especially when they occur in densely populated areas.

By studying natural disasters and their effects, scientists and engineers can devise ways to predict them, lessen their impact, possibly prevent them, and deal with their aftermath. These measures can help to minimize loss of life and property damage.

Investigation

1. Select a kind of natural disaster that has, at some point, affected the area in which you live.

2. Research the processes that caused the natural disaster that affected your community.

3. Use the internet or interview a local first-responder or emergency-preparedness official to find out how your community is prepared to deal with a similar natural disaster.

4. Research technologies that scientists and engineers have used to reduce the harmful effects of natural disasters like the one that occurred in your community.

Application of Science Practices

Answer

Record your interview notes and research in the following table.

Topic	Research and Interview Notes
Type of natural disaster	
Date of occurrence	
Cause of the disaster	
Consequences of the disaster	
Community preparedness for future disasters	
Technologies aimed at reducing future impact of a similar disaster	

Interpretation

Explain the relationship between the consequences of the natural disaster and your community's preparedness for future disasters.

_____

_____

Answer

What steps can your community take to avoid or lessen their effects of this type of natural disaster in the future?

_____

_____

Evidence

Identify specific strategies and technologies community officials are prepared to use to minimize or avoid the effects of another natural disaster.

_____

_____

Earth

The Earth is sometimes called a "big blue marble" or the "third rock from the Sun." Of course, Earth is not a marble or a rock; it is a dynamic place. Earthquakes and volcanoes shake and pummel people and structures. Massive storms take shape over oceans and build power and intensity over thousands of miles. The surface of the planet, its interior, and the surrounding atmosphere all play important roles in carving and changing the landscape.

Earth science is the study of the elements that make up the planet and their processes. In this chapter, you will study the different branches of Earth science to find out how Earth's processes interact.

In this chapter you will learn about:

Lesson 11.1: Geology
Geology is the study of the processes at work on and below Earth's surface. This lesson describes the structure of Earth and the movement of Earth's crust.

Lesson 11.2: Oceanography
More than 70 percent of Earth's surface is ocean. Investigate the features of the ocean floor, currents and waves, and the changing nature of other bodies of water.

Lesson 11.3: Meteorology
Have you got a soccer game or are you getting ready for a bike ride or a long walk? One aspect of Earth science will be of great interest to you—the weather. In this lesson, you will find out about the factors that cause the daily weather, how winds form, and the causes and behavior of storms.

Goal Setting

To help you set goals for learning more about Earth science as you read this chapter, complete a concept map. It may begin like the map below, but it is likely to grow much larger. Write the word *Earth* in a circle at the center of your map. Then draw circles around Earth. Use the circles to record important ideas and details from your reading.

You can also attach circles to circles beyond *Earth*. Make the concept map as large as you need to capture important information.

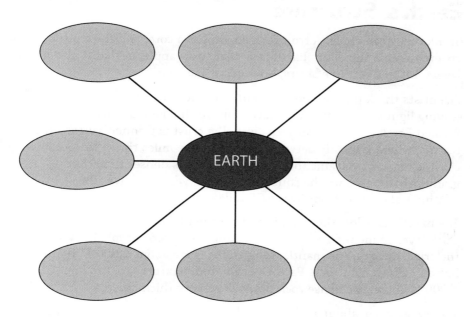

Geology

KEY CONCEPT: Earth is made of several layers. Rocks change form in a never-ending process called the rock cycle.

Have you ever piled sand on the beach to make a sand castle? If so, you know that such structures are temporary. Winds, rushing water, the actions of living things, or even a mild breeze can reshape the sand. Earth's mountains and other landforms may seem permanent. But like sand castles, they change over time—just more slowly.

Earth's Structure

Think of a fresh apple. A fruit formed around a core, and the fruit is covered with a thin skin. Earth is similar to an apple, in that it also has layers. Instead of three layers, however, it has four.

Scientists think that heavier elements sank toward Earth's center, leaving lighter elements in the layers above. The lightest elements formed Earth's crust, or surface layer. The crust is thinner beneath the oceans, where it is only between three and five miles thick. The oceanic crust is made of volcanic rock called basalt. It is denser than the granite that makes up the continental crust, where we live. The continental crust is from five to 25 miles thick.

The mantle is below the oceanic and continental crust. This 1,800-mile thick layer of melted rock flows constantly, like gooey asphalt. Where the crust rides on the mantle, temperatures are about 1,600°F. That is hot enough to melt rock. Farther down, temperatures soar to about 4,000°F. There, the dense rock flows like a very thick liquid.

The **outer core**, about 1,400 miles thick is made of melted iron and nickel, and temperatures range from 4,000°F to 9,000°F. The **inner core** extends about 800 more miles to Earth's center. It is also made of iron and nickel. Great pressures, however, force the iron and nickel of the inner core into a solid state.

Lesson Objectives

You will be able to
- Describe the structure of Earth
- Relate movement of Earth's crust to geologic activity
- Describe the three main types of rock and how they change in the rock cycle

Skills

- **Core Skill:** Integrate Text and Visuals
- **Core Skill:** Apply Scientific Models

Vocabulary

igneous
inner core
metamorphic
outer core
rock cycle
sedimentary

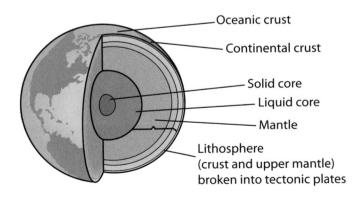

Oceanic crust

Continental crust

Solid core

Liquid core

Mantle

Lithosphere
(crust and upper mantle)
broken into tectonic plates

INTEGRATE TEXT AND VISUALS

A **visual** is a drawing, a diagram, a photograph, a graphic organizer, or a physical model. It is any tool that represents or helps explain a concept. Scientists often use visuals to help organize or represent data.

When visuals like the diagrams below accompany text, they often give new information, or they use images to interpret the text. Visuals can serve as tools for comprehending difficult concepts or for understanding processes and procedures.

To use visual tools effectively, examine them before reading. Then skim the text. Look for titles and subtitles that give clues to the text's central ideas. Examine each visual again. Ask yourself: *How does the visual relate to the text? What can I learn from it? How can I use it to better understand the text?*

Analyze the following diagram. Then read the text that follows. Explain how the text and diagram work together to help you understand the process of deformation.

BEFORE DEFORMATION
horizontal rock layers

AFTER DEFORMATION
folded rock layers

A **deformation** is a change in the shape of a rock caused by stress, or pressure. In one kind of stress, pressure pushes into a rock from opposite sides. Let a thick piece of fabric represent a rock layer. Now imagine placing your hands on either end of the fabric and sliding your hands together until they meet. The resulting stress squeezes the rock layers, causing deformation.

THINK ABOUT SCIENCE

Directions: Write a short response to the following question.

How is Earth's structure like an apple?

_____

_____

Movement of the Crust

The crust is fractured into several pieces, called **crustal plates** or **tectonic plates**. The plates move very slowly across the top of the mantle. Many plates contain both oceanic and continental crust.

About 250 million years ago, all of the world's continents were one large landmass called Pangaea. Over time, they began to break up and slowly move apart toward the positions they are in today. Their movement still continues, meaning the globe's appearance will continue to change.

PANGAEA

Mountain Building

Scientists think that many of Earth's most spectacular features resulted from the motion of the tectonic plates. Notice in the illustrations below what happens when tectonic plates meet. Some plates push up the crust, forming mountains.

Fold mountains are formed at the edge of a continent by two tectonic plates grinding into each other. As the two plates collide, the edges of the continents fold upward, forming large mountain ranges, such as the Appalachian Mountains.

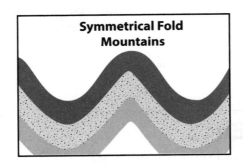

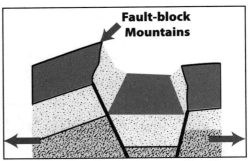

In other cases, huge blocks of crust are lifted, making high, steep mountains. These are called fault-block mountains. Examples of these kinds of mountains are found in East Africa and the West Coast of the United States. The mountains form when the crust cracks rather than folds. The cracks are called **faults**. On either side of the fault, blocks of rock move up or down in relation to each other. Sometimes the blocks subside, forming valleys between mountains.

Volcanoes

A **volcano** is an opening in Earth's crust, or the mountain that builds up around this opening. Volcanoes form in several ways. One way is by a process called **subduction**, in which one tectonic plate slips beneath another.

Because oceanic crust is denser than continental crust, the oceanic plate may slip beneath the continental plate, bringing both rocks and water into the mantle. In the mantle, rocks melt into a liquid-like form called **magma**, and the seawater turns into steam. This creates tremendous pressure underground. When the pressure becomes too high for the crust to withstand, a volcano erupts. The explosion can be extremely violent.

Mount St. Helens in Washington State formed in this way. Its eruption in 1980 was very explosive, taking off a part of the mountain and leveling the landscape around it.

Core Skill
Integrate Text
and Visuals

Models are visual tools. They come in a variety of forms. Some are diagrams and graphs. Others are mathematical explanations or three-dimensional structures.

Visuals can help you integrate technical information expressed in words with images or hands-on models that represent the same information. To use a visual effectively, be sure to look for titles and labels, and read captions or explanations that may be attached.

Look at the diagrams on this page. Ask yourself: How do the diagrams integrate, or build in, meaning to support the text?

FORMATION OF A VOLCANO

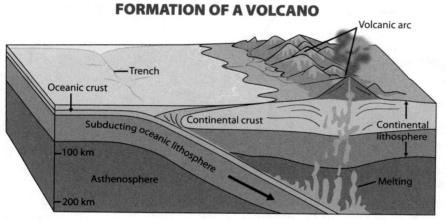

In a subduction zone, the oceanic crust slides under the continental crust.

Some volcanoes occur in mid-plate regions. These volcanoes typically form over **hot spots**, or breaks in Earth's crust where magma erupts. The Hawaiian Islands formed this way. Trace the chain of islands on a map, and you can see how the Pacific Plate has moved.

Eruptions on Hawaii tend to be relatively mild, producing small but steady flows of very fluid lava. **Lava** is magma that reaches Earth's surface.

PATH OF VOLCANIC ISLANDS

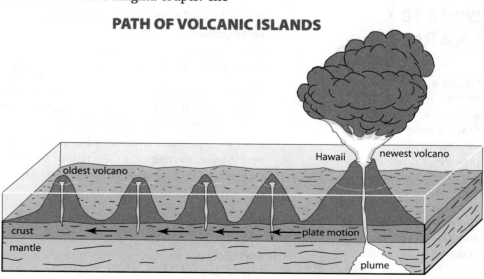

In their work, scientists follow an established process of scientific inquiry. Yet while this rigorous process is clearly established, it still demands that scientists have initiative, or the ability to take action and follow through on completing a task. For any inquiry to be completed successfully, scientists must also show self-direction, or the ability to monitor and improve their own performance.

How important are initiative and self-direction in other careers? Do clerks and IT specialists need these qualities? How about farm workers and small-business owners?

WRITE TO LEARN

Locate a map of your city or area. In a notebook, write several paragraphs that describe the kind of information the map provides. Include an explanation of what purpose the map has. If a friend needed directions to get from one place to another in your area, under what circumstances would the map be more helpful than your written description?

Earthquakes

An **earthquake** is the sudden shaking of the ground caused by movement of matter within Earth. Most earthquakes are very mild—too mild for people to feel or notice—but some are jarring.

Earthquakes typically form along the boundaries of tectonic plates. The map below shows these boundaries as thin black lines. Jagged lines mark areas of frequent earthquake activity.

PLATES AND QUAKES

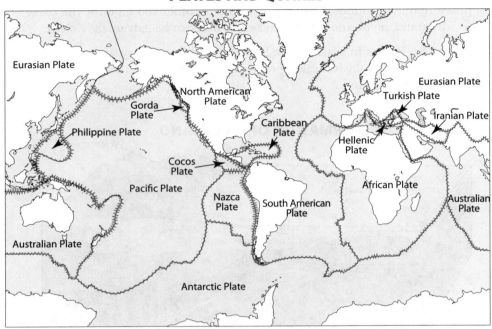

Many earthquakes occur where two tectonic plates slide past each other. The slowly moving plates strain the rocks that build up along the boundary. When the tension is too great, the rocks break, the tension is released, and an earthquake occurs.

In the United States, the San Andreas Fault is a common source of earthquakes. This fault runs along the length of California. Major earthquakes have struck California and the West Coast more than a dozen times in the past 100 years. Minor earthquakes happen more frequently.

Not all earthquakes occur at plate boundaries, however. Like volcanoes, some earthquakes occur in mid-plate regions. These earthquakes are far less understood. Scientists think that heat and pressure build up under the plate, possibly in areas where ancient faults exist. The stress may result in an earthquake. The city of New Madrid, Missouri, seems to be a center of mid-plate earthquake activity. Three violent quakes occurred there in the early 1800s.

Types of Rock

Earth's crust has three types of rock: igneous, sedimentary, and metamorphic. Melted rock from deep within Earth pushes up to fill cracks made when the tectonic plates move. This type of rock, called **igneous** from the Latin word for "fire," is molten rock that has cooled. Granite is an igneous rock. So is the melted rock that spews out of a volcano.

About 80 percent of the rock making up Earth's surface was formed by small particles of mineral or organic matter. Most of the deposits occurred on the ocean floor. As time passed, the loose material was pressed into solid rock, called **sedimentary** rock. Some types of sedimentary rocks were built up from the shells and skeletons of small sea animals. Some types were formed from the deposits of minerals that were dissolved in water. Other types of sedimentary rocks, such as sandstone, were formed through the erosion of other rocks. Sedimentary rocks usually lie in horizontal layers, but they may fold or undergo other changes if intense pressures are placed on them.

When intense heat and pressure are applied to igneous or sedimentary rock, the structure of the rock changes. These "changed" rocks are called **metamorphic** rock. There are many types of metamorphic rock. Marble is the changed form of limestone. Slate is the changed form of shale.

Some of the rocks in Earth's crust are considered quite valuable. Certain types of rocks, called **ores**, contain a metal or other mineral that is combined with other compounds. Ores can be processed to release their valuable metals, such as iron, copper, aluminum, and silver.

The Rock Cycle

If metamorphic rock is exposed to more heat and more pressure, the rock may melt. When it cools, it can once again be igneous rock. The igneous rock may weather into sediments that harden into sedimentary rock. The sedimentary rock may change into metamorphic rock. Scientists refer to the changes of rock from one form into another as the **rock cycle**. Steps of the rock cycle occur constantly, but usually very slowly.

Mineral

A **mineral** is a specific type of rock. Minerals have many properties, including luster (shininess), density, color, and hardness.

Identifying a mineral is not always easy, because many minerals share at least one or two properties. Minerals can be identified by their hardness. A useful test for hardness is a **scratch test**, in which one mineral is used to scratch another. The hardness of minerals is ranked on the Mohs scale, which ranges from 1 to 10. Talc, a soft powder, has a hardness of 1 on this scale. Diamond, the hardest natural substance, has a hardness of 10.

Core Skill
Apply Scientific Models

Models provide a variety of ways to visualize text. Examine the model of the rock cycle on this page. Then read or reread the text in "The Rock Cycle." Return to the model and describe how the model helps simplify the rock-cycle process. Then describe or create a new model that serves the same purpose but in a different way.

ROCK CYCLE

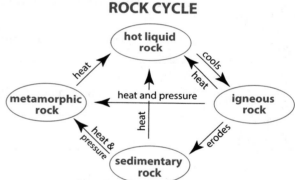

THINK ABOUT SCIENCE

Directions: Match the type of rock on the left with its description on the right.

_____ 1. igneous

_____ 2. sedimentary

_____ 3. metamorphic

A. rock made up of small particles of mineral or organic matter

B. rock that changes into another type of rock under heat and pressure

C. molten rock that has cooled

Weathering and Soil

Weathering is any process that breaks down rocks and other materials into smaller pieces. Different rocks weather into different materials. Many kinds of rocks weather and form soil. Weathering is often slow, taking thousands, even millions, of years. The process varies with rainfall, exposure to the air, and the actions of plants and animals.

The properties of soil depend on the materials they are made from. Some soils are rich in clay. They hold water well but drain poorly. In contrast, sandy soils hold water poorly and dry out quickly.

Humus, or decayed plant and animal matter, helps soil hold water and increases its fertility, or ability to support crops. Dark soil is usually rich in humus and is very fertile. Natural processes take about 500 years to produce one inch of topsoil from the soil layers beneath it.

Vocabulary Review

Directions: Complete each sentence correctly using one of the following terms:

igneous   inner core   outer core   rock cycle   sedimentary

1. A sedimentary rock may change into an igneous rock through the process of the
   _____.

2. A rock made up of many particles that are compressed and cemented together is a(n)
   _____ rock.

3. The Earth's _____ is composed of liquid nickel and iron.

4. A rock formed by cooled and hardened lava is a(n) _____ rock.

5. The pressures are so great on the _____, that it is composed of solid iron and nickel.

Skill Review

Directions: Choose the best answer to each question. Questions 1 and 2 refer to the diagram of the rock cycle on page 387.

1. What changes an igneous rock into a metamorphic rock?

   A. erosion
   B. cooling
   C. heat and pressure
   D. heat alone

2. What changes a metamorphic rock into an igneous rock?

   A. pressure
   B. heat and erosion
   C. weathering
   D. high temperatures

Skill Review (continued)

Question 3 refers to the map on page 386.

3. What is the relationship among tectonic plates, continents, and oceans?
    A. Each continent is a part of only one tectonic plate.
    B. Each ocean is a part of only one tectonic plate.
    C. Continents and oceans always meet at the border of two tectonic plates.
    D. Continents and oceans sometimes meet at the border of two tectonic plates.

Skill Practice

Directions: Choose the best answer to each question.

1. Which of these objects best models the layers of Earth?

    A. an apple, which has a core at its center
    B. a bowling ball, which is solid throughout
    C. a beach ball, which is filled with air
    D. an onion, which has thin layers

2. What events does the movement of tectonic plates help explain?

    A. earthquakes only
    B. earthquakes and volcanoes only
    C. mountain building only
    D. earthquakes, volcanoes, and mountain building

3. Where could you find recently formed igneous rocks?

    A. along the walls of a river canyon
    B. at an open gravel pit
    C. at the bottom of a lake or river
    D. along the slope of an active volcano

4. What type of rock forms deep underground and on the surface when lava cools?

    A. sedimentary
    B. igneous
    C. metamorphic
    D. ore

5. When a mountain is blasted for a new highway, curved layers of sedimentary rock are revealed. What happened first to form these layers?

    A. Tectonic forces squeezed the rock layers into a curved shape.
    B. Fragments of rock and minerals were squeezed into flat layers.
    C. Lava from a volcano flowed onto the rocks.
    D. Heat and pressure changed the rocks into metamorphic rock.

6. Which of these terms is used to describe large cracks in Earth's crust?

    A. plates
    B. blocks
    C. faults
    D. hot spots

Oceanography

KEY CONCEPT: Oceans cover most of Earth's surface. They affect Earth's climate and the shape of land.

Imagine being an explorer traveling across a vast plain. You reach a tall mountain range and then stop. You look down into a deep trench. You shine powerful electric lights into the depths and see strange and exotic creatures.

Could this happen to you? Well, it might, but not without a special ocean vehicle, because you are exploring an ocean floor. Some explorers call the oceans Earth's last frontier, because vast regions remain unexplored.

Earth's Oceans

Oceans cover about 70 percent of Earth's surface. The Pacific Ocean alone covers almost half the globe. The other two major oceans are the Atlantic Ocean and the Indian Ocean, and smaller oceans include the Antarctic Ocean and Arctic Ocean. All ocean waters are connected.

The ocean contains about 97 percent of Earth's water. Because ocean water is very salty, it cannot be used by humans and other land-dwelling living things. They need freshwater, which has very low salt content. Only 3 percent of Earth's water is freshwater, and much of that is trapped in frozen glaciers and icecaps.

Many living things thrive in the salt water, however. Among them are small organisms called **phytoplankton**, which are a kind of algae that live on the ocean's surface. They survive by performing photosynthesis, just as plants do. Phytoplankton form the base of the ocean food chain, and they release oxygen into the atmosphere.

OCEAN ZONES

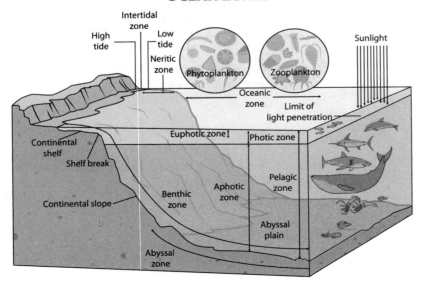

Ocean Zones

Temperature, depth, and brightness define different zones of the ocean. Different communities of living things are found in each zone.

The **intertidal zone** is the area between the high tide line and the low tide line on the shore. Organisms here must be able to survive underwater and in the open air. The **neritic zone** is the area from the shore to the point where the ocean floor takes a steep drop. More plants and animals live in this zone than in any other part of the ocean.

DETERMINE MEANING

Sometimes when you are reading, you come across an unfamiliar word. To understand the text, you need to know what the word means. You can use context clues to predict the word's meaning. That is, read the surrounding words and sentences. They may help you determine the word's meaning.

If context clues aren't sufficient, you can use a dictionary. Or the text you are reading may have a glossary. A **glossary** is an alphabetical list of words and their definitions, usually located at the back of a book. Glossaries normally include technical words that are used only in specific subjects, such as science.

It is likely that any special science words, or **jargon**, that you find in a text is also in the glossary. This is especially likely if the word is written in bold, or dark type.

Read the following passage. Underline any words you do not understand and would want to look up in a glossary.

There are photographs of every square kilometer of Earth's surface. Yet much of the ocean floor remains unexplored. The ocean's deepest waters are inhospitable to nearly all living things, humans included. Light from the surface does not penetrate to these depths. The water is extremely cold, and the pressure is enormous.

You may be unfamiliar with the words *inhospitable* and *penetrate*. The words are in plain type, so it is unlikely that you will find them in a glossary.

Use context clues to predict the meanings of the words as they are used in the passage. Then, use a dictionary to check your predictions.

The remainder of the ocean is described as the open ocean. From the surface to about 2,000 meters down is the **bathyal zone**. Beneath 2,000 meters is the **abyssal zone**, which is extremely dark and cold. Relatively few living things can survive in these conditions.

The Ocean Floor

Scientists classify the ocean floor into different regions. The **continental shelf** is a relatively flat region of the floor that extends from the edges of continents. Its width varies from place to place. At its edge is a steep drop called the **continental slope**. This marks the boundary between continental crust and oceanic crust. The wide, flat region of the floor in the center of the ocean is called the **abyssal plain**.

The ocean floor also has mountains. It has Earth's tallest and longest mountain range. This range is called the **mid-ocean ridge**. It spans the middle of the Atlantic Ocean, extends around Africa into the Indian Ocean, and then continues across the Pacific Ocean to North America. Unlike continental mountains, which formed when two plates collided, the mid-ocean ridge formed where two plates were moving apart.

The rocks of the ocean floor are quite young, especially near the mid-ocean ridge. This is due to a process called **seafloor spreading**. Magma moves upward in the space between the spreading plates near the mid-ocean ridge. The lava hardens to form new ocean floor, pushing older crust away. When these edges collide with a continental plate, the denser oceanic crust slips downward into the mantle, forming deep trenches.

OCEAN FLOOR

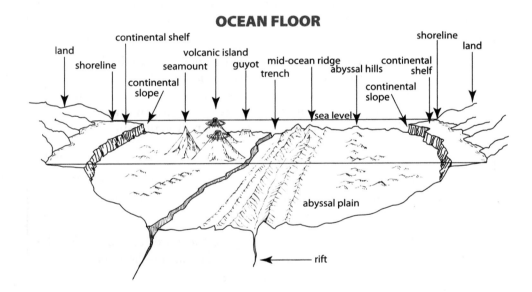

THINK ABOUT SCIENCE

Directions: Complete each sentence by filling in the correct term.

1. Ocean rocks near mid-ocean ridges are younger than other rocks on the ocean floor because of the process of _____.

2. The boundary between continental crust and oceanic crust is at the _____.

3. Earth's longest and tallest mountain range is the _____.

Currents

As the wind blows across the ocean, it creates surface currents. A **current** is a stream of water that moves through the ocean. Warm-water currents move water from warmer areas to cooler areas. Cold-water currents move water from cooler areas to warmer areas.

Most currents move in a circular **pattern**, or regular predictable movement, across the oceans, as shown below. The current that flows northward across the Atlantic Ocean is the Gulf Stream.

As the water moves past land, it affects the climate. Water temperatures don't change as quickly as land temperatures. In cold weather, the air over the ocean warms the land. In hot weather, the ocean cools the land.

Deep-water currents are caused by differences in the temperature and saltiness of ocean water, which affect density. Colder, saltier water is denser than warmer, less-salty water. Thus, cold water from polar regions moves toward warmer regions near the equator. There the water warms and rises. It flows back toward polar regions closer to the surface.

Core Skill
Determine
Central Ideas

The section "Currents" contains several paragraphs of text accompanied by a visual labeled "Circulation Patterns of Surface Currents." You can use the titles of the text and of the visual to help you determine the sections' central ideas. After reading the titles, scan the text to find words that may also serve as clues. You will notice, for example, that two words appear in bold type. This suggests that these words are important to the text and are probably related to the text's central ideas. Use context clues to determine their meanings, or use the glossary or a dictionary to find the meanings.

Combine all of the text and visual clues to determine the central idea of the section. State the central idea in your own words.

OCEAN CURRENTS
Circulation Patterns of Surface Currents

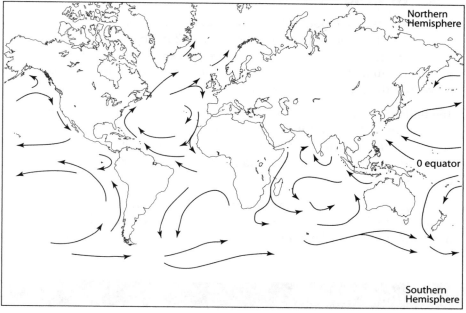

THINK ABOUT SCIENCE

Directions: Complete each sentence by circling the correct answer within the parentheses.

1. Surface currents are caused mainly by (wind, temperature).

2. (Cold-water, Warm-water) currents bring water from a colder area to a warmer area.

3. In hot weather, the ocean (heats, cools) the land.

4. (Deep-water, Surface) currents are caused by differences in the density of ocean water.

Waves

In addition to currents, winds produce the waves. Waves have two parts: the **crest** (high point) and the **trough** (low point). The height of a wave is measured from the top of the crest to the bottom of the trough. The **wavelength** is from the top of the crest to the top of the next crest.

MEASUREMENT OF A WAVE

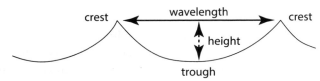

As waves approach shallow water near the shore, the bottom of the wave slows down as it rubs against the seafloor. The crest of the wave, however, continues at the same speed. The change in speed makes the crest curl farther and farther in front of the trough. Soon it breaks. Usually, the water then rushes back into the ocean, carrying sediments with it.

Lake to Land

Like rocks, lakes and ponds don't last forever. When a lake first forms, its water is clear and contains few organisms. But almost immediately, the lake begins filling with microscopic organisms, seeds, and bits of decayed plants and animals. Soon larger animals, such as insects, fish, and birds, will move into the water. Inevitably, debris collects at the bottom of the lake.

The buildup on a lake bottom is slow but steady—perhaps no more than one foot in 100 years. But eventually, plants take root in the sediments and speed the formation of new soil. Over time the lake becomes a **wetland**, such as a marsh or a swamp. As sediment continues to build up, the lake fills in completely.

The length of time a lake lasts depends on its size, water supply, and the local climate. The entire process is an example of **ecological succession**, in which one type of land and community develops into another.

Vocabulary Review

Directions: Write the letter of the correct definition in the blank before each term.

1. _____ current

2. _____ continental shelf

3. _____ mid-ocean ridge

4. _____ ecological succession

5. _____ pattern

A. flat region of the ocean floor that extends from the edges of continents

B. process by which a lake may change into land over time

C. stream of water at the surface or deep within the ocean

D. undersea mountain range found where new ocean floor is produced

E. a regular, predictable movement or change

Directions: Use the glossary at the back of this book to write a definition for each of the following words.

1. trough _____

2. intertidal zone _____

3. seafloor spreading _____

4. continental slope _____

5. wavelength _____

Directions: Organize the following steps in the correct order from 1–5 (1 being the first step) to show the correct stages in the ecological succession of a lake into land.

6. Write the correct number on the line before each step.
   A. _____ Plants take root in the sediment and help form soil.
   B. _____ Sediments build up so much that the land has dried up.
   C. _____ A melting glacier fills a large basin with water.
   D. _____ The lake becomes a marsh or swamp.
   E. _____ Living things move into the lake, and their remains collect on the bottom.

Skill Practice

Directions: Choose the best answer to each question.

1. What is one reason that phytoplankton cannot live in the abyssal zone of the ocean?
   A. The water is too salty.
   B. The water is too dark.
   C. The water is too bright.
   D. Phytoplankton cannot live in water.

2. What best describes the ocean floor?
   A. smooth and flat
   B. mountainous and hilly
   C. gentle and inactive
   D. different from region to region—much like the land

3. Where do ocean currents flow?
   A. on the surface only
   B. in deep waters only
   C. on the surface and in deep waters
   D. near the equator only

4. Where are the youngest rocks on the ocean floor found?
   A. on the continental shelf
   B. on the continental slope
   C. in the abyssal plain
   D. near the mid-ocean ridge

5. What have organisms in the abyssal zone adapted to live without?
   A. oxygen
   B. sunlight
   C. nutrients
   D. carbon dioxide

Meteorology

KEY CONCEPT: The atmosphere includes temperature, air pressure, and moisture content. These factors interact to cause weather.

Have you ever seen a hot-air balloon? Flames from a burner heat the air that fills the balloon. When the air inside is hot enough, the balloon will gently lift off and float across the sky. As long as the air remains hot, the balloon will stay aloft. When the burner is turned off, the air inside the balloon will cool, and the balloon will sink.

The behavior of air when it is warm and when it is cool affects more than hot-air balloons. Air temperature is one of the factors that determine the weather.

Earth's Atmosphere and Weather

Weather is the condition of the atmosphere at any given time and place. **Meteorology** is the study of Earth's atmosphere and the changes that produce weather. Meteorologists collect and analyze data to **forecast** (predict) how the weather will change.

The atmosphere is divided into five main layers. The **troposphere** extends from the ground to about 7.5 miles above Earth. This is the layer where we live and where most weather occurs. The **stratosphere** extends from the top of the troposphere to about 20 miles above Earth. A gas called ozone collects here. Ozone shields living things from the Sun's ultraviolet radiation.

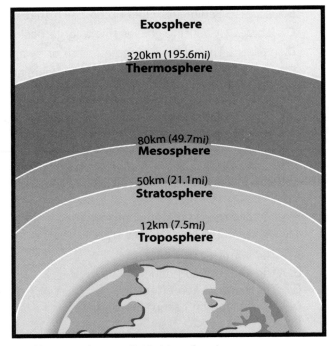

Earth's atmosphere is divided into four layers.

The **mesosphere** extends from the top of the stratosphere to about 50 miles above Earth. Above the mesosphere is the **thermosphere**. The outermost layer, called the **exosphere**, gradually diminishes into space.

Differences and Changes in the Weather

Weather can be different from day to day, even from hour to hour. Weather conditions are also different from place to place on Earth.

What causes differences and changes in the weather? One important factor is the Sun. The Sun heats Earth's surface unevenly. Places near the equator receive direct sunlight and thus are relatively warm year round. Places closer to the poles receive sunlight at a low angle and thus are quite cold. The angle of sunlight also changes with the seasons. In addition, Earth's land and water absorb heat at different rates.

Earth's surface warms the air above it. When hot air meets cold air, different kinds of weather can form.

Composition of the Atmosphere

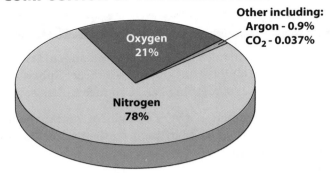

COMPOSITION OF THE ATMOSPHERE

Other gases such as carbon dioxide and argon make up 1% of Earth's atmosphere.

The atmosphere is a mix of gases: nitrogen (78%) and oxygen (21%). The remaining 1 percent is a mix of argon, carbon dioxide, water vapor, and other gases. These gases cycle between the atmosphere and the surface.

Moisture enters the atmosphere as water vapor, mainly through evaporation from the ocean. **Humidity** is a measure of the amount of moisture in the air. The humidity that the air can hold depends on temperature. Warm air can hold more moisture than cold air can. This explains why clouds often form when the temperature decreases. When the air cools, water vapor leaves it by condensing into liquid droplets.

The Greenhouse Effect

The levels of most atmospheric gases have remained constant for many years. However, scientists have observed that the levels of carbon dioxide in the atmosphere have been rising slowly for at least the past 50 years. Some scientists argue that the higher levels of carbon dioxide are contributing to an effect called the greenhouse effect.

The **greenhouse effect** is the trapping of heat within Earth's atmosphere. Certain gases, among them carbon dioxide, act like greenhouse walls to keep the atmosphere warm. Scientists are concerned that because carbon dioxide levels have been increasing, the greenhouse effect is increasing as well. An increase in the greenhouse effect could cause Earth's surface temperatures to gradually rise.

Air Pressure

Pressure is defined as the ratio of a force to the area on which it is applied. **Air pressure** is the pressure exerted by Earth's atmosphere above a specific region. Air pressure varies from place to place and from day to day, just like other weather factors. One reason is because warm air is lighter than cool air and exerts less pressure.

Air moves from places of high air pressure to places of low air pressure. This moving air is **wind**. One reason that winds are so common is because Earth is rotating, or turning, on its axis. The areas of high and low pressure move with Earth. In the Northern Hemisphere, areas of high pressure tend to move toward the right on the map, from west to east.

Thunderstorms

Thunderstorms begin when moist air from Earth's surface rises. When it reaches a layer of cool dry air, it condenses into tiny droplets of water. With enough moisture, puffy, white clouds form.

The moist air cools, releasing heat energy that causes a rapid updraft of air. More warm moist air is pulled up into the clouds. A puffy, white cloud may grow to cover several miles. Winds may flatten the top of the cloud into the familiar shape of a thunderstorm cloud.

Inside the cloud, water droplets combine. When they get big enough, they fall as rain. Hail forms when tiny bits of dust and ice collide with water droplets and freeze. They fall when they become too heavy for the wind to keep them in the air.

If conditions are right for the growth of one thundercloud, chances are that several thunderclouds will be produced. Thunderclouds often form in irregular rows, called **squall lines**.

When clouds are charged with electricity, lightning may leap from cloud to cloud, from cloud to ground, or from ground to cloud. As lightning moves through the air, the air instantly heats and expands rapidly. The result is the crack or rumble of thunder. Because light travels faster than sound, you see the lightning before you hear the thunder.

Tornadoes

If the wind in a thunderstorm begins to spin, a tornado may form. A tornado is a funnel-shaped cloud that can travel across the surface of Earth, destroying everything in its path. Tornadoes typically form in the spring over hot, flat lands.

The average forward speed of a tornado is 30 miles per hour, but sometimes they can reach speeds as fast as 70 miles per hour. The rotating column of air in a tornado picks up dirt, dust, and debris. This flying debris can cause great damage in highly populated areas. The average path of a tornado is 200 yards wide and 9 miles long. Fortunately, most tornadoes die out after a few minutes. They rely on updrafts of hot, humid air, which do not last long.

Hurricanes

Hurricanes form over warm oceans in late summer. As the warm water evaporates from the ocean surface, a tremendous amount of heat energy is pumped up into the atmosphere. The result is a widespread and powerful thunderstorm. If it begins to spin, a hurricane is born.

THINK ABOUT SCIENCE

Directions: Answer the questions below in the spaces provided.

1. _____ is caused by rapidly expanding air around a lightning bolt.

2. A(n) _____ forms when the winds spin faster and faster.

3. _____ form over warm ocean water in late summer.

When a hurricane reaches land, the wind may be strong enough to pull a tree out of the ground. Also it may sweep the water near shore into a wall of water called a **storm surge**. When the surge hits the beach, the ground can be washed away. Boats may be sunk or carried miles inland.

Like a tornado, hurricane winds spin around a low-pressure center called the **eye**. In the eye, the wind is calm and the sky is blue and clear. The ring of low pressure at the center of the storm can be more than 185 miles across. Moments later, however, the eye can shrink to about 30 miles across, and the winds will begin to swirl around it at hurricane force.

Hurricanes kill more people each year than all other storms combined. Fortunately, they can be predicted and tracked. Forecasting is the key to saving lives during a hurricane. Meteorologists try to find out as much as they can about the hurricane's strength and movement. Then they can try to plot where the hurricane will hit land. People in communities in the path of a hurricane are warned to evacuate, or leave to go to a safe place.

Vocabulary Review

Directions: Complete the sentences by using one of the terms below.

air pressure   forecast   greenhouse effect   humidity   meteorology   weather   wind

1. The force exerted by the weight of a column of air pressing down on Earth's surface is
   _____.

2. Meteorologists use data collected from around the world to _____ the weather.

3. _____ is the condition of the atmosphere at any given time or location.

4. The process in which gases in the atmosphere act as a barrier, preventing heat from escaping, is the _____.

5. _____ is the movement of air from an area of high pressure to an area of low pressure.

6. The study of Earth's atmosphere and the interactions that produce weather is
   _____.

7. The amount of moisture in the air is _____.

Skill Review

Directions: Find and underline the factual details that support the following weather forecast.

1. Today's weather will be sunny and mild with a chance of a thunderstorm in the evening. (1) The high temperature will be 87°F. (2) The humidity is 53%. (3) Golfers will be happy with today's weather. (4) Winds are from the south at 5 to 10 mph. (5) Air pressure is 29.98 in. (1014.5 mb). (6) Rainfall has been lower than normal this summer. (7) Tomorrow will bring more sunshine.

Skill Review (continued)

Directions: Read the passage below. Take note of the relationship among ideas to choose the correct answer to each question that follows.

> Ozone, a form of oxygen, collects naturally in Earth's atmosphere. It protects the surface from the Sun's ultraviolet (UV) radiation, which can cause skin cancer. In the 1970s, scientists were alarmed to discover a thinning of the ozone layer. The cause was the use of chemicals called CFCs (chlorofluorocarbons). CFCs, once used in spray cans and other products, destroy ozone when they escape into the atmosphere. Some refrigerants, such as Freon, also break down ozone. These chemicals were banned and have been replaced.

2. If scientists had not discovered the thinning of ozone in the 1970s, what most likely would have happened?

   A. Droughts would have become more common.
   B. Skin cancer would have become more common.
   C. Refrigerators would have become more expensive.
   D. Earth's weather would have been warmer and sunnier.

3. What lesson can be drawn from the events described in the passage?

   A. Some damage to the environment cannot be reversed.
   B. Skin cancer can be prevented and cured.
   C. Technology can have unexpected effects.
   D. Old technology is always better than new technology.

Skill Practice

Directions: Choose the best answer to each question.

1. Which of these gases makes up only a small part of Earth's atmosphere but has a large effect on Earth's surface?

   A. carbon dioxide
   B. oxygen
   C. nitrogen
   D. argon

2. What best describes the relationship between lightning and thunder?

   A. Lightning causes thunder.
   B. Thunder causes lightning.
   C. The same event causes both lightning and thunder.
   D. Lightning and thunder are not related.

3. What determines the size that a hailstone can become?

   A. the distance above the ground
   B. the increase in air temperature
   C. the level of humidity
   D. the strength and direction of the wind

4. How do meteorologists help save lives when a hurricane forms?

   A. by reducing the strength of the hurricane
   B. by preventing the hurricane
   C. by warning people about the hurricane
   D. by changing the direction of the hurricane

Directions: Choose the best answer to each question.

1. What best describes the frequency and location of volcanic eruptions today compared to when volcanoes first appeared on Earth?

   A. much less frequent; occur at predictable locations
   B. much less frequent; occur at random locations
   C. about as frequent; occur at regular intervals at known locations
   D. much more frequent; occur at known locations

2. A geologist finds a fish fossil in a rock layer. What rock forms the rock layer?

   A. metamorphic rock
   B. igneous rock
   C. sedimentary rock
   D. weathered rock, such as sand

3. In which direction is the seafloor spreading?

   A. along the paths of deep-water currents
   B. along the edges of continents
   C. inward toward the mid-ocean ridge
   D. outward from the mid-ocean ridge

4. In what layer of the atmosphere does weather occur?

   A. mesosphere
   B. stratosphere
   C. thermosphere
   D. troposphere

5. Wind is blowing steadily from east to west across a county. What can be concluded about the air above it?

   A. Air pressure is higher on the eastern side of the county.
   B. Air pressure is higher on the western side of the county.
   C. Humidity is higher on the eastern side of the county.
   D. Clouds are thicker on the eastern side of the county.

6. What pattern do water currents follow?

   A. from warm areas to cool areas
   B. from cool areas to warm areas
   C. from warm areas to other warm areas
   D. from cool areas to other cool areas

7. What is Earth's longest and tallest mountain range?

   A. the Applachian Mountains
   B. the mid-ocean ridge
   C. the Himalayan Mountains
   D. the Andes Mountains

8. What is the purpose of a scratch test?

   A. to measure the extent of weathering
   B. to determine a rate of erosion
   C. to identify a specific mineral
   D. to predict the likelihood of a volcanic explosion

Review

Directions: Choose the best answer to each question. Questions 9 and 10 refer to the following passage.

The strength of earthquakes is generally reported as a number on the Richter scale. The scale, named for the scientist who developed it, begins at 1. Each increase of one whole number means the earthquake's power has increased about 30 times.

The strongest quake ever recorded on the Richter scale was 8.9. That is about 12,000 times the energy released in an atom bomb. Two quakes share the 8.9 record. The first was in Ecuador in 1906; the second, near Japan in 1933. Both of these quakes occurred in sparsely populated areas and did not cause a major disaster. In contrast, a 1976 quake in China rated only an 8.0 but left most of the city of Tangshan in ruins. It also killed more than one-third of the population.

The strongest earthquake in the United States measured 8.4 on the Richter scale. It occurred near Prince William Sound in Alaska in 1964. The quake killed 131 people. It created a giant wave 50 feet high that destroyed the town of Kodiak. Tremors of the quake were felt in California, Hawaii, and Japan.

Another type of scale used to measure earthquakes is the modified Mercalli scale of earthquake intensity. On a scale of I (gentle tremor) to XII (total destruction), people's reactions to the earthquake rather than the strength of the tremor are measured.

A level of I is felt by birds, animals, and very few people. At level IV, the ground feels like a heavy truck is passing. At level VII, people run outdoors and find it difficult to stand up. Damage, however, is slight. At level IX, buildings shift off their foundations, and there is considerable damage and panic. At level XI, bridges are destroyed and fissures show in the ground. At level XII, damage is total. Lines of sight are distorted, objects are thrown up into the air, and actual waves in the ground's surface can be seen.

---

9. How much greater is the power of an earthquake that measures 8.0 on the Richter scale than the power of an earthquake that measures 6.0 on the Richter scale?

   A. 2 times
   B. 60 times
   C. 90 times
   D. 900 times

10. What factor or factors affect the amount of damage that an earthquake can cause?

   A. only the strength of an earthquake
   B. the strength and location of an earthquake
   C. only the location of an earthquake
   D. the scale used to measure the earthquake

Directions: Questions 11–13 refer to the following passage.

In about 5,000 BC, an underwater landslide of about 700 kilometers of rock and mud created the world's largest wave. The wave, traveling at 170 km per hour, was more than 300 meters high when it hit the Shetland Islands. Throughout history, giant waves, known as tsunamis, have caused devastation along many coastal areas.

A tsunami can occur whenever there is a sudden change in the seafloor. Earthquakes or volcanoes are frequent causes. Most tsunamis seem to start close to the plate boundaries at the edges of the Pacific Ocean.

At sea, a tsunami might be less than a meter tall. From aboard ship, the wave might not even be noticed. The wavelength, however, may be greater than 200 km, and it may be traveling at 700 km per hour. As the wave approaches land, it slows down. Like passengers in a car crash, the water in the wave keeps moving even when the wave "puts on the brakes." The rushing water pushes into itself, quickly building into a high wall.

If the trough of the tsunami reaches land first, water near the shore may suddenly rush out to sea. If the crest reaches land first, there will be a sudden increase in water level. The tsunami crest may be 50 meters high. It crashes against the land with enormous power.

At 5:36 p.m. on March 27, 1964, an earthquake in Alaska caused a tsunami in the Pacific Ocean. The wave reached California four hours later. Although it caused a great deal of property damage, only a few people were killed. This was due mainly to the timing of the tsunami's arrival.

Today, in many parts of the ocean, early warning systems have been installed to help detect tsunamis. Yet tsunamis can still cause great loss of human life, especially in places without these systems. On December 26, 2004, a tremendous tsunami in the Indian Ocean killed more than 200,000 people. None of them knew a tsunami was coming until they saw the wave with their own eyes.

11. Based on the passage, which three adjectives best describe a tsunami?

   A. rapid, frequent, manageable
   B. gradual, predictable, enjoyable
   C. sudden, powerful, destructive
   D. immediate, electric, violent

12. When the tsunami of 1964 struck California, what most likely accounted for no one being injured?

   A. an early warning system that detected the tsunami
   B. people were away from the shoreline when it struck
   C. the weakness of the tsunami
   D. the low population of California

13. If an early warning system had been in place, predict how the tsunami of 2004 most likely would have been different.

   _____

   _____

   _____

   _____

   _____

   _____

   _____

   _____

   _____

Directions: Questions 14 and 15 refer to the following passage.

Water moves back and forth between Earth's surface and atmosphere. Water evaporates into the air from oceans, rivers, and moist land. The amount of water vapor the air can hold depends on the temperature of the air. Warm air can hold more water than cool air.

Water in the air eventually falls back to Earth's surface. This precipitation happens when the air can't hold the water anymore. Sometimes warm, moist air rises and cools off, or it may meet a mass of cold air. The cold air forces the warm air up, and the warm air cools off. The cooler air can't hold all the water. The water vapor turns into the water drops that make up clouds. If the water drops get too heavy, the water falls as precipitation.

Will the precipitation be rain, snow, or sleet? This, too, depends in part on temperature. The following table shows which precipitation is likely to reach Earth's surface at a particular temperature.

Temperature of Clouds	Temperature at Ground	Precipitation
below 32°F	up to 37°F	snow
below 32°F	37°F–39°F	sleet
any temp.	above 39°F	rain

14. One day the temperature in the clouds is 30°F and is 33°F on the ground. If precipitation falls, what will it most likely be?

   A. sleet
   B. snow
   C. rain
   D. hail

15. Snow is falling on a cold winter morning,. One hour later, the snow has stopped, but sleet has started to fall instead. Explain where the temperature changed and how it has changed.

Check Your Understanding

On the following chart, circle the number of any item you answered incorrectly. Next to each group of item numbers, you will see the pages you can review to learn how to answer the items correctly. Pay particular attention to reviewing those lessons in which you missed half or more of the questions.

Chapter 11 Review

Lesson	Item Number	Review Pages
Geology	2, 8, 9, 10	382–389
Oceanography	1, 3, 6, 7, 11, 12, 13	390–395
Meteorology	4, 5, 14, 15	396–401

Application of Science Practices

CHAPTER 11: EARTH

Question

How do meteorologists measure wind speed?

Background Concepts

An anemometer is a device that meteorologists use to measure wind speed. These devices are commonly seen at weather stations and airports.

When pushed by wind, the cups of an anemometer revolve, or turn.

Investigation

Materials required

5 three-ounce paper, plastic, or
   Styrofoam™ cups, labeled 1, 2, 3, 4, 5
2 soda straws
pushpin
marker or pen

scissors
stapler or tape
pencil with eraser on top
3-speed fan
watch or timer

Assembly

1. Use scissors to punch one hole on the side of each of Cups 1, 2, 3, and 4, about $\frac{1}{2}$ inch below the rim. Set the cups aside.

2. Punch a hole about $\frac{1}{4}$ inch below the rim of Cup 5. Push the pencil all the way through the cup's opposite side. Remove the pencil and move it 90 degrees from one of the holes. Push the pencil all the way through the cup to make two more holes. All four holes should be equally spaced.

3. Push the pencil through the bottom of Cup 5 until it is level with the rim. Set it aside.

4. Push a straw through the hole in Cup 1. Fold the end of the straw inside the cup and staple or tape it to the cup.

5. Now push the part of the straw coming out of Cup 1 through two opposite holes in Cup 5. Once the straw goes through both holes, push it into the hole of Cup 2. Tape or staple the end of the straw inside the cup.

6. Repeat the previous step using Cups 3 and 4.

7. Arrange the four cups so that the straws are perpendicular to each other and so the open ends of the cups all face the same direction, either clockwise or counterclockwise.

8. Use the pushpin to attach the straws and cups to the pencil's eraser.

Application of Science Practices

Procedure

1. Mark a spot 3 to 5 feet away from the fan so you remain the same distance in every trial. Set up the fan, turn it on high speed, and move to the spot.

2. Hold or prop the anemometer in the path of breeze from the fan. Count the number of times that the anemometer makes a complete revolution in one minute. To make counting easier, count the number of times Cup 1 moves directly in front of you. Use revolutions per minute, or rpm, as the units of measurement. Record the measurement in the table below.

3. Adjust the speed of the fan to medium. Count the number of times that the anemometer makes a complete revolution in one minute. Record the measurement in the table.

4. Adjust the speed of the fan to low. Count the number of times that the anemometer makes a complete revolution in one minute. Record the measurement in the table.

Fan Speed	Wind Speed (rpm)			
	Reading #1	Reading #2	Reading #3	Average
High				
Medium				
Low				

Interpretation

Compute the average wind speed for each fan setting. How do the fan speed and wind speed in rpm vary with respect to each other?

Answer

Imagine that the fan has two more speeds—medium-low and medium-high. Predict the approximate wind speed in rpm for these fan speeds.

_____   _____

Evidence

Which speed measurements did you use to make each prediction?

_____

_____

The Cosmos

Do you enjoy watching movies about "outer space" or reading science fiction books? The enormous expanse of the cosmos, or universe, has fascinated our collective imagination for thousands of years.

Over the centuries, astronomers and other scientists have learned more and more about Earth in relation to the solar system and universe. This chapter describes our planet's place in our solar system and galaxy, and discusses scientists' explanation for the origin of the universe itself.

In this chapter you will learn about:

Lesson 12.1: Earth's Origins
Dating technologies have helped scientists estimate that Earth formed about 4.6 billion years ago. Find out about the origin of the Moon and learn about the early days on Earth and the conditions that made it habitable.

Lesson 12.2: Origins of the Universe
The Sun, Earth, and the planets of our solar system are just a small part of the vast universe. In this lesson you will learn how the universe began and the elements that combined to form Earth.

Lesson 12.3: The Milky Way and the Solar System
Earth's immediate "neighbors" are the other planets in our solar system. The planets create a neighborhood in the Milky Way galaxy. Learn more about the Milky Way, the planets, and other space bodies in the solar system.

Lesson 12.4: Earth and the Moon
Why is it summer in one hemisphere while it is winter in the other? This lesson explains the seasons, tides, the phases of the Moon, and the characteristics of Earth that make it habitable.

Goal Setting

There are many reasons to study space science. You might want to

- know more about Earth, the Moon, the solar system, or the universe.

- learn more about phenomena such as tides and eclipses.

- understand why the Moon appears to go through a regular cycle of changes.

Think about your reasons for wanting to study space science. What do you want to learn? Write your questions below. Leave a space after each question. Then, as you read, return to the chart to write the answers. If you don't find all of the answers you are looking for, use print or online resources to find more information.

_____

_____

_____

_____

_____

_____

_____

_____

_____

_____

_____

_____

Lesson Objectives

You will be able to
- Describe the unique characteristics of Earth that allow it to sustain life

- Sequence events in the development of Earth and the Moon

Skills

- **Reading Skill:** Understand Science Texts

- **Core Skill:** Identify Hypotheses

Vocabulary

comprehension
habitable
mantle
nebula

Reading Skill
Understand Science Texts

Science texts often use high-level words that sometimes get in the way of **comprehension**, or understanding. If you find yourself confused about what you are reading, use these strategies to better comprehend the text:

- Reread the passage.

- Read more slowly.

- Look for descriptions, examples, and other context clues.

KEY CONCEPT: Earth, which formed 4.6 billion years ago, has unique characteristics that allow it to support life.

On a chilly evening, a campfire will keep you warm. However, if you sit too close to the campfire, you will soon get too hot. But if you sit too far away from the campfire, you will soon be too cold. To be comfortable, you sit the ideal distance from the fire.

Earth is the ideal distance from the Sun. If Earth were much closer to or farther away from the Sun, life could not exist.

Earth and Its Origins

Dating Earth

Scientists use different techniques to measure **geologic time**, or Earth's history. One technique is radioactive dating. Some chemical elements are unstable. They break down, or decay, automatically, producing more stable atoms. Most rocks and minerals have tiny amounts of these radioactive elements, and their decay occurs at constant rates. By measuring the quantity of products that form during decay, and comparing this quantity to remaining amounts of the parent element, scientists can calculate a rock's age. Scientists estimate that Earth is about 4.6 billion years old.

Billions of years ago, a star exploded, expelling huge amounts of gas and dust. The debris crashed into another cloud of dust and gas, or a **nebula**. The energy in the explosion was enormous. The dust, cooked by high temperatures, formed droplets. Gravity pulled droplets together, making larger lumps. Temperatures in a central lump grew, until the matter ignited, forming the Sun. The remaining lumps grew, forming early planets.

Early Earth had no water, no atmosphere, and no land. Matter mixed together until temperatures rose sufficiently to melt iron. The iron formed Earth's **core**. Temperatures cooled, producing three primary layers. As magma at the surface cooled, it formed a **crust**. Beneath the crust, molten, plastic-like magma formed the **mantle**.

The Moon

About 4.5 billion years ago, a body the size of Mars slammed into Earth. It pushed the angle of Earth's tilt on its axis from zero to 23.5 degrees, resulting in the seasons. The collision left material in Earth's crust. It also caused a huge amount of material to fly into space, where gravity pulled it together to form the Moon, which began orbiting Earth. Scientists studying Moon dust have found material like that found on Earth's surface and in Earth's mantle.

UNDERSTAND SCIENCE TEXTS

Scientists share their findings in a variety of ways, including writing articles and books on the topics they study. Although scientists write for other scientists, others read about science, too.

There are some strategies you can use to understand the science texts you read.

1. Scan the text to look for titles and subtitles. These will help you identify the main topic of the text.

2. Skim the text to find special science words, or jargon. Use context clues to determine their meaning and rewrite the meanings in your own words.

3. Examine visuals, such as diagrams and graphs. Read labels and captions to determine the purpose of each visual.

4. Predict and record the text's main idea.

5. Pause after reading each paragraph to summarize the content in your own words. Determine if the summaries you write support your predictions, or if you need to revise your predictions.

6. Refer frequently to the visuals to determine their relevance to what you are reading.

7. If something doesn't make sense to you, look for additional information to help you make meaning of what you read.

8. Build concept maps to explain relationships among the ideas and details you read about.

Apply these strategies as you read this lesson. Build a concept map to help you determine important ideas and details.

THINK ABOUT SCIENCE

Directions: Answer the question in the space provided.

How did the Moon form?

_____

_____

WRITE TO LEARN

Locate a science article online. Apply the strategies you learned for understanding science texts. Describe the strategies you found most helpful. Explain why you found these particular strategies effective.

Before scientists plan an investigation, they read about their topic. They learn what other scientists know about their topic and what remains to be learned. They collect information, make observations, and eventually form a **hypothesis**, or possible answer to a scientific question. Then, they design an investigation that will help them find evidence that either supports or rejects their hypothesis.

Over time, scientists have formed hypotheses about the source of Earth's water. As you read about Earth's history, summarize the current explanation of how Earth got its water.

Conditions for Life

Life has unique properties that allow life to thrive. For example, Earth's gravity holds on to light gases to create an **atmosphere**, or layer of mixed gases that surrounds the planet. As Earth cooled, volcanoes spewed gases into the air, forming an early atmosphere of carbon dioxide, water, and ammonia, a source of nitrogen.

Oxygen and Water

About one billion years after Earth formed, primitive organisms began using photosynthesis to manufacture their own food. The process released oxygen, which entered the atmosphere. Levels of different gases have changed over Earth's history. Today, the atmosphere is composed of about 78 percent nitrogen, 21 percent oxygen, and less than one percent each of carbon dioxide, water, and argon. Earth's atmosphere helps moderate temperatures on Earth, resulting in a range of temperatures that help make the planet **habitable**, or capable of supporting life.

Almost all living things need water, and water is the main ingredient in most life forms, including humans. Today, about three-fourths of Earth's surface is covered in water. Scientists continue to look for evidence to explain how it appeared. Recent research indicates that it probably came from millions of water-rich asteroids that slammed against the planet during its early formation.

Change Continues

Earth continues to change. The process of **weathering** physically or chemically breaks down rock. Water, wind, and gravity cause **erosion**, or the movement of soil or sediment from one location to another. Climates have changed over time and continue to change now.

Vocabulary Review

Directions: Complete the sentences using one of the following words:

habitable   mantle   nebula

1. The _____ is a layer of hot, molten rock within Earth's interior.

2. Earth formed from a _____, or cloud of gas or dust in space.

3. The characteristics of Earth, such as its temperature, atmosphere, and water, make it _____ for living things.

Directions: Read the following passage. Then answer the questions below.

> Scientists once considered several theories to explain how the Moon formed. Some argued that the Moon was captured by Earth's gravity. Others thought it formed from the same nebula as Earth. Today, evidence from the *Apollo* missions supports another explanation.
>
> Analysis of Moon rock shows that the Moon is made of the same material as Earth's crust (surface layer) and mantle, with some exceptions. Unlike Earth, the Moon has little or no iron. These facts suggest that the Moon formed as a result of a giant impact with young Earth. The impact broke away a large amount of vaporized material. It is thought that gravity pulled this material together to form the Moon.

1. What happened during the *Apollo* missions that led scientists to develop a new theory about the origin of the Moon?
   A. Astronauts walked on the Moon.
   B. Scientists calculated the Moon's orbit very precisely.
   C. Astronauts collected Moon rocks and brought them to Earth.
   D. Astronauts took close-up photos of the Moon.

Directions: For items 2–5, place an *F* before each statement that is a fact and an *H* before each statement that is a hypothesis.

_____ 2. The Moon was captured by Earth's gravity.

_____ 3. Moon rocks are similar in composition to materials in Earth's crust and mantle.

_____ 4. The Moon formed from the same nebula as Earth.

_____ 5. Unlike Earth, the Moon has little or no iron.

Skill Practice

Directions: Choose the best answer to each question.

1. Which phrase best describes how the young Earth compares to Earth today?
   A. very similar
   B. very similar, but significantly hotter
   C. somewhat similar, but with fewer living things
   D. very different

2. Which evidence best supports the theory that the Moon was once a part of Earth?
   A. The Moon lacks an atmosphere.
   B. The Moon is relatively close to Earth.
   C. The Moon is made of rocks that astronauts can walk across easily.
   D. Moon rocks are similar to Earth rocks.

3. What do the formations of Earth and the Moon have in common?
   Both formations
   A. involved gravity pulling matter together.
   B. involved a huge collision.
   C. occurred at the same time.
   D. occurred at cold temperatures.

4. Which was most significant in the formation of the current atmosphere on Earth?
   A. outgassing by volcanoes
   B. rise in plants that undergo photosynthesis
   C. formation of Earth from a nebula
   D. formation of a solid crust

Origins of the Universe

KEY CONCEPT: According to the big bang theory, the universe began with an explosion of matter and energy from an extremely small and dense particle. The universe has been expanding ever since. Reactions that occur during the life cycle of a star form the elements found on Earth.

As people grow and develop, they progress through many stages of the life cycle, including infant, child, teenager, and adult. Changes in the life of a human being continue until death occurs.

Although a star is not a living thing, scientists describe a star in terms of a life cycle that lasts billions of years. Millions of stars are "born" and "die" every day.

21st Century Skill
Critical Thinking and Problem Solving

The term *rigor* is often associated with science. **Rigor** is a commitment to careful planning, methodical procedures, control of factors beyond those being tested, precise measurements, and thoughtful analysis of results. It is a structure in which science is conducted, and it is accepted by members of the scientific community.

In a notebook, write down what you think someone who applies rigor to their daily efforts needs to do.

Origins of the Universe

The universe includes everything that exists in space—stars, planets, asteroids, comets, dust, gas, and matter that has not been observed. Other than the Sun, the closest star is Proxima Centauri, which is 4.3 light-years away. A **light-year** is the distance that light travels in one year. Light travels about 186,000 miles per second, or about 5,865,696,000,000 miles in one year!

Earth belongs to the Milky Way galaxy, which is 100,000 light-years wide. The Milky Way is only one of millions of galaxies in the universe.

The Big Bang

How did the universe and all of its contents form? According to one model, it began with a big bang. The big bang theory says that all of the matter and energy of the universe was once packed into an extremely tiny and dense particle. About 14 billion years ago, this particle suddenly exploded, sending an extremely hot "soup" of subatomic particles expanding outward. This early universe cooled as it expanded. As temperatures cooled, particles combined, eventually forming atoms, the building blocks of matter.

Gravity caused clouds of matter to clump together, forming the first galaxies. Later, as matter condensed and temperatures rose, matter ignited, forming stars. Our Sun is only about 5 billion years old.

An Expanding Universe

The night sky appears still, yet a discovery by Edward Hubble in 1929 indicated the opposite. The universe, Hubble learned, has continued to expand at a tremendous speed since the big bang. Galaxies beyond the Milky Way are moving farther and farther away, at speeds proportional to their distances from our own galaxy. Research continues, as scientists want to learn whether the universe's expansion will continue, or whether it will stop and eventually collapse upon itself.

DETERMINE THE CONCLUSIONS OF A TEXT

A conclusion is a summarizing statement or argument ordinarily found at the end of a text. It restates the main points of a text, leaving readers with an understanding of the text's most important information or points.

Although a conclusion is a component of good writing, an author may not make the conclusion clear. Instead, the reader is left with the task of identifying and summarizing important details to create a meaningful conclusion.

A conclusion is built on facts, or details relevant to a text's main idea. The facts are evidence. Relevant facts provide accurate information that supports the text's central ideas.

Read the following passage. Identify the conclusion scientists reached about the nature of the enormous area of swirling matter that appears around the brown dwarf.

> Researchers using the Hubble and Spitzer Space Telescopes focused on a small star, called a brown dwarf. Light from the brown dwarf dimmed and brightened as the star completed a rotation once each 90 minutes. Scientists observed changes in brightness levels when they viewed the brown dwarf using different wavelengths of light. Earth-sized patches of material swirling around the star produced the variations in brightness. Different wavelengths of light penetrate different layers of the patches, helping scientists determine that the patches are made of hot grains of sand, liquid drops of iron, and other unusual chemicals.

A scientific model is based on observations that are explained completely, yet as simply as possible. In science, the principle of "Occam's razor" applies to explanations. The principle is named after William of Ockham, a 14th-century friar in England. Since William's time, mathematicians and scientists have applied the principle to mean that if there are two explanations for the same set of observations, the simpler explanation is better. Occam's razor is applied to theories when many facts are unknown or cannot be known.

Do you think scientists no longer look for facts that prove theories such as the big bang theory? Why or why not?

THINK ABOUT SCIENCE

Directions: Complete each sentence by filling in the space provided.

1. The Milky Way galaxy is 100,000 _____ wide.

2. The _____ theory states that the universe is the result of an enormous and powerful explosion.

3. The universe has been _____ since the big bang.

An **assumption** is a
statement that is
accepted as true or as
certain, but without any
definite proof. As you
read about the big bang,
you should recognize
the unstated assumption
in the text, namely,
that scientists have the
ability to estimate vast
periods of time and huge
distances. What other
unstated assumptions do
you see in this lesson?

WRITE TO LEARN

Read a news article to
evaluate the strength of
its conclusion. Check, for
example, to see if the
author summarizes the
article's main points.
Write a paragraph
describing the strengths
and weakness of the
author's conclusion. If
there is no conclusion,
write one.

The "Life" of a Star

Stars are not alive. Like living things, however, a star exists for a limited
time. A star forms, changes over time, and eventually either darkens or
explodes. This process has happened throughout the history of the
universe, and it continues today.

A star forms from a nebula, a vast cloud of dust and gases in space. Most
of the gas is hydrogen, the smallest and simplest element. Very slowly,
the force of gravity pulls the dust and gases together to fill a much
smaller space. The increased density of matter results in an increase in
temperature. Eventually, the matter is hot enough for nuclear fusion to
occur. In nuclear fusion reactions, the cores of atoms combine and release
energy. This energy is released as light and heat. A star is "born."

Element Factories

Earth is made of about 90 elements including carbon, oxygen, iron, gold,
and lead. Have you ever wondered how these elements were made? They
were forged inside ancient stars.

Hydrogen is by far the most common element in young stars, and it is the
"fuel" that powers them. Inside a star, hydrogen nuclei fuse together to
form helium, the next-largest element. This process can continue for
billions of years. As a result, a star's supply of hydrogen decreases, while
its supply of helium increases. Helium eventually begins to take part in
fusion reactions. This leads to the formation of lithium, beryllium, and
other elements. Over time, a small star is able to form carbon, the sixth-
heaviest element. Very large stars can make elements as heavy as iron.

Not all stars are hot enough to make elements heavier than iron. Small
and medium-sized stars, such as our Sun, are believed to fade to cold
objects called black dwarfs. But much larger stars come to an end in a
tremendous explosion called a **supernova**. The energy of a supernova is
great enough to form the heaviest elements that exist in nature.

Vocabulary Review

Directions: Complete the sentences below using one of the following
words:

assumption   light-year   nebula   supernova

1. The distance that light travels in one year is a _____.

2. A star forms from a cloud of dust and gases called a _____.

3. An unstated _____ may be that all stars share the same
chemical composition.

4. A _____ is an enormous explosion in which a star
breaks apart.

Directions: Read the passage below. Then answer the question that follows.

> All stars begin in the same way, as a protostar. A protostar forms from a spinning cloud of gas within a nebula. Over time, a protostar may develop into a main-sequence star, which is the average-size star. These stars eventually burn out. A larger protostar follows a different path. It forms a massive star that eventually expands to become a supergiant. A supergiant dies suddenly in a massive explosion called a supernova.

1. Which sentence best states the main idea of this passage?
   A. A nebula is a giant cloud of dust in space.
   B. The mass of a protostar determines its future.
   C. The life cycle of every star is unique in some way.
   D. Stars have a life cycle similar to that of living things.

Directions: Read the following passage. Then answer the questions that follow.

> The big bang theory states that the universe began with a sudden explosion about 14 billion years ago. All of the matter and energy of the universe was released at that moment. By applying the big bang theory and the laws of physics, scientists have been able to construct a history of the universe. According to this history, the forces of gravity and electrical attraction helped form atoms, then galaxies and stars, and eventually Earth and the solar system.

2. When scientists constructed a history of the universe, what assumption did they make about the laws of physics?
   A. The laws have changed slowly over time.
   B. The laws were the same in the past as they are now.
   C. The laws explain why the big bang occurred.
   D. The laws can predict both the future and the past.

3. Which of these is the most likely reason the author mentioned Earth in the passage?
   A. Earth is the likely cause of the big bang.
   B. Earth formed soon after the big bang.
   C. Earth has a central position in the universe.
   D. Earth is a planet familiar to the reader.

Skill Practice

Directions: Choose the best answer to each question.

1. Which property of a star indicates its age?
   A. composition of elements
   B. size
   C. mass
   D. distance from Earth

2. Based on the information about the history of the universe, when could a planet similar to Earth have started to form?
   A. soon after the big bang
   B. when the first stars had formed
   C. when the first nebulas had formed
   D. after the first supernovas

The Milky Way and the Solar System

Lesson Objectives

You will be able to

- Describe the Milky Way galaxy
- Identify the objects that make up the solar system
- Understand the definition of a planet

Skills

- **Reading Skill:** Analyze Author's Purpose
- **Core Skill:** Evaluate Conclusions

Vocabulary

asteroid
bias
comet
criteria
galaxy
satellite
solar system

KEY CONCEPT: Earth is one of eight planets that orbit the Sun in the solar system. Other objects in the solar system include asteroids, dwarf planets, and comets.

How often do you gaze at the night sky? Have you observed the Moon's phases? Have you identified and tracked constellations, or familiar patterns of stars? There is much to see, and even more you cannot see.

The Milky Way Galaxy

A **galaxy** is a large group of stars. Almost everything you can see in the night sky without the aid of a telescope is part of the Milky Way galaxy. A **solar system** is a collection of planets and their moons in orbit around a star. Our solar system, with its star, the Sun, is located in the Milky Way galaxy, as are about 200 to 400 billion other stars. The galaxy is in the shape of a giant spiral, with a bulge in the center and arms that extend from the center to create the pattern of a pinwheel. It acquired its shape by "eating" other galaxies, which it continues to do today.

The Milky Way is approximately 120,000 light-years from side to side. A light-year is the distance that light travels in one year, or about 6 trillion miles. The Sun is located about 28,000 light-years from the center of the Milky Way. It revolves around the galaxy's center at a speed of about one-half million miles per hour. The galaxy is so large, however, that the Sun has completed only 25 orbits in its lifetime.

What scientists can see within the Milky Way is far less than what they cannot see. Most of the Milky Way is made of dark matter, or matter that reacts gravitationally with the matter scientists can observe.

An extremely large black hole, about 14 million miles across, is at the center of the galaxy. A **black hole** is a region left behind after a tremendous explosion of a massive star. Such a star would be ten or more times more massive than the Sun. After the explosion, the gravity of the core left behind is so strong that neither matter nor light can escape.

The Milky Way is one many galaxies in the universe. Some have fewer stars, about 10 billion stars in all. Others have closer to 100 trillion stars. The oldest and largest galaxies are typically elliptical galaxies, meaning they have oval shapes. The closest galaxies to the Milky Way are irregular galaxies. Their stars are arranged in asymmetric, or irregular, shapes.

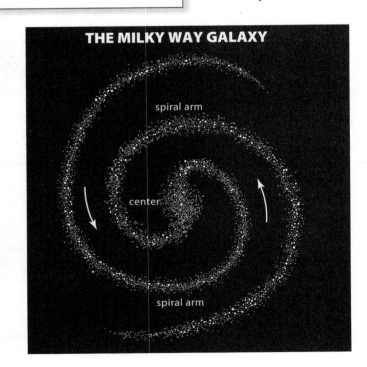

THE MILKY WAY GALAXY

spiral arm

center

spiral arm

ANALYZE AUTHOR'S PURPOSE

All authors write for a purpose, or reason. Some write to inform, or teach. Others write to entertain, and others write to persuade, or convince their readers of the truth of an idea or to encourage a specific action. It is important to determine an author's purpose for writing. That understanding will help you understand a text more fully.

When authors write to inform, they give information about a topic. If they write to entertain, they choose language and content that you will find funny, dramatic, or simply interesting. If they write to persuade, they may present a specific **bias**. In other words, they may show only one side of an argument. To support that argument, they may include only those facts they find helpful.

Authors may have more than one purpose also.

To understand the author's purpose for writing, ask yourself: *What is the author writing about? Is the author giving information? Is the author writing to entertain? Is the author trying to persuade me to think a certain way?*

Read the following passage. What is the author's purpose?

> Have you met your neighbors yet? Neither is very sociable, but you can glimpse them in the night sky.
>
> Mars is sometimes as close as 55 million kilometers from your front door. It has a rocky, lifeless surface and a thin, oxygen-free atmosphere. Your other neighbor, Venus, has a thick atmosphere that would crush you instantly if your blood did not boil first. If you are invited for coffee on either planet, think twice before accepting.

This passage discusses the two planets closest to Earth. The author presents information about both planets in a humorous way. The author's purpose is to inform and entertain.

THINK ABOUT SCIENCE

Directions: Complete each sentence below.

1. A _____ is a large group of stars.

2. The Milky Way galaxy contains approximately _____ stars and is about _____ light-years across.

3. Three types of galaxies found in the universe include spiral galaxies, _____, and _____.

The Solar System

The solar system consists of the Sun, eight planets, many moons, and small objects such as asteroids, dwarf planets, and comets. At the center of the solar system is the Sun. The huge mass of the Sun—and the gravity produced by that mass—keeps all the objects in the solar system in orbit around it.

Earth and its moon, like the other planets and their moons, revolve around the Sun in a regular and predictable motion. Due to this motion, we can experience day and night, the seasons of the year, phases of the Moon, and eclipses. The Sun also is the major source of energy that powers Earth's winds, seasons, and ocean currents.

OUR SOLAR SYSTEM

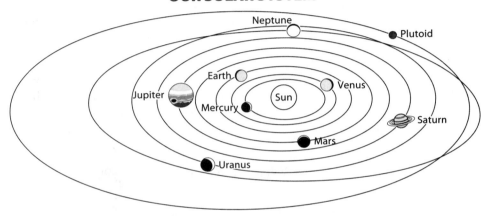

The Sun is a **star**, a huge ball of a gas-like substance that radiates heat and light. The Sun is the closest star to Earth. Although a million Earths could fit into the Sun, the Sun is only about one-fourth as dense as Earth.

At the Sun's center, temperatures are believed to be about 15 million °C. It is in this core that hydrogen fuses into helium and the other elements in nuclear fusion reactions. The energy produced during these reactions radiates out of the core as light and heat.

THINK ABOUT SCIENCE

Directions: Complete each sentence below.

1. The force that holds the planets in orbit around the Sun is _____.

2. The solar system consists of _____ planets, asteroids, dwarf planets, and comets.

3. The Sun is the closest _____ to Earth.

The Planets

A **planet** is a large object that orbits the Sun in a nearly circular path. A planet is massive enough to have its gravity pull it into a round or nearly round shape. Planets do not produce their own light. However, they can be seen from Earth because they reflect light from the Sun.

Most planets have their own **satellites**, or moons, orbiting them like miniature solar systems. The planets that are closest to the Sun are similar in size and chemical makeup. Mercury—the planet closest to the Sun—together with Venus, Earth, and Mars are called the **inner planets**. The inner planets are solid with rocky surfaces. Mercury and Venus have no moons, whereas Earth has one and Mars has two.

The four **outer planets**—Jupiter, Saturn, Uranus, and Neptune—are also the largest planets. These planets are called gas giants because they are made mostly of gas. Most scientists agree that the gas giants have small cores made of hot, melted metal. The cores are surrounded by thick layers of liquid and gas.

Each gas giant has many moons. Jupiter and Saturn have more than 50 moons each. Saturn is well known for its huge, bright rings, but the other gas giants have rings as well. All four of the gas giants were visited by the *Voyager* space probes. The probes took pictures of each planet and some of its moons.

Asteroids, Comets, and Dwarf Planets

Between the inner planets and outer planets is a belt of rocky objects called **asteroids**. Like the planets, asteroids orbit the Sun. No one is certain of the origin of these asteroids. Many astronomers, however, think they are fragments remaining from the formation of the solar system, which was 4.6 billion years ago.

Every so often, the gravitational tug from another planet or object will disturb the orbit of an asteroid and push it away from the main asteroid belt. When this happens, an asteroid may collide with a planet or moon. An asteroid is thought to have struck Earth about 65 million years ago and changed the climate so much that dinosaurs could not survive.

Reading Skill
Analyze Author's Purpose

Authors always have a purpose, or reason, for writing. You should be aware of the author's purpose as you read. The text you are reading now is informational text, meaning it is intended to teach or explain. There are signs to look for when you identify informational text. Such texts usually contain facts, and authors avoid repeating the same facts. Instead, they present information clearly, beginning by introducing a topic and then developing it in a logical, precise way.

Examine the text in "The Planets." What clues help you identify this text as informational?

THINK ABOUT SCIENCE

Directions: Answer each of the following questions.

1. How are the inner planets different from the outer planets?

   _____

   _____

2. What are asteroids? How do scientists think they formed?

   _____

   _____

WRITE TO LEARN

Think of a science topic that interests you. In a notebook, write three paragraphs about it. Write the first paragraph to inform, the second one to entertain, and the third one to persuade. Keep your purpose for writing in mind so that someone reading your paragraphs will clearly understand your purpose.

Comets are small objects made of dust and frozen gas that orbit the Sun. A comet consists of a head that looks like a dirty snowball and a long tail that extends far behind it that appears as it nears the sun. They come from the outer edges of the solar system and become visible from Earth only when they approach the Sun and reflect sunlight. Halley's comet is one of the most famous comets. It approaches Earth every 76 years.

Pluto once was considered the ninth planet of the solar system. Pluto was thought to be unusual because it was quite small and its orbit was unlike orbits of other planets. As scientists began identifying similar objects beyond Pluto, they reconsidered Pluto's classification. They defined new **criteria**, or rules or measures for judging, the definition of a planet. The new criteria disqualified Pluto. In 2006, Pluto was reclassified as a dwarf planet. The former asteroid Ceres and another space object similar to Pluto called Eris have also been classified as dwarf planets.

Vocabulary Review

Directions: Match each word on the left with its description on the right.

1. _____ satellite

2. _____ criteria

3. _____ solar system

4. _____ comet

5. _____ galaxy

6. _____ asteroid

A. objects that orbit the Sun

B. huge collection of stars, dust, and gas

C. any object orbiting another object

D. rocky objects found mainly in a belt between the inner and outer planets

E. rules or measures for judging something

F. object made of dust and frozen gas

Skill Review

Directions: Read the following passage. Then answer the questions that follow.

Apophis is a large asteroid that is approaching Earth. In 2004, scientists estimated that Apophis had a 1 in 37 chance of striking our planet. Today, after much further study, scientists now predict that Apophis will not strike Earth, but it will come close. On April 13, 2029, Apophis will pass Earth at a distance ten times closer than the Moon. If a large asteroid ever did strike Earth, it would be similar to the blast of one hundred nuclear bombs.

1. What is the author's purpose for writing this passage?

   A. to persuade readers to become a scientist

   B. to entertain readers with a science fiction story

   C. to inform readers about Apophis and the dangers of asteroid strikes

   D. to explain how scientists change their minds

2. How did scientists change their prediction about Apophis after 2004?

   A. They concluded that Apophis would strike the Moon.

   B. They used new evidence to make their prediction more accurate.

   C. They wanted to do more research before making a prediction.

   D. They revised their prediction to include the damage Apophis would cause.

Skill Review (continued)

Directions: Read the following passage. Then answer the questions that follow.

The Hubble Space Telescope provides clear, beautiful images of objects in space. These images have helped shape our views of the universe. Now NASA wants to spend billions of dollars on new, more powerful space telescopes. While such projects may be worthy, the time has arrived to spend our resources elsewhere. Before we explore space even further, we should improve the health and well-being of the citizens of Earth.

3. The author would most likely support NASA's plans for new space telescopes when

   A. public funding becomes available.
   B. the public agrees that new telescopes are worthwhile.
   C. living conditions improve for people on Earth.
   D. space telescopes could be proven to take clear images.

4. Which phrases from the passage show the author's bias?

   A. "clear, beautiful images"
   B. "spend billions of dollars"
   C. "new, more powerful space telescopes"
   D. "spend our resources elsewhere"

Skill Practice

Directions: Choose the best answer to each question.

1. To circumnavigate means to travel around the edge of something. Why could a spacecraft circumnavigate the Milky Way, but not travel through its center?

   A. The Milky Way has a black hole at its center.
   B. The Milky Way is too large to travel across.
   C. The Milky Way has too many stars in its center.
   D. The Milky Way is a spiral galaxy.

2. What property do all eight planets have in common?

   A. a rocky surface
   B. at least one moon
   C. a nearly circular orbit
   D. layers of liquid and gas

3. What new evidence led scientists to reclassify Pluto as a dwarf planet?

   A. new measurements of Pluto's size and mass
   B. new criteria for defining a planet
   C. the discovery of an asteroid that is larger than Pluto
   D. the discovery of many objects similar to Pluto

Earth and the Moon

KEY CONCEPT: Earth is in constant motion. It turns on its axis, causing the cycle of day and night. Earth's tilt on its axis and its movement around the Sun results in Earth's seasons. Earth's distance from the Sun makes it habitable, a place for living things.

Do you remember Goldilocks? The fairy tale character found porridge that wasn't too hot or too cold. It was "just right." Scientists used to say Earth was just right, too. They said Earth was in the Goldilocks Zone, neither too hot or too cold. But in recent years scientists have discovered some very extreme conditions that support life. If Earth is in the Goldilocks Zone, the zone is much wider than we thought.

Lesson Objectives

You will be able to

• Relate Earth's motion to day and night and to the seasons

• Discuss the characteristics that make Earth habitable for living things

• Identify the interactions between the Earth, Sun, and Moon that cause the phases of the Moon and tides

Skills

• **Reading Skill:** Cite Textual Evidence

• **Core Skill:** Apply Scientific Models

Vocabulary

habitable
interactions
phase
revolution
rotation
tides

Earth's Journey

Earth is the third planet from the Sun in our solar system. Every 24 hours, Earth completes one **rotation**, or turn, on its axis. The Sun shines on only half of Earth at a time. As each point on Earth turns toward the Sun, it begins its day. As each point turns away from the Sun, night begins.

In science, a **revolution** is a complete journey around the Sun. It takes $365\frac{1}{4}$ days for Earth to complete one revolution. The fraction of one day explains why a full day is added to the calendar every fourth year.

Earth travels at a speed of about 67,000 miles per hour. As it revolves, or orbits the Sun, Earth tilts on its axis. One hemisphere, or half of Earth, tilts toward the Sun, while the other hemisphere tilts away, as shown in the illustration. Earth's tilt as it revolves produces the seasons.

When Earth tilts away from the Sun, the Sun's rays strike less directly, and the result is winter. When Earth tilts toward the Sun, the Sun's rays are more direct, resulting in summer.

When the Northern Hemisphere tilts toward the Sun, summer occurs

TILT OF EARTH'S AXIS

Sun's rays in June

Equator

Sun's rays in December

across Earth's northern half, but the opposite season occurs in the Southern Hemisphere. Then, when the Northern Hemisphere resumes its tilt away from the Sun, winter occurs across Earth's northern half, and summer occurs in the Southern Hemisphere.

CITE TEXTUAL EVIDENCE

Authors provide evidence to support important ideas in their texts. You can cite that evidence when you clarify or communicate those main ideas. You cite evidence when you quote, refer to, or direct someone's attention to a specific piece of information in a text.

Citing evidence is much like producing a map. You show the route your mind followed as you made sense of a text. Others can use your map, meaning the evidence you cited, to evaluate your analysis of a text and any conclusions that you share.

When asked to cite evidence of an opinion, a conclusion, or a summary, be prepared to point to specific information in a text that supports your statements.

Read the following passage. Determine which sentences provide evidence for the main idea that seasons begin at precise moments.

> (1) Search the pages of the calendar, and you will find a date in June labeled as the first day of summer. (2) A date in December is labeled as the first day of winter. (3) In astronomy, however, these seasons begin at very precise moments during Earth's orbit around the Sun. (4) A **solstice** marks the beginning of both summer and winter. (5) A solstice occurs twice each year, when the north pole points as directly as possible toward the Sun, and again six months later, when the south pole points toward the Sun.

Sentences 4 and 5 provide textual evidence that support the main idea that seasons change in relation to precise positions of the Earth as it orbits around the Sun.

21st Century Skill
Communication

Communication is the transfer or exchange of information between individuals. Among scientists, effective communication includes precision, or exactness, in their shared explanations of research. This precision ensures that an experiment will be repeatable, an important aspect of science.

Precise explanations become especially important when many scientists work together. For example, the joint European/U.S. Asteroid Impact and Deflection Assessment mission (AIDA) involves scientists from several countries. This project is working to develop a trial mission to intercept an asteroid. Scientists estimate that there are billions of asteroids; a direct hit from an asteroid could lead to millions of human deaths. The AIDA scientists will share their explanations and findings. This information will help others develop a plan to deflect or destroy an asteroid headed toward Earth.

In your notebook, write an example of a time when you gave or received an explanation that was less than precise. What were the consequences?

THINK ABOUT SCIENCE

Directions: Complete each sentence by underlining the correct word.

1. Earth makes one complete (revolution, rotation) every 24 hours.

2. One complete (revolution, rotation) around the Sun takes $365\frac{1}{4}$ days.

3. Winter occurs in the Northern Hemisphere when the (northern, southern) half of Earth is tilted away from the Sun.

When authors present important ideas, they also provide evidence—information to support those ideas. When you analyze a text, you can cite evidence to support your analysis.

The text on this page describes conditions on Earth that make it **habitable**, or able to support life. What evidence can you cite that supports this central idea?

WRITE TO LEARN

Think of a sport or hobby that you especially enjoy. In a notebook, write two paragraphs describing or explaining the sport or hobby. Use at least three details in each paragraph to give more information about your main idea.

A Habitable Planet

It is impossible to know exactly how many living things reside on Earth, but it is possible to explain what makes the planet so habitable. Earth's distance from the Sun is about 150 million kilometers (93 million miles). This distance is critical for several reasons. One, it limits the range of temperatures on Earth's surface. The planet is neither too hot, like its neighbor Venus, nor too cold like its opposite neighbor, Mars.

The limited range of surface temperatures also makes it possible for water to exist in the liquid state. Almost all living things depend on liquid water for survival. Plus, water is a solvent, meaning it is able to dissolve the substances essential for life processes.

Life needs energy, too. The Sun's energy, in the form of sunlight, remains relatively constant. Plants depend on the energy in sunlight to conduct photosynthesis. Most forms of life are sustained either directly or indirectly by the Sun.

Earth's atmosphere also makes the planet habitable. Earth's gravity keeps a blanket of gases surrounding the planet. Nitrogen represents 78 percent of those gases, and oxgyen occupies 21 percent. Most living things, even those that live in water, need oxygen to survive. In addition, a form of oxygen called ozone collects in the upper atmosphere. This gas helps protect Earth from the Sun's harmful ultraviolet (UV) rays.

THINK ABOUT SCIENCE

Directions: Answer each question below.

List three reasons that Earth is able to support life.

_____

The Moon

The Moon is a dusty, rocky ball with no atmosphere. Past asteroid impacts have left craters on its surface. Because there is no weather, and therefore no erosion on the Moon, the craters have stayed almost the same for billions of years.

Earth and the Moon exert a pull on each other, or gravity. Some scientists think that without the Moon's gravity, Earth's axis might wobble back and forth. Living things might not be able to adapt to the severe changes in weather and climate that this would cause.

When you see the light of the Moon, you are seeing reflected sunlight. The Moon itself generates no light. When you see the Moon's changing shapes, you see the results of **interactions** between Earth, the Moon, and the Sun. These shapes are called the phases of the Moon. In astronomy, a **phase** is a particular appearance of the Moon or planet at any given time.

Interactions are the actions of things that create an effect on each thing. When the Moon is between Earth and the Sun, no sunlight is reflected, and you cannot see it. This phase is called a new moon. When Earth moves between the Moon and the Sun, the full side of the Moon reflects sunlight. This phase is called a full moon. You see varying amounts of reflected sunlight as the Moon revolves around Earth each month, completing a lunar cycle.

PHASES OF THE MOON AS SEEN FROM EARTH

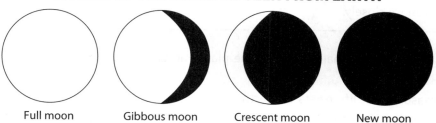

| Full moon | Gibbous moon | Crescent moon | New moon |

At times during Earth's travels, the planet casts a shadow over the Moon. This shadow darkens some or all of the Moon's surface, creating a lunar eclipse. The Moon can also cast a shadow when it moves between Earth and the Sun, creating a solar eclipse.

Tides

Gravity exists within the Sun-Earth-Moon system. Both the Sun and Moon pull on Earth, creating **tides**, but the Moon's influence is greater because it is closer to Earth. The Sun is about 93 million miles from Earth, but the Moon is less than 240,000 miles. On the side of Earth facing the Moon, gravitational force pulls water into a bulge called high tide. Inertia, which works to keep water moving in a straight line away from Earth, overcomes gravity on the opposite side of Earth, producing another high tide. As Earth and the Moon rotate, the location of high tides changes.

Vocabulary Review

Directions: Complete each sentence using one of the words below.

interactions   phases   revolution   rotation   tide

1. _____ between the Earth, Sun, and Moon create different appearances of the Moon.

2. The Earth makes one complete _____ around the Sun about once every $365\frac{1}{4}$ days.

3. Earth makes one _____ on its axis every 24 hours.

4. A(n) _____ is the rise and fall of the oceans due to the Moon's gravitational pull on Earth.

5. _____ of the Moon include a full, gibbous, crescent, and new moon.

Directions: Read the passage below. Then answer the question that follows.

> Scientists have discovered a planet in orbit around a star named Gliese 581. This planet might support life. This planet is fifty percent larger than Earth and five times as massive. It orbits its star about fifteen times closer than Earth orbits the Sun. However, Gliese 581 is fifty times dimmer than our Sun. Because this star is much cooler, a life-supporting planet could orbit it more closely.

1. How do details in the passage support the main idea that the planet around Gliese 581 might support life?

   A. The details identify the reasons that life exists on Earth.
   B. The details compare the planet and Gliese 581 to Earth and the Sun.
   C. The details describe the formation of life-supporting planets.
   D. The details explain how a planet orbits a star.

Directions: Read the passage below. Then answer the questions that follow.

> Space probes to Venus have photographed structures that might be the remains of coastlines. For this reason and other reasons, scientists think that Venus may once have had oceans and an atmosphere. Life may have existed there. Then rising levels of carbon dioxide caused the greenhouse effect to increase dramatically. The surface temperatures increased to 467°C (862°F), where they remain today. If any water had once gathered on Venus, it long since has boiled away.

2. What caused the temperatures on Venus to rise to such high levels?

   A. the boiling away of the oceans
   B. rising levels of carbon dioxide
   C. visits by space probes
   D. distance from the Sun

3. Which detail would best support the main idea of the passage?

   A. the chemical formula for carbon dioxide
   B. the names of the space probes to Venus
   C. a description of Earth's oceans and atmosphere
   D. an explanation of the greenhouse effect

Skill Practice

Directions: Choose the best answer to each question.

1. Australia is located in the Southern Hemisphere. When does Australian winter occur?

   A. June to September
   B. December to March
   C. September to December
   D. March to June

2. The moon is in orbit around Earth. What occurs when the Moon is directly between Earth and the Sun in its orbit?

   A. a full moon
   B. a new moon
   C. high tides
   D. low tides

3. In 1969, astronauts planted a flag on the Moon's surface. From the information provided and your own knowledge, what might you conclude about the condition of this flag today?

   A. It has been changed by the actions of water and air.
   B. It may have suffered rips and tears.
   C. It has become buried under lava or dust.
   D. It has changed very little or not at all.

Directions: Choose the best answer to each question.

1. Which measurement of distance is the longest?

   A. 1,000 kilometers
   B. 100 miles
   C. 1 light-year
   D. 1 billion miles

2. The big bang theory is widely accepted within the scientific community. What fact about the universe supports this theory?

   The Universe

   A. is expanding.
   B. is made of matter and energy.
   C. contains galaxies, planets, and stars.
   D. is about 14 billion years old.

3. The eight planets are classified into two groups. What are the groups based on?

   A. the shapes of their orbits
   B. their ability to support life
   C. their date of discovery
   D. their distance from the Sun

4. If a full Moon is observed one night in the Northern Hemisphere, what phase of the Moon will be observed in the Southern Hemisphere?

   A. full moon
   B. gibbous moon
   C. new moon
   D. crescent moon

5. The energy in a supernova is sufficient to make

   A. heavy elements such as iron
   B. a universe of dust and gases
   C. a new star
   D. a nebula

Directions: Question 6 refers to the following passage.

The temperature of the surface of Venus is 475°C. The weight of its atmosphere is 100 times greater than that of Earth, and clouds of sulfuric acid cover the sky. In addition, carbon dioxide and other gases in Venus's atmosphere trap the heat from sunlight. This runaway "greenhouse effect" makes the planet's surface extremely hot. Life as we know it could not survive on Venus.

6. Which of the following statements best summarizes the information in the passage?

   A. Although hidden from view, Venus has many features similar to Earth.
   B. Trapped sunlight spreads over the surface of Venus.
   C. The thick atmosphere and high temperatures of Venus make it an unlikely home for living things.
   D. Venus could be habitable without the greenhouse effect.

Review

Directions: Questions 7 and 8 refer to the following passage.

Sunspots are dark, relatively cool (4,500°C) areas on the Sun's surface that are regions of intense magnetic activity. The number of sunspots increases and decreases at fairly regular intervals. Whenever the number of sunspots is highest, solar flares erupt on the Sun's surface and shoot out particles and radiation. When they reach Earth's atmosphere, they may interfere with radio transmissions and electrical power.

7. What happens when solar flares erupt on the surface of the Sun?

   A. The number of sunspots decreases.
   B. Solar matter and energy are sent into space.
   C. Sunspots become cooler and smaller.
   D. The Sun becomes even hotter than normal.

8. According to the passage, when are radio broadcasts likely to be interrupted by static?

   A. when sunspots become cool
   B. when the sunspot cycle is at its highest
   C. when magnetic activity on the Sun decreases
   D. when flares explode on Earth

Answer each question completely.

9. List six steps that explain the formation of the solar system, including the Sun and planets.

   _____
   _____
   _____
   _____
   _____
   _____
   _____
   _____
   _____
   _____

10. Some scientists refer to Earth as being in the Goldilocks Zone. Explain what they mean.

   _____
   _____
   _____
   _____
   _____
   _____
   _____
   _____
   _____

11. Write an argument justifying Pluto's reclassification as a dwarf planet.

_____

_____

_____

_____

_____

_____

_____

_____

12. Describe the conclusion that Edward Hubble formed in 1929 that endures today. Then explain why scientists continue to collect information related to Hubble's observations.

_____

_____

_____

_____

_____

_____

_____

_____

13. Draw and label a diagram to explain the summer and winter solstices in the Northern Hemisphere.

14. Gravity is a force that exists among all objects in the universe. The force of gravity depends on the masses of both objects and the distance between them. When two objects approach each other, the attraction of their gravity affects their motions.

If an object in the solar system dramatically changed its orbit around the Sun, which explanation could account for this change?

A. The planet changed in volume, but not mass.
B. An unidentified object is passing close to the planet.
C. The force of gravity is not strong enough to hold the planet in orbit.
D. The planet is moving quickly away from the Sun.

15. Given your understanding of the factors that determine the strength of a gravitational force, explain why the Moon has more influence over Earth's tides than the Sun does.

_____

_____

_____

_____

_____

_____

_____

_____

_____

_____

Check Your Understanding

On the following chart, circle the number of any item you answered incorrectly. Next to each group of item numbers, you will see the pages you can review to learn how to answer the items correctly. Pay particular attention to reviewing those lessons in which you missed half or more of the questions.

Chapter 12 Review

Lesson	Item Number	Review Pages
Earth's Origins	1, 2	410–413
Origins of the Universe	5, 12	414–417
The Milky Way and the Solar System	3, 7, 8, 9, 11	418–423
Earth and the Moon	4, 6, 10, 13, 14, 15	424–429

CHAPTER 12: THE COSMOS

Question

What is the relationship between the sizes of craters and the meteorites that created them?

Background Concepts

Meteoroids are pieces of rock and debris. When they fall through a planet's or a moon's atmosphere, friction heats them, causing them to radiate light. These flaming objects are called meteors, or shooting stars. Once they hit the surface of a planet or moon, they are called meteorites.

Meteors can be tiny or up to about 33 feet wide. Millions of meteors fall through Earth's atmosphere every day. Most burn up before reaching Earth's surface. When meteorites strike Earth, they produce impact craters.

Investigation

Materials required

- 5 round or nearly round objects of various sizes, from small to large, to represent meteors. Possible examples include walnuts, gumballs, marbles, apples, oranges, golf balls, tennis balls, baseballs, and rubber balls.

- a cardboard box at least 24 in. long × 18 in. wide × 24 in. high

- one 5 lb. bag of flour

- ruler

1. If possible, do this experiment outdoors on a clear day with little or no wind. If not, line the floor with newspaper to make cleanup easy.

2. Place the box on the floor. Pour flour into the box until about two inches of flour cover the bottom of the box. Shake the box from side to side to ensure that the flour is evenly distributed.

3. List the objects that will serve as meteorites in the table.

4. Use a ruler to measure the approximate diameter of each meteor. Record the measurements in the table.

5. Arrange the meteors from small to large.

6. Hold the meteor just above the top edge of the box. Then release it. Remove the meteor, being careful not to disturb the impact crater.

7. Move to a different location around the edge of the box and repeat Step 6. Drop the meteor from the same height as you dropped it the first time.

8. Move again. Choose a location that will prevent impact craters from overlapping. Holding the meteorite from the same height as before, drop it.

Application of Science Practices

9. Use a ruler to measure the diameter of each impact crater. Record the measurements in the table.

10. Shake the box from side to side to repeat the procedure using the next largest meteor.

11. Repeat Steps 6 through 10 until you have dropped each meteor three times.

Meteor Mass (oz.)	Meteor Diameter (in.)	Size of Crater #1 (in.)	Size of Crater #2 (in.)	Size of Crater #3 (in.)	Average Crater Size (in.)

Interpretation

Determine the average impact crater diameter for each meteorite.

_____

_____

Answer

Describe the relationship between the mass of the meteorite and the diameter of the impact crater.

_____

_____

Evidence

What measurements would you expect to get if you dropped each meteor a fourth time?

_____

_____

Science

Directions: The Science Posttest consists of forty-five multiple-choice and short-answer questions. Choose the best answer to each multiple-choice question, and write a complete response to each short-answer question. If a question seems to be too difficult, do not spend too much time on it. Work ahead and come back to it later when you can think it through carefully.

When you have completed the test, check your work with the answers and explanations on pages 452–454. Use the chart on page 454 to determine which areas you need to review.

1. An osprey, or sea eagle, is an ocean-dwelling bird. Two alleles, or forms of a gene, determine the osprey's beak size. Large beaks (B) are dominant over small beaks (b).

   The Punnett square shows the possible offspring of a cross between a male and female osprey. In this cross, both parents have the gene combination Bb.

	B	b
B	BB	Bb
b	Bb	bb

   Based on the information in the passage and the diagram, what percentage of the ospreys' offspring will have large beaks?

   A. 25%
   B. 50%
   C. 75%
   D. 100%

2. A molecule called DNA carries coded information. The cell uses this information to make proteins. Decoding DNA involves two important steps. In transcription, the DNA codes for another code-carrying molecule called RNA. In translation, the code in RNA is used to make proteins.

   Which best describes the role of RNA in the process of making proteins?

   A. It plays the role of a byproduct.
   B. It plays an intermediate role.
   C. It plays the role of an end product.
   D. It plays the role of a warehouse.

Science

3. Ants are fast, strong animals with an outstanding sense of smell. They are social animals that live in colonies and work together to carry out complex tasks, such as finding food.

   Scientists have investigated how ants communicate with each other to carry out food-finding tasks. One possibility is that they use chemical signals. According to this model, an ant that discovers food lays down a tiny chemical trail as it returns to the nest. Then other ants can follow the trail.

   Which statement explains why scientists are confident that ants communicate with one another?

   A. Ants have a simple language.
   B. Ants have moving body parts.
   C. Ants have a well-developed sense of smell.
   D. Ants work together to perform complex tasks.

4. Green plants and a few other living things make their food through the process of photosynthesis. This process, which occurs in pigment-filled cell parts called chloroplasts, is powered by the energy of sunlight.

   To perform photosynthesis, plants take in carbon dioxide gas and water. The materials are converted into a compound called glucose, a simple sugar. Oxygen is released as a byproduct of the process.

   As the passage describes, what do plants need to perform photosynthesis?

   A. carbon dioxide, water, and glucose
   B. carbon dioxide, oxygen, sunlight
   C. carbon dioxide, water, and sunlight
   D. oxygen, water, sunlight, glucose

Directions: Questions 5 and 6 refer to the diagram and passage below.

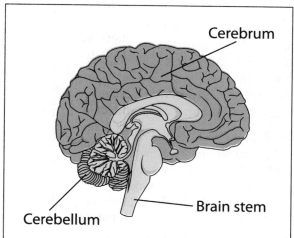

The human brain is divided into three main parts. Each part has a specific job to do.

The brain stem controls basic body functions. It regulates heartbeat, breathing, and other jobs not under conscious control. The cerebellum coordinates balance and movement. The cerebrum is the large, outer part of the brain that appears to be folded tightly. It receives information from the senses, controls conscious movement, and allows the brain to process data, to reason, and to feel emotions.

5. Which part of the brain helped you understand the passage?

   A. brain stem
   B. cerebellum
   C. cerebrum
   D. spinal cord

6. Which action requires the use of the brain stem?

   A. running up and down stairs
   B. maintaining blood pressure
   C. smelling a flower
   D. using a screwdriver

Science

Directions: Questions 7 and 8 refer to the passage below.

Several decades ago, chemists discovered a new form of the element carbon. This substance, called buckminsterfullerene, is named after the American engineer Buckminster Fuller. It is made of carbon atoms joined into the shape of a soccer ball.

"Buckyballs," as chemists call them, have interesting properties. First, when a buckyball collides with a silicon surface, it does not break apart. Instead, it bounces back. Second, buckyballs can trap other molecules within their round structure. Third, at cold temperatures, buckyballs lose almost all of their resistance to the flow of electric current. They become superconductors.

---

7. What property other than shape makes buckyballs similar to soccer balls?

   A. size
   B. usefulness
   C. superconductivity
   D. being bouncy

8. Which statement explains the reason buckyballs might be useful as carriers of drugs in the bloodstream?

   A. Buckyballs could trap drug molecules inside them.
   B. Buckyballs would flow quickly in the blood.
   C. Buckyballs would carry electricity easily.
   D. Buckyballs could dissolve the drug molecules.

9. One of the most celebrated landmarks in India is the Taj Mahal. On the Yamuna River, the faithful reflection of this monument in the waterway is a favorite image for tourists and photographers.

   Unfortunately, the building and waterway sit in stark contrast to each other. While the Taj is considered the finest example of Mughal architecture, the Yamuna has become one of the most polluted rivers in India.

   Based on the information in the passage, what can you conclude about the pollution of the Yamuna River?

   A. The pollution is not visible on the water's surface.
   B. The pollution comes from agriculture.
   C. The pollution has lowered the value of the Taj Mahal.
   D. The pollution is concentrated at the river's edge.

10. A car is traveling north across a distance of 250 miles. Its speed is 47 mph. What is the car's velocity?

    A. 47 miles per hour
    B. 250 miles at 47 miles per hour
    C. 250 miles traveling west
    D. 47 miles per hour traveling north

Science

Directions: Questions 11 and 12 refer to the following passage.

> In 1991, scientists saved the lives of tens of thousands of people. These scientists were geologists, and their tool was an improved method of forecasting volcanic eruptions.
>
> Mount Pinatubo in the Philippines was dormant for 600 years. Then scientists observed unexpected changes near the volcano's top. They began studying the volcano and noted several minor earthquakes occurring deep below it. In addition, emissions of sulfur dioxide gas were increasing.
>
> As earthquake activity moved closer to the top of the volcano, scientists became convinced that magma was moving under the mountain. They sent out a general warning, and about 60,000 people were evacuated. A short time later, the volcano exploded in several large eruptions, sending hot, molten rock across the landscape. More than 800 people were killed, but many more were saved.

11. What were the first signs that Mount Pinatubo was becoming active?

    A. shaking of the ground above it
    B. decreased sulfur dioxide emissions
    C. earthquake activity below the mountain
    D. people fleeing from their homes

12. Based on the passage, how could magma be described?

    A. earthquake activity
    B. molten rock
    C. sulfur dioxide
    D. metals from below the crust

Directions: Questions 13 and 14 refer to the information and diagram below.

> During photosynthesis plants take in carbon dioxide and release oxygen. During respiration, animals take in oxygen and release carbon dioxide.
>
> The diagram illustrates gas exchange in plants.

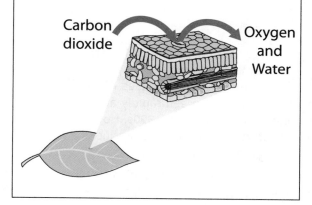

13. What should be added to the diagram to make it more useful in explaining gas exchange during respiration?

    A. one arrow pointing to the leaf
    B. two arrows pointing to and from the leaf
    C. an animal and two arrows pointing to and from it
    D. an animal and an arrow pointing from it to the leaf

14. Carbon dioxide contains both carbon and oxygen. As the passage and diagram suggest, photosynthesis acts to remove which elements from air?

    A. carbon only
    B. oxygen only
    C. carbon and oxygen
    D. carbon, oxygen, and hydrogen

Science

15. The chemical formula for photosynthesis is written as:

$$6CO_2 + 6H_2O \xrightarrow{\text{light}} C_6H_{12}O_6 + 6O_2$$

What kind of reaction does this formula represent?

A. a form of radioactive decay
B. an endothermic reaction
C. an ectothermic reaction
D. an acid-base reaction

16. Before the 1980s, few people had heard about the human immunodeficiency virus (HIV). This is the virus that causes AIDS, a deadly disease that spread rapidly across the world in the 1980s and 1990s. How did AIDS get started? Researchers are looking for the origin of the virus.

The likely origin of HIV is the simian immunodeficiency virus (SIM). This causes a disease similar to AIDS in chimpanzees and other nonhuman primates. Strong evidence suggests that SIM crossed species, infected a human, and changed form to become HIV.

What best describes the relationship between HIV and SIM?

A. identical viruses that infect different species
B. similar diseases among different species
C. similar viruses that infect the same species
D. similar viruses that infect different species

Directions: Questions 17 and 18 refer to the passage and diagram below.

Double Stranded Chromosome with Two Chromatids

Genes, which code for inherited traits, exist on chromosomes. Most human cells have 23 pairs of chromosomes, or 46 chromosomes in all. When these cells divide to make new body cells, the chromosomes split and duplicate. Each chromosome and its "sister" is called a chromatid.

Chromatids are critical for cell division. New cells other than sex cells receive one chromatid from each pair of sister chromatids.

17. According to the passage, what is the purpose of chromatids?

A. They repair damaged chromosomes.
B. They change genetic codes.
C. They control cell activities.
D. They are necessary for cell division.

18. In the diagram, what could the dark and light bars represent?

A. traits of the cell
B. chromatids
C. genes
D. new cells

Science

19. A solution is made by thoroughly mixing two substances together. The substance in the lesser amount is called the solute. The solute is mixed into what is called the solvent. What kind of substance is the sugar in a pitcher of lemonade?

   A. solute
   B. solvent
   C. mixture
   D. solution

Directions: Questions 20 and 21 refer to the passage and illustration below.

Visible light is the light that people are able to see, such as light from the Sun. Although sunlight appears colorless, it is actually a mixture of colors. Each color has a different wavelength. A wavelength is a measure of the distance between successive crests or successive troughs of a wave.

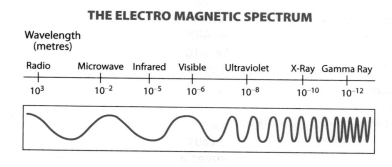

The wavelengths in visible light range from violet (the shortest wavelength) to deep red (the longest wavelength).

Visible light is only a small part of the electromagnetic spectrum. Other forms of light that cannot be seen include radio, ultraviolet, and x-rays.

While humans cannot see these invisible parts of the spectrum, they have found numerous ways to use them. Doctors use gamma rays, for example, to treat cancer and to locate disorders inside the body. Ultraviolet rays are used in sunlamps. Infrared rays are used to treat skin diseases and to dry paint on cars. Radio waves, like the ones in the illustration, are used in radio and television broadcasting.

20. Which conclusion about electromagnetic waves does the passage support?

   A. Broadcasters can use visible light to enhance communication.
   B. Pigments in matter determine wavelengths.
   C. All wavelengths share the same properties.
   D. Visibility is unrelated to usefulness.

21. Which common observation provides the best evidence that sunlight is a mixture of colors?

   A. A rainbow appears in the sky after a rainstorm.
   B. Leaves change color during the autumn.
   C. Leaves appear green in sunlight during spring and summer.
   D. Colored spotlights at theaters can combine to form colorless light.

Science

Directions: Questions 22 and 23 refer to the passage below.

The reactions of nuclear fusion power the Sun and all other stars. Here on Earth, nuclear fusion might someday provide a tremendous source of energy. Many researchers are trying to stage and control fusion reactions. One of the problems they face, however, is that fusion only occurs at extremely high temperatures. No known container is able to hold a fusion reaction.

In 1989 two chemists in Utah announced that they had observed nuclear fusion at room temperature, a process called cold fusion. Many scientists doubted that this was possible. Suspicion grew when other scientists were unable to repeat the experiment. Several scientists were injured and one was killed in laboratory explosions related to cold-fusion testing. Today, most scientists dismiss any experimental claims of cold fusion.

22. Which statement describes the future of controlled nuclear fusion as a source of energy?

   A. Current research makes it almost certain that nuclear fusion will one day be a source of energy.

   B. The use of nuclear fusion as a source of energy depends on whether scientists can replicate the original cold fusion experiment.

   C. The use of nuclear fusion as a source of energy is impossible because it occurs only in stars of a certain density.

   D. The use of nuclear fusion as a source of energy is possible if serious obstacles are overcome.

23. What is the best evidence that the Utah chemists had not really observed cold fusion?

   A. the doubts of other scientists at the time

   B. the inability of other scientists to repeat the experiment

   C. the injuries and deaths of scientists studying cold fusion

   D. the conclusion today that cold fusion is not possible

24. Acceleration describes a change in the velocity of an object. An object that increases its speed is accelerating. However, an object that decreases its speed is also accelerating. This is called deceleration, and the direction of the acceleration is opposite to the direction of motion.

The graph below shows the acceleration of four different cars.

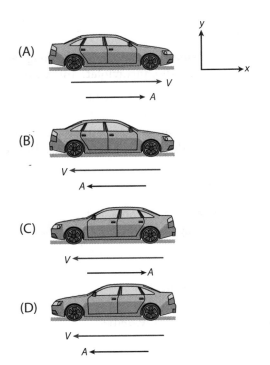

Which car is decelerating?

   A. A   **C.** C

   B. B   **D.** D

Science

Directions: Questions 25 and 26 refer to the passage and illustration below and to the right.

Lightning is a discharge of the buildup of electrical charges during a thunderstorm. As the storm cloud develops, interactions of charged particles produce an intense electrical field within the cloud. A large positive charge is usually concentrated in the frozen upper layers of the cloud. A large negative charge, along with a smaller positive area, build up in the lower portions of the cloud.

Normally, Earth's surface is negatively charged with respect to the atmosphere. However, as a thunderstorm passes over the ground, the negative charge in the base of the storm cloud induces a positive charge on the ground below. This ground charge follows the storm like an electrical shadow, growing stronger as the negative charge increases in the cloud.

The attraction between positive and negative charges makes the positive ground current flow up buildings, trees, and other elevated objects in an effort to establish a flow of current. Air is a poor conductor of electricity, so it insulates the cloud and ground charges. By preventing a flow of current, air allows a huge difference in electrical charge to build up.

Lightning strikes when the difference between the positive and negative charges—the electrical potential—is great enough to overcome the resistance of the insulating air. The electrical potential can be as much as 100 million volts. Lightning strikes may proceed from cloud to cloud, cloud to ground, or in certain circumstances, from ground to cloud.

25. What normally prevents the flow of electricity between the negative charge in the base of the storm cloud and the positive charges on the ground?

   A. air
   B. water vapor
   C. raindrops
   D. ice crystals

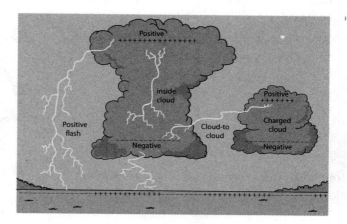

26. What is one consequence of air's insulating property?

   A. A large difference in positive and negative charges builds up between the clouds and the ground.
   B. The electrical charges within clouds reverse position.
   C. Objects on the ground become negatively charged.
   D. Lightning is restricted to cloud-to-cloud rather than cloud-to-ground paths.

Science

27. Before tools were developed to measure changes in the atmosphere, people had to rely on observation to describe the weather. They often used folklore and proverbs to account for and predict atmospheric changes. Many proverbs were set to rhyme so they could be easily remembered. One example is "When the dew is on the grass, rain will never come to pass."

Which atmospheric processes are referred to in this proverb?

A. evaporation and condensation
B. condensation and precipitation
C. precipitation and transpiration
D. precipitation and evaporation

Directions: Question 28 refers to the diagram below.

28. A hydrocarbon is a compound made of carbon (C) and hydrogen (H). The diagram shows the structures for three hydrocarbons. Each line, or dash, represents a chemical bond between two atoms. Double dashes represent two bonds.

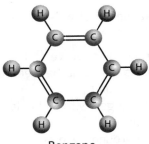

Benzene

As shown in the diagram, what rule applies to the bonds that carbon atoms form?

A. A carbon atom bonds only with other carbon atoms.
B. A carbon atom bonds only with hydrogen atoms.
C. A carbon atom bonds to carbon atoms and hydrogen atoms.
D. A carbon atom forms bonds with other atoms under specific circumstances.

Directions: Question 29 refers to the passage and diagram below.

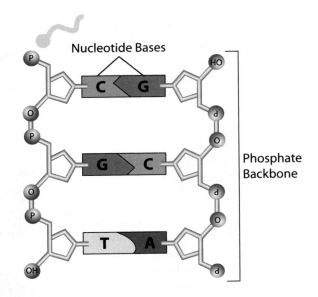

A DNA molecule is made of four simpler molecular units called nucleotides. The nucleotides, or bases, are abbreviated A, G, C, and T. The letters shown in the diagram represent a sequence of nucleotides on a DNA molecule. Specific sequences of nucleotides determine the traits of living organisms. In other words, they are genetic codes.

Cells use DNA to make proteins, and proteins control all cell activities. Inside DNA, the nucleotides are repeated billions of times in a variety of combinations. The order of the nucleotides determines which proteins are made.

29. What best describes the role of the nucleotide sequence of DNA?

A. a key to a lock
B. a coded message
C. a shield against enemies
D. an identity tag

30. Geneticists often use karyotypes to portray the molecular characteristics of an individual, whatever species that individual may belong to. A karyotype is an organized profile of an individual's chromosomes. The chromosomes are arranged and numbered by size, from largest to smallest. This way scientists can evaluate the size, shape, and number of chromosomes in a cell. This kind of genetic profiling is limited to eukaryotes, or multicellular organisms.

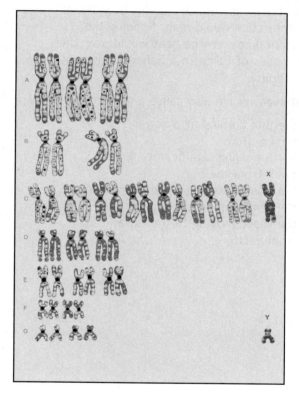

Which specific factor shows that this karyotype identifies a human?

A. the presence of sex chromosomes
B. the number of chromosomes
C. the distinct shape of the chromosomes
D. the size of the chromosomes

Directions: Questions 31 and 32 refer to the passage below.

> Radiation can kill or damage cells within an organ. If enough cells die, the organ dies. If crucial organs die, the organism dies. Thus, death is a possible consequence of radiation.
>
> To be immediately lethal, the body's overall exposure to radiation must exceed 1,000 rems over a brief period of time, meaning minutes or hours. Such exposures occurred in Hiroshima and Nagasaki during bombings in World War II. A dose of 500 rems, delivered at one time to the whole body, causes death in about 50 percent of cases.
>
> In the 100 to 500 rems range, radiation sickness occurs, and some individuals die. At lower radiation levels, the consequences are harder to predict and detect.
>
> Low radiation doses often cause genetic changes, or mutations, within cells. If changes occur to body cells, such as blood, bone, or skin cells, they are called somatic mutations. When these cells replicate, the cellular changes affect the organism itself. If mutations occur in reproductive cells, however, they can be transmitted to future generations.

31. Based on the information in the passage, what is a rem?

A. an abbreviation for "rapid eye movement"
B. the loudness of a sound
C. a type of poisoning
D. a measured unit, or dose, of radiation

32. Based on the information in the passage, what is a somatic mutation?

A. an injury to a part of the body
B. organ failure
C. changes within a reproductive cell
D. damage to a nonreproductive cell

Science

33. On Earth, life of every kind—from a tiny virus to the largest whale, from blue-green algae to giant sequoia, and even humans—is fully and precariously dependent on the Sun. All life on Earth originated and evolved in the Sun's light and warmth.

Before machines were introduced in the Industrial Revolution, animals provided power and transportation. How was this work style dependent on energy from the Sun?

A. The Sun dried the ground, making it a solid surface to work on.

B. Uneven heating by the Sun created breezes that cooled the animals.

C. When focused on a point, the Sun's energy could be harnessed for burning wood.

D. The animals derived their energy from plants, which used the Sun's energy for their growth.

34. Life on Earth depends on a precise amount of energy that it receives from the Sun, which is only 150 million kilometers away. For example, a drop of 200 degrees in the Sun's 10,000°F surface temperature would initiate a global glacial advance. The new ice age would turn our habitable blue planet into a giant, uninhabitable snowball in perhaps fifty years.

In contrast, a 200 degree rise in the temperature of the Sun would cause severe global warming. The ice caps of Greenland and Antarctica would melt, flooding the continental coasts. The heat would sear the landmasses of Earth to a Saharan stillness and sterility.

What event would also cause a new ice age?

A. the Sun burning at a hotter temperature

B. Earth moving significantly farther away from the Sun

C. the Sun expanding while maintaining the same temperature

D. the Moon moving farther away from Earth

Science

35. The fossil fuels we depend on today—oil, gas, and coal—took hundreds of millions of years to form. Nuclear energy comes from radioactive materials within Earth's rocks. These energy sources are nonrenewable.

The Sun drives the water cycle, and thus is the source of hydroelectric power. Tidal energy is caused by the gravitational pull of the Moon. The Sun causes winds to blow. Geothermal energy comes from the high temperatures of Earth's interior. Biomass is waste from once-living matter. These energy sources are renewable.

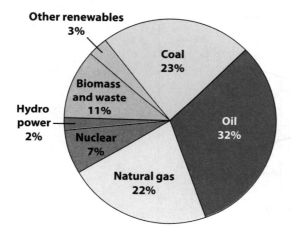

Based on the information and the graph, what percentage of energy sources used in one year came from renewable energy sources?

A. 2%
B. 16%
C. 23%
D. 84%

36. In the state of Wyoming, you can find at least two hundred clues to the nature of Earth's interior. The clues are geysers. A geyser is a hole in the ground that erupts in a jet of hot water and steam. The water is heated deep underground, which shows that Earth's interior is very hot.

The most famous geyser is "Old Faithful" in Yellowstone National Park. This geyser earned its name because it erupts just about every hour. It has been doing this for at least a hundred years.

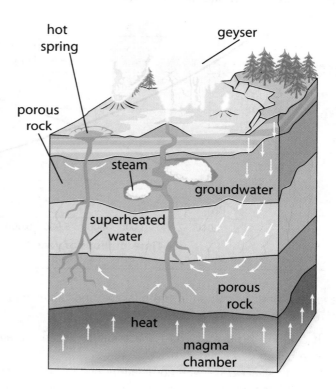

Based on the information in the passage, what does a geyser do that offers clues to Earth's interior?

A. extends deep into the ground
B. erupts very regularly
C. erupts much like a volcano
D. emits water that was heated underground

Science

Question: Question 37 refers to the passage and diagram below.

Soon after he defined the structure of the atom, British physicist Lord Rutherford suggested the use of radioactivity as a tool for measuring geologic time. The year was 1905, a jumping off point for research on the nature and behavior of atoms. By 1950, scientists had developed and refined several techniques for dating rocks and other Earth materials.

The basis of these techniques is simple, although the laboratory procedures are not. The radioactive decay of certain elements, such as uranium-238 and carbon-14, provides an "atomic clock." The clock allows scientists to accurately measure the age of almost any substance over a few hundred years old. The table lists a few examples and their ages.

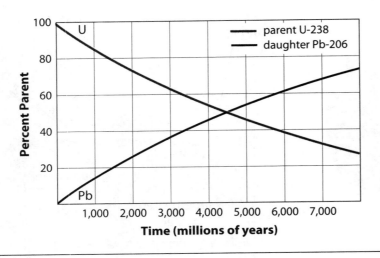

37. Based on the information in the passage and table, what can you conclude about the usefulness of radioactive dating?

A. It can only be used to date materials between the ages of two thousand and one billion years old.

B. It is most accurate when uranium is used in the laboratory procedure.

C. It will become even more accurate as our knowledge of atomic behavior improves.

D. It cannot be used to measure the age of quartz, gneiss, or other nongranitic rocks.

Science

Question: Questions 38 and 39 refer to the passage and diagram below.

Socks removed from a hot dryer often cling together. In the dark, you may even notice a tiny spark when the socks are pulled apart. The cause of this phenomenon is static electricity.

As the socks rub together in the dryer, electrons are rubbed from one atom to another. The result is atoms that are no longer electrically neutral. They have either a positive or a negative electrical charge. The socks stick together because opposite electrical charges attract.

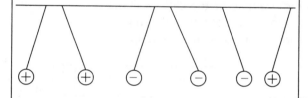

38. Which situation illustrates the principle of static electricity?

   A. walking across a rug and feeling a shock when you touch a metal doorknob
   B. sliding down a slope of loose gravel after you take a small step
   C. replacing flashlight batteries
   D. heating a pot of water and watching the steam escape from the surface as the water boils

39. Based on the passage, which statement best summarizes the behavior of electrically charged atoms?

   A. Electrically charged atoms will attract each other when they gain electrons.
   B. Electrically charged atoms will attract each other when they lose electrons.
   C. Electrically charged atoms will attract each other when they have the same electrical charge.
   D. Electrically charged atoms will attract each other when they have opposite electrical charges.

Science

40. Scientists recognize there are millions of species of living things on Earth. Exactly how many millions is another matter, however. Regardless of how many different species inhabit the planet, they share common characteristics. For example, they are made of one or more cells, and cells have structures for performing tasks. Even a single-celled amoeba can capture food, break down the food to release energy, grow, reproduce, respond to stimuli, and adapt to its environment.

While an amoeba carries out life functions within a single cell, multicellular organisms have complex and specialized systems for activities such as getting energy from food. For example, humans have a digestive system made of different kinds of tissues that combine to form organs, including the stomach, small intestine, liver, pancreas, and large intestine.

Which statement best summarizes the concept in the passage?

A. Body structures vary from organism to organism.
B. All living things have similar requirements.
C. There are many kinds of organisms.
D. The level of complexity of an organism is determined by its digestive system.

Directions: Questions 41 and 42 refer to the passage below.

> For three decades, scientists have recorded lower traffic-related death rates for cliff swallows, a kind of bird, despite increasing numbers of vehicles on the road. Scientists studying the birds have found a difference between cliff swallows in the general population and those whose bodies they find along roadways. Most of the dead birds have slightly longer wings than most cliff swallows, perhaps making them less able to turn away quickly from oncoming vehicles.

41. Use the example of the cliff swallows to relate genetic variations within a population and adaptation to environmental change.

_____

_____

_____

_____

_____

_____

_____

42. Explain why genetic variation forms the foundation of natural selection.

_____

_____

_____

_____

_____

_____

_____

Science

Directions: Question 43 refers to the illustration and passage below.

The DNA molecule has two strands, which resemble the two poles of a ladder. Each strand is connected to nucleotides, or bases, and the bases on the two strands join in pairs. The base pairs are similar to the steps or rungs of a ladder.

In the process of DNA replication, the strands of the DNA molecule split apart along the rungs, like a ladder divided lengthwise. The chart shows the steps of DNA replication.

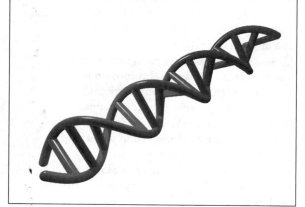

43. After a DNA molecule is replicated, where are its original strands?

   A. broken apart into individual bases
   B. rearranged within the original DNA molecule
   C. attached to existing DNA molecules
   D. divided into two new DNA molecules

Directions: Questions 44 and 45 refer to the illustration and passage below.

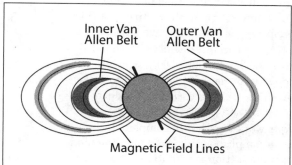

Zones of high-energy particles circle Earth, three hundred miles above Earth's surface. These zones, which are held in place by Earth's magnetic field, are named the Van Allen belts after James Van Allen, who discovered them in 1958. The belts protect Earth from blasts of dangerous x-rays and ultraviolet rays that erupt from the Sun.

The Van Allen belts also funnel incoming particles toward Earth's north and south poles. This creates auroras, or brilliant displays of shimmering colored light.

44. Based on the passage information, what does the round object in the center of the diagram represent?

   A. auroras
   B. Earth's magnetic field
   C. radiation
   D. Earth

45. Based on the information in the passage, what phenomenon could scientists not explain until the Van Allen belts were discovered?

   A. Earth's magnetic field
   B. bursts of radiation from the Sun
   C. auroras forming near Earth's poles
   D. x-rays and ultraviolet rays

Answer Key

1. **C.** Because B is dominant over b, the genotypes BB and Bb will each result in a bird with a large beak. The only genotype that will result in a small beak is bb. The Punnett square shows that $\frac{3}{4}$ of the offspring, or 75%, will have large beaks.

2. **B.** RNA plays an intermediate, or "middle" role. RNA is made by DNA, which carries coded information, and is used to make proteins, which are the end products of transcription and translation.

3. **D.** Ants work together to complete tasks. To accomplish such tasks as finding food, communication is essential.

4. **C.** Plants need carbon dioxide, water, and sunlight for photosynthesis. The process yields oxygen and glucose.

5. **C.** According to the passage, the cerebrum controls the ability to think and reason, skills necessary for reading.

6. **B.** The passage states that the brain stem controls the basic body functions that are not under conscious control. Blood pressure, which is related to heart rate, is not under conscious control.

7. **D.** Buckyballs can bounce off a silicon surface, much like a soccer ball can bounce off the ground.

8. **A.** The passage describes how buckyballs can trap molecules. The molecules could be those of drugs that the buckyballs carry through the blood.

9. **A.** Because the Taj Mahal is reflected faithfully by the river, the river's surface remains clear and clean. The pollution in the river is not visible on the surface.

10. **D.** Velocity is defined as both speed and direction. The car is traveling 47 miles per hour northward.

11. **C.** The passage states that when scientists began studying the volcano, several minor earthquakes were occurring deep below it.

12. **B.** The passage states that magma began moving under the mountain, and it describes the hot, molten rock from the volcano's eruption. The molten rock is magma that reached the surface.

13. **C.** The passage describes how during respiration, animals perform gas exchange in the reverse direction of photosynthesis. Adding an animal to the diagram and labeling arrows to and from the animal could show this process.

14. **A.** During photosynthesis, plants take in carbon dioxide from the air and release oxygen into the air. The carbon in carbon dioxide is taken into the plant.

15. **B.** In plants, the process of photosynthesis usually occurs in the leaves, where chloroplasts capture the energy in sunlight to start a series of chemical reactions that lead to the production of sugar, or food for the plant, and oxygen, which is released as waste. When energy is required for a chemical reaction to occur, the reaction is endothermic.

16. **D.** HIV and SIM are the names of viruses, not the diseases they cause. The two viruses cause similar diseases in different species: humans and nonhuman primates.

17. **D.** As described in the second paragraph, a pair of chromatids is needed for cell division, in which each new cell receives one of the chromatids.

18. **C.** As described in the first paragraph of the passage, genes reside on chromosomes, and genes control traits. The black and white bars most likely represent genes.

19. **A.** The lemonade is a solution made of sugar—the solute—dissolved in water and lemon juice—the solvent.

20. **D.** Visibility is unrelated to usefulness. Other parts of the electromagnetic spectrum are used in medicine, communications, and industry.

21. **A.** A rainbow forms when sunlight shines after a rainstorm. The colors of the rainbow show that sunlight is made of many colors.

22. **D.** The passage identifies problems associated with staging and controlling fusion reactions, which include containing matter at extreme temperatures. Fusion can be an energy source only if these problems are overcome.

Answer Key

23. **B.** While the opinions and conclusions of scientists are important, they are not the same as evidence, which is direct observation of the natural world. The inability of other scientists to replicate, or reproduce, the results of the original cold fusion experiment is strong evidence that the original claim was false.

24. **C.** Car C is the only one that shows deceleration, as a force is applied in the opposite direction of the car's velocity, thus changing the car's velocity.

25. **A.** The third paragraph states that air is a poor conductor and insulates the cloud and ground charges. This prevents a flow of current until the electrical potential becomes great enough to overcome the resistance of the air.

26. **A.** A large difference in positive and negative charges builds up between the clouds and the ground.

27. **B.** Dew is a form of condensation that occurs when water vapor in the air cools and adheres to a surface, forming liquid droplets. Rain is a form of precipitation.

28. **C.** In a hydrocarbon, carbon bonds with carbon and hydrogen. In the diagram of the hydrocarbon called benzene, each carbon atom has a total of four bonds to other atoms.

29. **B.** As the passage describes, the cell uses DNA to make proteins. The order of the four nucleotides—A, T, G, and C—determines exactly how these proteins are made. In this way, DNA acts like a coded message that the cell is able to interpret.

30. **B.** The karyotype represented in the diagram has 23 pairs of chromosomes, or 46 chromosomes in all. Only humans have 46 chromosomes in their body cells (other than sex cells).

31. **D.** Although the passage does not mention what a rem is, you can infer that it is a unit of measure. Because the passage is about radiation, it is logical to infer that a rem is a measure of a radiation dose.

32. **D.** The passage identifies mutations within blood, bone, and skin cells as examples of somatic mutations. Only the organism itself is affected by the reproduction of these cells. To the contrary, mutations in reproductive cells can be transmitted to future generations.

33. **D.** The energy the animals used in their work came from the plants they ate, such as oats, corn, barley, and hay. Plants use the energy in sunlight for photosynthesis, in which they produce their own food.

34. **B.** Earth depends on the Sun for its energy. If Earth moved farther from the Sun, less of the Sun's energy would reach it. Consequently, Earth's temperatures would fall, and liquid water would freeze into ice.

35. **B.** $2\% + 11\% + 3\% = 16\%$

36. **D.** A geyser erupts water that is very hot, which is a clue that Earth's interior has a high temperature.

37. **C.** The table provides just a few examples of radioactive dating results, so few generalizations can be drawn from the table alone. The passage implies that radioactive dating techniques improved over time, as scientists gained more knowledge of atomic behavior. Therefore, we can conclude that this trend will continue, and techniques will become even more refined as more knowledge is gained of how radioactive elements decay in natural materials.

38. **A.** Rubbing your feet across a rug will remove electrons from the rug and add them to your feet. A static charge builds up. When you touch a metal doorknob, the electrons jump from your body to the doorknob. You experience that jump as a small shock.

39. **D.** The last sentence states that socks stick together because opposite electrical charges attract.

Evaluation Chart

40. B. All living things have similar requirements. Whether an organism is composed of only one cell or many, it carries out the same functions of life.

41. Long before the death rate of cliff swallows decreased, genetic variations, or differences, existed in the population. There were birds with wings that were slightly shorter than average and birds with wings slightly longer than average. When humans built roads for vehicle travel, they introduced a change to the environment. Many cliff swallows that flew in the path of cars couldn't turn away from traffic fast enough to avoid being struck and killed. However, the birds with slightly shorter than average wings could change direction more swiftly than their longer-winged relatives. Consequently, they lived longer. They reproduced, passing the trait for shorter wings to offspring. In time, as the trait continued to be advantageous, it spread through the population, until more birds had shorter wings.

Genetic variations are differences in the genetic makeup of individuals within a population of the same species. These variations always exist and make it possible for organisms to adapt to changes in their environments.

42. Genetic variation is the foundation of natural selection. Genetic differences must exist within a population because environments are always changing. When a change occurs, organisms that happen to have genetic differences that enable them to adapt to the change, survive. Those without the traits they need to adapt, die.

43. D. Each strand serves as a template, or model, for a new strand. When DNA replication is complete, each of the original strands is in one of the two new DNA molecules.

44. D. The round object or sphere between the two sets of Van Allen belts is Earth.

45. C. As described in the second paragraph, the Van Allen belt steers radiation from the Sun toward Earth's poles, thus causing auroras. Without knowledge of the Van Allen belt, auroras could not be explained.

Check Your Understanding

On the following chart, circle the number of any item you answered incorrectly. Next to each group of item numbers, you will see the pages you can study to learn how to answer the items correctly. Review the content and skills in the areas in which you missed half or more of the items.

Unit	Item Number	Review Pages
Life Science	1, 2, 3, 4, 5, 6, 9, 13, 14, 16, 17, 18, 29, 30, 31, 32, 40, 41, 42, 43	13–234
Physical Science	7, 8, 10, 15, 19, 20, 21, 22, 23, 24, 28, 31, 33, 38, 39	235–354
Earth and Space Science	11, 12, 25, 26, 27, 34, 35, 36, 37, 44, 45	355–435

Answer Key

CHAPTER 1 Human Body and Health

Lesson 1.1

Think About Science, page 16
1. marrow
2. bones

Think About Science, page 21
1. Exercise benefits all of the major systems of the human body and helps lead to a longer, healthier life.
2. People should exercise throughout their lives because exercise increases bone density, size, and strength, even in older adults.
3. Yoga and tai chi help increase flexibility and range of motion.

Vocabulary Review, page 22
1. B.   4. F.
2. E.   5. C.
3. A.   6. D.

Skill Review, page 22
1. C.
2. One muscle, the bicep, pulls the bone of the lower arm up; a second muscle, the tricep pulls the bone of the lower arm down.

Skill Practice, page 23
1. D. The other choices describe the digestive, nervous, and muscular systems.
2. A. The shoulder and hip joints both allow flexible movement in many directions.
3. C. Ligaments are tough strands of tissue that connect bones.
4. B. All muscles contract, and they work in pairs to control movement.
5. A. The flow of blood through arteries is not controlled by skeletal or cardiac muscles and can't be conciously controlled.
6. D. Ends of bones are padded with cartilage, which acts as a shock absorber.

Lesson 1.2

Think About Science, page 24
Answers will vary but should explain that food enters the mouth, is ground into small pieces by teeth and broken down further by saliva. It travels down the esophagus to the stomach where it is broken down further by churning and acids. It then travels to the small intestine where nutrients diffuse into the bloodstream. Undigested food moves into the large intestine where water is removed and solid wastes exit via the anus.

Think About Science, page 27
1. kidneys
2. liver
3. platelets
4. red
5. ventricles
6. lungs

Vocabulary Review, page 28
1. C.
2. B.
3. D.
4. A.

Skill Review, page 29
1. The circulatory system carries waste products, such as carbon dioxide, away from cells and to the lungs of the respiratory system where it is released. Blood carries other waste products through the kidneys where the wastes are removed and excreted in urine.
2. The digestive system breaks down food into particles that can be used by cells. Nutrients are absorbed into the bloodstream in the small intestine and carried to cells throughout the body.

Skill Practice, page 29
1. B. Inside the lungs, oxygen and carbon dioxide are the two gases that travel between the blood and the air.
2. C. Red blood cells carry oxygen from the lungs to cells throughout the body. White blood cells fight off germs, platelets help to form clots to stop bleeding, and plasma is the fluid that helps the blood cells circulate in the body.
3. C. Food is broken down by the digestive system and provides the energy that cells need.

Lesson 1.3

Think About Science, page 33
1. testes
2. ovaries
3. fertilized

Vocabulary Review, page 34
1. menstrual cycle
2. hormones
3. labor
4. fetus

Skill Review, page 35
1. B.
2. B.
3. Sensory nerves from the eyes travel to the brain with information about the pencil. The brain processes this information, then sends instructions through motor nerves through the spinal cord, eventually reaching muscles in the hands, arms, legs, and back. They act to grab the pencil.

Skill Practice, page 35
1. A. Each of the answers involves moving a foot. Jerking a foot from a sharp tack, however, is a reflex action that does not involve the brain, so the path of nerves is shortest.
2. B. Both of these systems act to send messages between body parts. The nervous system acts quickly and uses nerves, while the endocrine system acts relatively slowly and uses chemicals.
3. D. The uterine lining and an unfertilized egg are shed in a process called menstruation during which the muscles of the uterus contract. During a pregnancy, the uterine lining is maintained.
4. C. The placenta is the organ that connects an embryo to the uterus and allows nutrients and oxygen to be passed from the mother to the embryo.

Lesson 1.4

Think About Science, page 39
1. C.
2. E.
3. F.
4. D.
5. A.
6. B.

Think About Science, page 41
1. A balanced diet provides the body with the variety of nutrients it needs to stay healthy.
2. "Empty" calories are calories that provide few nutrients.
3. The different parts of the food guide represent the five groups of food that promote good health: fruits and vegetables, grains, proteins, dairy, and oils.

Think About Science, page 43
1. Over-the-counter drugs can be obtained without a doctor's prescription. Prescription drugs require a prescription from a physician, nurse, or other health care provider, and may only be purchased at a pharmacy.
2. People should consult a health care provider about an illness when the symptoms of an illness do not improve or when the illness seems very serious.
3. Bacteria become resistant to antibiotics over time. New antibiotics must be developed to replace ineffective ones.

Vocabulary Review, page 43
1. prescription
2. well-balanced
3. over-the-counter
4. symptoms
5. immunity
6. calorie

Answer Key

(Lesson 1.4, cont.)

Skill Review, pages 44–45

1. Answers will vary. Students should clearly explain which resource provides more information and why one format is easier to understand than another. For example, if they refer to a video, they may point out that visuals make it easier to understand how the drugs are delivered.

2. Answers will vary. Students should explain that nursing students would want information with a greater deal of technical detail. For example, nursing students might need to know which drugs should be administered topically, orally, or intravenously.

3. Nurses need to be flexible and adaptable in their careers because they need to keep up with advancements as new drugs and new delivery methods are developed.

4. **A.** Sunlight includes a broad range of electomagnetic wavelengths. Ultraviolet (UV) radiation is part of sunlight.

5. **A.** Because the passage talks about prevention of skin cancer, the most likely author would be a doctor who studies skin cancer.

6. **C.**

7. **B.**

Skill Practice, page 45

1. **D.** All infectious diseases are caused by germs and can affect adults and children, but only some are sexually transmitted.

2. **A.** Grains, fruits, and vegetables provide a great deal of carbohydrates. Meats, fish, eggs, and cheese provide more fat, protein, and amino acids, which are the building blocks of protein.

3. Answers will vary. Students should detail all the organs and major parts within the system and give evidence to support their statements about keeping the system healthy.

Chapter 1 Review, pages 46–49

1. **D.** Bone breaks from falls are much more common in older adults than in children. Bones tend to weaken and become more brittle with age, which is why older adults often take calcium supplements or increase their intake of milk and dairy products.

2. **D.** Bones provide structure and support to the human body, and allow the body to stand and move.

3. **B.** Fixed joints hold the bones of the skull tightly together. They do not allow movement but do help the skull protect the brain inside it.

4. **D.** Ligaments hold two bones together at a joint. Tearing a ligament may allow the bones to move out of place.

5. **A.** Like an absorbent bath towel, the lining of the small intestine picks up broken-down food particles and carries them into the blood.

6. **A.** Smooth muscles are under involuntary control, meaning you cannot decide when or how to use them. Only skeletal muscles, which are the muscles that move bones, are voluntary.

7. **A.** The kidneys have the job of removing wastes from the blood. When they fail, wastes build up. (Note that healthy kidneys can make urine that is either watery or thick.)

8. **B.** the cardiovascular system

9. **A.** The respiratory system takes in oxygen the air and transfers it to the blood so that it can be delivered to cells throughout the body.

10. **D.** A variety of forms of cancer may strike almost any organ of the body. The early warning signs of cancer include those listed here.

11. **C.** AIDS and other sexually-transmitted diseases (STDs) are typically acquired through sexual contact. Avoiding direct contact with the disease-causing agents is the best way to avoid contracting the diseases. Note that antibiotics are not effective against viruses, such as the virus that causes AIDS.

(Chapter 1 Review, cont.)

12. **C.** Calories are the unit used to measure the energy content of food. Although carbohydrates often provide the bulk of the energy in a food product, other nutrients provide energy as well.

13. **D.** Antibiotics, such as penicillin, help the body fight bacterial infections. These drugs are ineffective against other types of infections and against noninfectious diseases.

14. **A.** muscle coordination and balance

15. **D.** Drinking alcohol while pregnant can harm a fetus.

16. **B.** Characteristics of FAS can occur as a child grows older.

17. **B.** Body systems equipped to deal with alcohol are not fully formed.

18. **B.** The normal human body has two functional kidneys but can function with only one.

19. **D.** As the passage infers, a donor may help up to fifty people by donating a wide variety of organs, which means that the donor no longer needs the organs. While living donors often donate a kidney or part of a liver, these donations would help only a small number of patients.

Application of Science Practices

Interpretation

Examples of nutrients that are critical to more than one system include calcium, vitamin D, and fiber (from complex carbohydrates).

Examples of diseases that affect more than one system of the human body include diabetes, vitamin deficiencies, and cancer. The fact that diseases can affect more than one system indicates that body systems work together.

Answer

Nutrients affect almost every body system. Here are some examples: Carbohydrates are the only fuel that the brain (part of the nervous system) can use. Proteins are needed to build and repair the muscular system. The skeletal system requires calcium and vitamin D.

The nutrients that we must get from our food include some minerals, amino acids, and fatty acids. The human body can make many of the nutrients needed in small amounts, but all nutrients can also be found in the foods we eat.

Evidence

Examples of evidence include the connection between calcium and the health of the skeletal system; the connection between water and the health of the excretory system; and the connection between vitamin A and the nervous system health.

Answer Key

CHAPTER 2 Life Functions and Energy Intake

Lesson 2.1

Think About Science, page 57
1. C.
2. B.
3. D.
4. E.
5. A.

Think About Science, page 58
1. sunlight
2. oxygen
3. carbon dioxide
4. water

Think About Science, page 59
1. carpel
2. genetic information

Vocabulary Review, page 60
1. A.
2. F.
3. D.
4. E.
5. C.
6. B.

Skill Review, page 60
1. The arrows show the movement of energy or materials into the plant, such as sunlight, oxygen, carbon dioxide, and water.
2. Labels or a caption would help explain what the arrows represent.
3. A. Hummingbirds are examples of pollinators. They spread pollen by flying from flower to flower.
4. D. As the passage states, these flowers attract hummingbirds, and hummingbirds visit tube-shaped flowers for their nectar.

Skill Practice, page 61
1. A. During respiration, the lungs take in the oxygen from the atmosphere. Photosynthesis by plants and algae is the source of this oxygen. Although all human activities depend on oxygen, respiration is the activity most directly affected.
2. D. As mentioned in the last paragraph of the passage, the Venus flytrap responds to touch. The passage does not mention the stimuli listed in the other answer choices.

3. A. Plants do not respond to a stimulus as quickly as animals do. Typically, a plant must be observed over the course of a day or longer to study its responses.

Lesson 2.2

Think About Science, page 64
In photosynthesis, water and carbon dioxide are combined in the presence of sunlight to yield sugar (glucose) and oxygen. The opposite occurs in cellular respiration, as the molecular bonds of sugar molecules are broken to create molecules of ATP, energy-rich molecules. The process also produces carbon dioxide and water as waste.

Think About Science, page 65
The purpose of cellular respiration is to break the molecular bonds of food molecules to release energy, which is then stored in molecules of ATP.

Think About Science, page 66
Special bacteria digested molecules of jet fuel to get the energy and nutrients they needed for growth and survival. These molecules were broken down during cellular respiration. Consequently, bacterial cellular respiration eliminated the jet fuel from the water supply.

Vocabulary Review, page 68
1. initiative
2. glycolysis
3. procedure
4. aerobic
5. cellular respiration
6. mitochondria
7. process

Skill Review, page 68
1. A molecule of <u>glucose</u> enters a cell.
2. The energy locked in two molecules of <u>ATP</u> is used to start the process.
3. A six-carbon sugar molecule is split into two <u>three</u>-carbon sugar molecules.
4. The three-carbon sugar molecules are altered to become <u>pyruvate</u>.
5. If <u>oxygen</u> is present, the pyruvate molecules move into the mitochondria.
6. <u>Carbon dioxide</u> and <u>water</u> are given off as waste.

(Lesson 2.2, cont.)

Skill Practice, page 69

1. D.
2. D.
3. The process of photosynthesis:
   carbon dioxide + water + sunlight →
   sugar + oxygen
   The process of cellular respiration:
   sugar + oxygen →
   energy + heat + carbon dioxide + water
4. Glycolysis, which occurs in a cell's cytosol, is the first part of the process of cellular respiration. During the process, the energy from two molecules of ATP are used to break a six-carbon molecule of sugar into two three-carbon molecules of sugar. Chemical changes alter those molecules, changing them into pyruvate. If oxygen is present, the pyruvate enter the mitochondria, where the process of cellular respiration is completed.
5. At the end of glycolysis, less than 25 percent of the energy trapped in the glucose molecule has been stored in molecules of ATP. For the remaining 75 percent to be converted to ATP, more chemical reactions in the mitochondria's cytosol and cristae are necessary. Ultimately, the process yields about 38 molecules of ATP.

Lesson 2.3

Think About Science, page 71
Your muscle cells couldn't get enough oxygen for cellular respiration to occur, so lactic acid fermentation began. The process produces lactate as a waste product. This waste product builds up in your muscle cells, causing a burning sensation. The sensation disappears after your breathing returns to normal and your muscle cells get the oxygen they need to resume cellular respiration.

Think About Science, page 73
Photosynthesis helps plants such as corn to grow. Corn and other plants with high sugar content are taken to an ethanol processing plant, where alcohol fermentation produces ethanol. Gasoline is added to the ethanol to produce a fuel for operating cars and other machinery, producing carbon dioxide as a waste product. Carbon dioxide is a necessary ingredient in photosynthesis.

Vocabulary Review, page 74

1. anaerobic
2. research
3. fermentation
4. accountability
5. productivity

Skill Review, pages 74–75

1. A.
2. B.
3. A.
4. A.
5. You may have observed that Bag 1 inflated the most. Bag 3 may have inflated, too, but Bag 2 probably did not, and Bag 4 definitely did not.
6. Bags inflated because yeast in the bags were breaking down the sugar for energy, but no oxygen was present. Without oxygen, yeast cells use the process of fermentation to get the energy they need. The process produces carbon dioxide as a waste product. This gas causes the bags to inflate.

Skill Practice, pages 76–77

1. Possible answers for Lactic Acid Fermentation: Produces lactic acid, which changes to lactate, a waste product during fermentation; occurs in muscle cells, where lactate builds up, making muscle cells less effective and causing a burning sensation

   Possible answers for Alcohol Fermentation: Occurs in plant and yeast cells; produces an alcohol called ethanol and carbon dioxide as a waste product

   Possible answers for the intersection: Processes break down sugar molecules for energy; occur without the presence of oxygen
2. Fermentation occurs so that cells can break down sugars to get the energy they need.
3. ethanol and carbon dioxide
4. Cooler water slows the fermentation process, causing less gas to form than forms in the same amount of time in warm water.
5. The yeast would use the sugar in the honey as a food source.
6. By following the same practices, scientists are able to reproduce other scientists' work to confirm results.
7. Records of actions and measurements make it possible to analyze results more accurately and repeat investigations to test those results.

Answer Key

Chapter 2 Review, pages 78–83

1. **B.** They anchor the plant and absorb nutrients for growth. Roots grow into the soil, where they can absorb water and nutrients necessary to allow the plant to grow.

2. **C.** The glycolysis process results in the production of ATP molecules, high-energy molecules that cells use to do work.

3. **D.** Mitochondria are called powerhouses because the process of respiration is completed here.

4. **C.** The sticky fluid on the pistil, or female reproductive structure, helps capture pollen grains carried in the wind, in rain, or on the bodies of passing animals.

5. **C.** Respiration is an aerobic process, meaning that oxgen must be present for the process to occur.

6. **A.** The process of cellular respiration breaks down sugar molecules to release energy stored in their chemical bonds.

7. **A.** The presence of enzymes, or proteins, speeds the respiration process.

8. **B.** If insufficient oxygen reaches muscle cells, the process of lactic acid fermentation replaces glycolysis. The process leads to the production of lactate, which can make muscles feel fatigued.

9. **A.** During fermentation, yeast cells change glucose into ethyl alcohol, or ethanol, and carbon dioxide is released as a waste product.

10. **B.** Glycolysis occurs in the presence of oxygen. Fermentation occurs in the absence of oxygen.

11. **A.** Alcohol fermentation occurs in plant and yeast cells, while lactic acid fermentation occurs in animal tissues.

12. **A.** The process of photosynthesis produces glucose, a sugar, and oxygen, which is released as a waste.

13. **B.** Pollen are male reproductive cells. They carry half of a plant's genetic information.

14. **C.** A stomate is an opening in a plant leaf that allows gases to move in and out of leaf cells.

15. **A.** Flowers attract pollinators and contain the reproductive parts of a plant.

   B. Cells in green leaves contain chloroplasts, the sites of food manufacturing in the plant. Inside the chloroplasts, the pigment chlorophyll captures the energy in sunlight that makes photosynthesis possible.

   C. The stem provides structure for a plant. It also has tissues that transport nutrients and water up and down the plant.

   D. The root anchors a plant in the ground. It also absorbs water and nutrients from the soil.

   E. Root hairs are threadlike structures that provide more surface area, allowing the root system to absorb greater quantities of water and nutrients.

16. **A.** Insects like bees serve as pollinators. They are attracted by a flower's colors, markings, or scents. They feed on nectar, and during the process, pollen stick to body parts and hair. When insects travel to another plant of the same kind, pollen falls from their bodies onto the new plant's reproductive structures.

   B. Birds contribute to plant reproduction when pollen attaches to feathers as birds settle on flowers.

   C. Pollen sticks to animal fur as animals walk by. Later, the pollen falls off the animals. If it falls upon new plants, it can fertilize eggs.

   D. Wind carries pollen from plant to plant.

   E. Pollen can be carried across plants in rain drops.

17. A sample answer might describe photosynthesis as the process by which a plant combines carbon dioxide and water in the presence of sunlight to create glucose, a sugar, and oxygen, a waste product. The energy in the original bonds of carbon dioxide and water are stored in the bonds of a sugar molecule. The plant uses the sugar it produces or stores it for later use, when it is broken down during the process of cellular respiration. The respiration process is a chemical reaction in which sugar reacts with oxygen to produce energy in the form of high-energy ATP molecules, or adenosine triphosphate, carbon dioxide, and water. So, photosynthesis creates sugar molecules and cellular respiration breaks them down, making them opposite reactions.

18. A flowchart showing the stages of cellular respiration would begin in the cytosol, or the cell's cytoplasm. There, two molecules of ATP provide the energy necessary to split a molecule of 6-carbon glucose into two molecules of 3-carbon pyruvate. This process is called glycolysis, or the splitting of sugar. Action then moves to the mitochondria, where pyruvate undergoes further chemical changes that include the movement of electrons. These electrons shuttle through the inner folds of a mitochondrion's membrane, where an enzyme contributes to the production of about 38 molecules of ATP.

(Chapter 2 Review, cont.)

19. An argument in favor of bioremediation might suggest that using living organisms that feed naturally on pollutants such as spilled oil serves the environment well because it eliminates the need for chemical or technical intervention. Plus, bacteria used in bioremediation give off carbon dioxide and water as wastes, the same wastes that cells give off during respiration.

20. Fermentation is the process of glycolysis, or breaking down sugar molecules, in the absence of oxygen. Sugar must be added to the water to provide yeast food that they can break down for the purpose of creating energy molecules, a process that releases carbon dioxide, which would inflate the balloon.

Application of Science Practices

Interpretation
1. The water in the jar with the plant changed from purple to pink. There was no color change in the jar with only water.
2. The color change indicates the presence of an acid.

Answer
1. Carbon dioxide and water are produced during plant respiration.
2. Water is neutral, so water would remain purple.
3. If plant respiration produced and alkaline substance, the water would turn from purple to green.

Evidence
If the color change was caused by something in the water, the color change would have happened shortly after the cabbage water was added. The color change took time to occur, indicating that the cause of the change was the production of acid by plant respiration

Answer Key

CHAPTER 3 Ecosystems

Lesson 3.1

Think About Science, page 88
1. biosphere
2. habitat
3. community
4. niche

Think About Science, page 90
1. producers
2. If there were no decomposers, dead material would not be decomposed and necessary nutrients would not be returned to the soil.
3. If the number of grasshoppers decreased, the voles would have to find another food source, or their numbers might decrease due to lack of food. This, in turn, could reduce the number of hawks.

Think About Science, page 92
1. F.
2. A.
3. E.
4. D.
5. C.
6. B.

Vocabulary Review, page 93
1. ecosystem
2. biosphere
3. interact
4. food chain
5. ecosystem
6. biome

Skill Review, page 94
1. D. The passage disproves the idea that the tundra is a desert. Rather, it is home to many living things, although not as many or as varied as elsewhere.
2. C. A community is defined as all the living things in an area, not merely some of them.
3. D. The sharks prey on seals. If people did the same, then the sharks would suffer.
4. D. If sharks prey on seals more aggressively, then the seal population would likely drop as a result.

Skill Practice, page 95
1. B. The passage describes this phenomenon. The other effects are not described.

2. D. The passage mentions that the rabbits have no natural predators in Australia. A predator would control the population of rabbits.
3. A. This is the situation described in the passage.

Lesson 3.2

Think About Science, page 97
Equilibrium doesn't occur in a single moment. Balance is achieved only over time, and even then there will be slight increases and decreases in population numbers.

Think About Science, page 99
The reindeer were isolated on an island, where they relied on lichens for food. They consumed the food faster than lichens could grow back in such a harsh environment, and there was no way for the reindeer to leave the island to find food elsewhere. Consequently, the population fell dramatically as reindeer died of starvation and disease.

A possible graph could look like this:

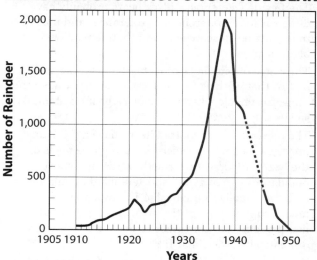

REINDEER POPULATION ON ST. PAUL ISLAND

Vocabulary Review, page 100
1. population
2. habitat
3. limiting factor
4. jargon
5. carrying capacity
6. equilibrium
7. exceed

(Lesson 3.2, cont.)

Skill Review, pages 100–101

1. B.
2. C.
3. Limiting factors can include food, water, space, and shelter.
4. You might ask how the highway construction project will affect: 1) the pond's water level; 2) the pond's water quality; 3) the presence or growth of trees used for food and building lodges; 4) noise levels that may affect beavers' reproductive behavior; and 5) the movement of predator species closer to the pond.
5. Equilibrium is a balance between available resources and wildlife population levels.

   When a population is in equilibrium, numbers move slightly above and below a habitat's carrying capacity.

   Carrying capacity is the highest population that a habitat can support.

   Both factors are influenced by population levels and resources.

Skill Practice, page 101

1. The female Okefenokee black bears probably have different food preferences or tolerances and must move farther in search of food than female Osceola black bears.
2. The range of female Osceola black bears was about half of that of their Okefenokee counterparts, so the range for female Okefenokee black bears is about 61 square kilometers (60.6).
3. The mongoose and spotted lizard have a predator-prey relationship, in which the mongoose feeds on the lizard. At first, the growth of the mongoose population soared as the spotted lizard population declined equally dramatically. The mongoose now occupies 100% of the area, and the spotted lizard population continues to decline but not as dramatically as it did at first.
4. You might ask: 1) Why did the rate of decline for spotted lizard spotted lizards change after about 1890? 2) What factors other than mongoose affect the spotted lizard population?

Lesson 3.3

Think About Science, page 103

The relationship must be beneficial to both organisms.

Think About Science, page 105

Foodborne and waterborne parasites are endoparasites because they live and reproduce within the tissues and organs of infected human and animal hosts.

Think About Science, page 106

The burs stick to the fur of a passing animal, allowing the seed to travel and be dispersed in a new area in which to grow. The animal is not affected, and the bur plant is able to reproduce.

Vocabulary Review, page 107

1. C.
2. E.
3. D.
4. B.
5. A.

Skill Review, pages 107–108

1.

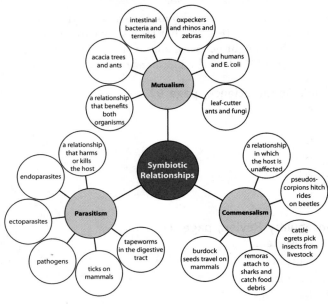

POSSIBLE CONCEPT MAP

2. The barnacle has access to more water and thus more food as it travels on the whale. Barnacles can migrate to different areas by traveling on the whale.
3. You can conclude that barnacles are selective in choosing which type of whale they attach to, depending on the benefit the whale can provide for the barnacles.

Skill Practice, pages 108–109

1. The question could be what different kinds of chemical signals do different whales produce?
2. Barnacles attach mainly to the head or fins of whales because this is where the water flows consistently and therefore, the barnacle have access to more food.

Answer Key

(Skill Practice, pages 108–109, cont.)

3. A whale's territory is vast, and whales are often on the move, making study difficult. In the water, barnacle larvae are very small and are often mixed in with other types of larvae.

4. The fungus and the plant have a mutualistic relationship because the fungus supplies nutrients to the orchid, and fungus eats food the plant makes and stores in its roots. This kind of relationship is similar to the relationship between the oxpecker and rhinos and zebras. The oxpecker feeds on ticks and other parasites or harmful organisms that attach to the skin of rhinos and zebras. As the oxpecker eats, it rids rhinos and zebras of pests.

5. The ants and the fungus have a mutualistic relationship. The ants farm the fungi gardens, and the fungi provide a constant source of food for the ants. Their relationship is mutually beneficial. Another example of a mutually beneficial relationship is the one between acacia tree and ants. The acacia tree attracts animals that strip the leaves from its branches, which slows its growth. However, the acacia tree also grows thorns at the base of its leaves. The thorns provide a home for the ants. The ants bite animals that attempt to eat the acacia. This deters predators, thus allowing the acacia tree to continue growing at a normal rate.

6. The birds help pollinate the plants, giving the plants opportunities to reproduce.

7. The nectar mites and the hummingbirds might have a parasitic or commensal relationship, as the nectar mites take the nectar from the birds, therefore taking the hummingbird's food source.

8. Before you could decide, you would need to determine if nectar mites harm the hummingbirds. If the birds aren't harmed, then the relationship is commensal. If they are harmed, the relationship is parasitic.

Lesson 3.4

Think About Science, page 112
Answers will vary.

Sample answers:
Advantages
Some plants release their seeds.
Burned vegetation adds nutrients to the soil.
Microbial activity in the soil increases.

Scavengers come in search of food.
Seeds are released.

Disadvantages
Water-repellant soil can lead to erosion.
Small animals, insects, and sick or old organisms may die.
Displaced species may not be able to return immediately.

Think About Science, page 113
Humans are constantly moving into new environments. Like other invasive species, they compete with native species for resources, including space. Also, like invasive species, they change the environment, making it difficult or impossible for native species to remain.

Vocabulary Review, page 115
1. threatened
2. abiotic
3. invasive species
4. fragmentation
5. biotic
6. destruction
7. biodiversity
8. degradation
9. endangered

Skill Review, pages 115–116
1. Humans brought rabbits to Australia. Some of the rabbits were released into the wild. They did very well in their new habitat, and they spread quickly. The rabbits were so successful that they ate their way across the continent, leaving places bare of vegetation.
2. Wild European rabbits eat the bark off of trees. They also eat so many seeds that plants cannot reproduce fast enough to meet herbivore demand. At first this degrades habitats but can ultimately destroy them.
3. Poisons introduced into the rabbit population will not simply disappear after they kill the rabbits. They may be spread to predators, decomposers, and other species in the environment, causing unintended consequences.
4. What measures are you taking to ensure that more invasive species are not introduced to Australia? Have you taken steps to make sure that native Australian species are not transported to other locations where they might become invasive?

(Lesson 3.4, cont.)

Skill Practice, pages 116–117

1. A new insect called the glassy-winged sharpshooter has arrived in California, and it carries bacteria that causes disease. Unfortunately, there is no cure for the disease and no effective means for destroying the insect population across the area. One of the insect's preferred foods is grape plants, so check your plants regularly for signs of the insect or disease.

2. Destroy the insects and destroy the diseased plants so that more insects don't ingest the bacteria and spread it to more of the crop.

3. Global economies promote the distribution of goods from one corner of the world to another. This has the unintended consequence of also spreading insects and other pests around the world.

4.

Kind of Loss	Definition	Example
Degradation	A habitat that has become unfit for life because of pollution or invasive species	European rabbits in Australia
Fragmentation	The breaking up of a habitat into smaller pieces that may not be able to sustain native populations	Floods or fires separating previously joined habitats
Destruction	The complete loss or removal of habitat	Human construction of cities

Lesson 3.5

Think About Science, page 121

1. Death rate decreased due to improvements in health care, agriculture, and sanitation; birth rates increased.

2. Nonrenewable resources cannot be replaced in our lifetime; renewable resources can be replaced within an average lifetime.

Think About Science, page 122

1. C.
2. D.
3. B.
4. A.

Think About Science, page 123

1. Climate is the average weather conditions from year to year. Daily weather describes the meteorological conditions for a location at just one point in time.

2. Global climate change is the change being observed in climates around the world, including rising temperatures across Earth.

Vocabulary Review, page 124

1. natural resources
2. pollution
3. speculation
4. conservation

Skill Review, page 124

1. B.
2. D.
3. Yes. Examples include the burning of fossil fuels, which can cause smog, acid rain, and global warming.
4. Acid rain can mix in with the lake water, as could pollutants from farms, factories, or power plants. Airborne pollutants could come from anywhere on Earth.

Skill Practice, page 125

1. A. DDT and HCH are two of the dangerous chemicals that were found in breast milk. The chemicals were used to grow food.

2. A. The goal of the study was to show how the chemicals affected the general public through food consumption, not people who were directly exposed to them.

3. D. The results of the study show that women were acquiring PCB from the ecosystem where they lived.

Chapter 3 Review, pages 126–129

1. B. Ecologists would be most likely to investigate Emerald ash borers and the way they are changing forests in North America. Ecology is the study of how organisms interact with, or relate to, one another and with the world around them.

2. B. A tundra is similar to a "cold desert" because it has harsh, dry conditions, though it is not actually a desert.

3. C. A habitat has a limited number of resources that organisms in that habitat must share. This number of resources determines a habitat's carrying capacity.

Answer Key

(Chapter 3 Review, cont.)

4. **A.** A bee pollinating a flower is an example of mutualism, because both the bee and the flower benefit from the interaction. The bee obtains food, and the flower has its pollen carried to other flowers.

5. **D.** In commensalism, the host is neither helped or harmed by the presence of the other organisms. An example of a commensal relationship is a remora that attaches to a shark and eats food debris leftover from the shark eating.

6. **B.** Sunlight is an abiotic factor. It is not living and it does not come from living materials. A dead tree is a biotic factor, even though it is no longer alive.

7. **B.** The future of an endangered species depends on whether the factors that have caused its numbers to dwindle still impact the species. Species that are endangered by human development, hunting, or overharvesting may be able to return to larger numbers if humans take care to protect the species so that it does not go extinct.

8. **B.** The Minamata tragedy was caused by the mercury poisoning of fish in the ocean. People were injured or died after eating poisoned fish.

9. **C.** People ate the contaminated fish and shellfish and were affected by the poisons.

10. **A.** The seagulls were killed by the mercury poisoning in the fish they had eaten.

11. You answer should include the following information: A desert is characterized by a lack of rainfall. Tundras are extremely cold, and ice and snow can cover the ground for a long time. Grasslands get more rainfall than deserts but not enough to support trees. Tropical forests have mild weather and abundant sunlight and rain. Temperate forests have distinct seasons in which temperature and rainfall vary seasonally. Oceans are aquatic biomes in which most of the plants and animals live near the surface. Freshwater areas include lakes, ponds, streams, and rivers.

12. A predator-prey relationship is an example of competition. A predator hunts for, captures, and eats all or part of its prey. A parasite-host relationship is an example of symbiosis. Like a predator, a parasite usually harms its host. Unlike a predator, a parasite does not usually bring about the death of its host.

13. A limiting factor is any factor that limits the growth, the numbers of, or the distribution of organisms within an ecosystem. Limiting factors can include food, water, shelter, and space. Limiting factors reduce a habitat's carrying capacity if there are not enough resources for organisms within a population to survive.

14. Your sketch should include the major characteristics of the biome you chose. Biotic factors should include the animals and plants present; don't forget to include once-living materials, too, in your biotic factors. Abiotic factors should include things like sunlight, precipitation, soil, and rocks.

15. **Renewable**
    cattle
    crops
    sunlight
    trees
    wind
    Nonrenewable
    coal
    minerals
    natural gas
    oil
    topsoil

Application of Science Practices

Critical Thinking

1. The effect on the ecosystem depends on the living component that you removed. If you remove a producer, the consumers may starve. If you remove a consumer, the producers may overpopulate to the point that there's not enough food for them. If you remove a detritivore, wastes may build up that contaminate the ecosystem.

2. The effect on the living components depend on the nonliving component that you removed. For example, if you removed dirt than this would leave no place for plants to grow, which would affect those organisms that depend on grass for food. If you removed air than all of the living organisms would die.

Evidence

Describe how your evidence supported your prediction. If your evidence did not support your prediction, explain why.

Answer Key

CHAPTER 4 Foundations of Life

Lesson 4.1

Think About Science, page 135
1. C.
2. D.
3. E.
4. B.
5. A.

Think About Science, page 136
The cell membrane and nucleus are parts of both plant cells and animal cells.

Think About Science, page 137
1. Mitochondria trap the energy from food and release it to the cell.
2. A plant cell has a stiff structure called a cell wall around its outer edge, giving a defined shape. Animal cells do not have a cell wall so they do not have a rigid structure.
3. Chlorophyll is found in the chloroplasts of plant cells. It is only useful in plants that trap energy from the Sun to make food.

Vocabulary Review, page 138
1. function
2. nucleus
3. diffusion
4. cells

Skill Review, pages 138–139
1. C. Titles (A) and (D) are too specific, and title (B) is too general to describe what the passage is about.
2. A. Sentences 2, 4, and 7 give specific details about the shape and function of different kinds of cells.
3. At first, the particles are more concentrated on the outside of the membrane than the inside. They will gradually diffuse from the outside to the inside until the concentrations are equal.

Skill Practice, page 139
1. B. A plant cell's chloroplasts are the site of photosynthesis, or food production. The other answer choices are structures found in both a plant cell and an animal cell.
2. D. The strength of the cell wall helps a plant cell maintain its shape. Releasing energy (A) is accomplished by mitochondria, while trapping energy (B) is accomplished by chloroplasts.

3. D. Proteins are assembled on ribosomes. The endoplasmic reticulum transports and stores substances; the nucleolus makes ribosomes; and vacuoles digest and store food.
4. A. Iodine is being transported from an area with less iodine concentration (the seawater) to an area with greater iodine concentration (the sea plants). This type of transport requires energy, and is called active transport.

Lesson 4.2

Think About Science, page 141
A microbe is an organism that is not visible without the aid of a microscope.

Think About Science, page 142
1. B
2. A
3. P
4. B

Think About Science, page 143
1. because they are not complete cells capable of reproducing on their own
2. tissue, organ, organ system

Think About Science, page 145
1. virus
2. single cell
3. cell wall
4. some
5. multicell
6. yes
7. no
8. no
9. in nucleus

Vocabulary Review, page 146
1. decomposers
2. microbe
3. organ
4. thrive

Answer Key

(Lesson 4.2, cont.)

Skill Review, page 146–147
1. headings and boldface words
2. viruses, immunity, vaccine, inactivated
3. research done to discover a polio vaccine
4. it compares and contrasts viruses and bacteria
5. no; "can be used to make yogurt" appears in the part of the diagram that is labeled "Bacteria," so it is a difference between viruses and bacteria
6. things that are true about viruses and bacteria

Skill Practice, page 147
1. **A.** Vaccinations prepare the body to fight viruses. Antibiotics fight bacteria, not viruses, while doctor visits and hand washing are of limited use in stopping a virus.
2. **A.** The chart lists "red bumps" as a symptom of chicken pox. These red bumps are a type of rash.
3. **B.** Of the diseases listed in the chart, only strep throat is caused by bacteria.

Lesson 4.3

Think About Science, page 149
Vertebrates have a backbone. Invertebrates do not have a backbone.

Think About Science, page 150
Statements 1, 2, and 3 apply to earthworms.

Think About Science, page 152
1. Mollusks include snails, slugs, squids, octopuses, clams, and oysters.
2. Shedding an exoskeleton enables an arthropod to grow.
3. All insects have six legs and three body regions: a head, a thorax, and an abdomen.

Think About Science, page 153
1. egg, larva, pupa, adult
2. In three-stage metamorphosis, the larva and pupa stages are skipped. Instead, the eggs hatch into nymphs.

Vocabulary Review, page 154
1. C.
2. A.
3. D.
4. E.
5. B.

Skill Review, page 154
1. **C.** The passage discusses how sponges take in food without moving around.
2. **A.** The passage included this sentence, but it relates the least to the main idea.
3. Grasshopper eggs are deposited in soil. ⟶ A nymph hatches from the egg. ⟶ The nymph develops into an adult. ⟶ The adult grasshopper continues to grow and molt.

Skill Practice, page 155
1. **A.** All of the specific details mentioned in the passage are examples of how different types of bees do different tasks, and how they cooperate with one another to keep the colony alive.
2. **A.** The worker bees communicate through dance. This helps them all learn where to find food, rather than having to search for it individually.
3. **D.** Earthworms have very small, simple brains that control their movements and other activities.
4. **C.** Molting is the process by which an arthropod sheds its old exoskeleton so that a new, larger one can grow in its place. Because an exoskeleton is not alive and cannot grow, molting allows the arthropod to grow larger.
5. **B.** All insects have three body regions, six legs, a pair of antennae, and a mouth. Some, but not all, have wings.

Lesson 4.4

Think About Science, page 157
An instinctive behavior is automatic and does not have to be taught. A learned behavior is one that is acquired through observing or direct teaching.

Think About Science, page 158
1. Its body temperature rises.
2. Fish populations can shrink due to over-fishing or due to pollution.

Think About Science, page 159
1. egg, tadpole, adult
2. lungs, skin, respiration
3. eggs, water, reproduce

(Lesson 4.4, cont.)

Think About Science, page 161
1. C.
2. D.
3. A.
4. B.

Think About Science, page 162
   All birds are warm-blooded and have feathers, two legs, and two wings.

Think About Science, page 163
1. Mothers feed their young with milk produced in special glands.
2. Mammals have hair.
3. Mammals are highly intelligent.

Vocabulary Review, page 164
1. mammals
2. amphibians
3. instincts
4. respond
5. reflex

Skill Review, page 164
1. reptiles
2. cartilaginous
3. mammals
4. vertebrate
5. amphibians

Skill Practice, page 165
1. B.   This is mentioned in the last paragraph.
2. C.   Koko displays the ability to think. She also displays emotion and the ability to plan. It can be inferred that Koko likes to communicate with Penny.

Chapter 4 Review, pages 166–171
1. B.   It contains all of the materials a cell needs to function.
2. C.   small blocks that snap together
3. C.   Archaea, bacteria, protists, fungi, plants, and animals
4. A.   Unlike an animal cell, a plant cell is surrounded by a cell wall and contains chloroplasts. The structures listed in the other answer choices are common to both plants and animals.

5. D.   A virus is made of genetic material (either DNA or RNA) surrounded by a protein. It is not made of cells. Although viruses cause disease and need hosts to reproduce, these qualities also apply to many living things.
6. A.   Sponges are the simplest of all invertebrates.
7. D.   Active transport requires energy, but diffusion does not.
8. A.   Metamorphosis is a dramatic change in body form. Many insects undergo metamorphosis, as do amphibians.
9. C.   Molting is the process in which an arthropod sheds its exoskeleton, or outer shell-like covering. The exoskeleton is nonliving and must be shed before the arthropod can grow larger.
10. B.   In large plants, roots and stems carry water from the soil to the leaves. Mosses lack roots and stems, so their leaves soak up water directly from the ground. This limits the size of the moss plant.
11. D.   Birds and mammals are warm-blooded.
12. B.   Birds assist in plant reproduction by carrying pollen from plant to plant and also by releasing seeds in their waste.
13. A.   An amphibian's skin is thin, making it easy for the animal to lose body moisture. Therefore, amphibians live in or near moisture.
14. C.   Once an invertebrate attaches itself to a solid object, such as a rock, it is no longer capable of moving. Thus, it captures food from water as the water passes through filter structures.
15. D.   Without a microscope it would have been difficult to examine the bacteria that caused different diseases.
16. C.   As mentioned in the third paragraph, a rabies shot is an example of a vaccination. It protects the dog from acquiring the disease.
17. C.   As mentioned in the fourth paragraph, the process of pasteurization kills microbes in foods.
18. Archaebacteria thrive in harsh conditions. The human digestive system relies on chemicals, including acids, to help break down food. Arhaebacteria are able to survive, even in acidic environments.

(Chapter 4 Review, cont.)

19. Mosses have no specialized cells that allow water and the nutrients in water to be transported through the body. With no internal transportation system, these organisms depend on diffusion, or the flow of water in and out of individual cells. The diffusion process is slow. Consequently, these organisms do not grow to be large.

20. Some protists have characteristics of both plants and animals. They are capable of making their own food, like plants. But when light is unavailable, these organisms capture food, making them more similar to animals.

21. Your diagram should include a cell wall around the organism, with a cell membrane directly inside the wall. A flagellum, used for movement, appears at one end of the organism's cell. A nucleus appears inside the cell. Euglena are unusual in that they contain chlorphyll, which they use to make their own food, but they also have a mouth-like structure for capturing and eating other organisms.

22. You may have said that fungi are fed by releasing a chemical that digests organic material around them. They feed on plants and animals, both dead and alive. Some fungi reproduce by growing a new cell, called a bud, or by releasing reproductive cells called spores. Plants contain chlorophyll and make their own food. They reproduce sexually. Both fungi and plants are rooted in one place, and most are multicellular.

Application of Science Practices

Interpretation
The organelles that appear in plant cells but not animal cells are the cell wall and chloroplasts.

Answer
Be sure to include as many organelles in your models as possible. The purpose of a cell wall is to provide rigid support for a plant cell. The purpose of the cell membrane, found in both the plant cell and the animal cell, is to hold the cell together and serve as a "gate" for the passage of substances. The purpose of the nucleus, found in both plant and animal cells, is the control of all activities in the cell.

Plant and animal cells have some differing organelles because plants have different requirements for function and survival than animals do.

Evidence
Plants, rooted into the ground, do not move as animals do. The cell walls that only plants have help plant stems have a rigid structure that helps them grow and remain upright. Because plants are producers, they require chloroplasts that enable them to create food from other substances.

CHAPTER 5 Heredity

Lesson 5.1

Think About Science, page 177
1. the study of how traits are inherited
2. Possible answer: Why do plants have differences, such as growing tall or short, or growing yellow peas or green peas?

Think About Science, page 179
1. A.
2. D.
3. B.
4. E.
5. C.

Think About Science, page 181
1. thymine
2. 46 chromosomes
3. separating
4. mutation

Vocabulary Review, page 182
1. genetics
2. recessive
3. genes
4. dominant
5. trait
6. chromosome

Skill Review, page 182
1. B.
2. A.
3. Someone who carries an allele for a trait but does not show a form of it.
4. A pedigree traces the history of a trait inherited by a family over several generations.

Skill Practice, page 183
1. B. Plants that have the hybrid combination of alleles (Tt) are tall, not short. Therefore, the trait for tallness (T) is dominant, and the trait for shortness (t) is recessive.
2. D. Because tallness is dominant over shortness, a plant needs only one allele for tallness to grow tall. Its allele combination may be TT or Tt.

3. B. The recessive gene and its harmful consequences are described in the second paragraph of the passage.
4. B. The damage that PKU can cause is described at the end of the second paragraph, while the harmful diet is described in the third paragraph. By placing PKU children on the special diet, the harm caused by the disease can be avoided.

Lesson 5.2

Think About Science, page 187
9: wet earwax, freckled
3: wet earwax, unfreckled
3: dry earwax, freckled
1: dry earwax, unfreckled

Think About Science, page 188
The cross is between $X^C X^c$ and $X^C Y$. Outcomes include: $X^C X^c$; $X^C X^c$; $X^C Y$; and $X^c Y$.

Think About Science, page 189
Possible phenotypes are: a red flower, a pink flower, a white flower.

Think About Science, page 190
The cross is between $I^A I^B$ and $I^O I^O$. From left to right in row 1, the combinations are $I^A I^O$ and $I^B I^O$. From left to right in row 2, the combinations are $I^A I^O$ and $I^B I^O$.] There is a 50 percent chance that the offspring will be Type A and an equal chance that they will be Type B.

Vocabulary Review, page 192
1. Punnett square
2. allele
3. heredity
4. phenotype
5. offspring
6. genotype

Answer Key

(Lesson 5.2, cont.)

Skill Review, page 192

1. The heterozygous carnation shows an example of incomplete dominance, in which neither allele is completely dominant over the other. The result is a combined phenotype.

2.

	R	W
R	RR	RW
W	RW	WW

There is a 25 percent chance that the offspring will be red, a 25 percent chance it will be white, and a 50 percent chance it will be pink.

3.

	R	W
W	RW	WW
W	RW	WW

There is a 50 percent chance that the offspring will be pink and an equal chance that they will be white.

4. Both parents are either IAIo or IBIo.

Skill Practice, page 193

1.

	R	R
W	RW	RW
W	RW	RW

All of the possible allele combinations will be the same— RW. Because these alleles are incompletely dominant, all of the offspring will be pink.

2. Because the condition is recessive and most often affects males, the genetic information appears on the X sex chromosome. Male offspring receive only one X chromosome, from the mother. If the gene resulting in the disorder exists on the X chromosome he inherits, the genotype will be expressed. Because the trait is recessive, a woman can carry the allele for the disorder without expressing it.

3. Answers will vary. Sample answer: You might first determine if the mother has hemophilia, indicating a genotype of X^hX^h. If she does not have the disease, but her father had it, she is a carrier.

   If the mother is not a carrier, all of their children would be free of the disease, but any female children born to the couple would be carriers.

   If the mother is a carrier, there is a 25 percent chance that she will have a male child with the disease.

4. You may have used a table like the one below, a full Punnett square, or a shortened Punnett square, since one parent could contribute only "al" to the offspring's genotype.

	AL	Al	aL	al
al	AaLl	Aall	aaLl	aall

5. There are four possible genotypes: AaLl; Aall; aaLl; and aall. There is an equal chance of each combination occurring in the F1 generation.

6. The offspring would have banded-color short hair; banded-color long hair; one color, short hair; or one color long hair.

Chapter 5 Review, pages 194–195

1. **B.** Capital letters are used to identify dominant traits, while lowercase letters represent recessive traits. An organism that is homozygous dominant for tallness (T) has the genotype TT.

2. **D.** Most cells in the human body have 46 chromosomes.

3. **C.** An allele is a form of a gene. For example, say the trait for freckled skin is controlled by a single pair of genes. Those genes may take different forms, one being for unfreckled skin and one being for freckled skin. These forms of the gene controlling freckling are alleles.

4. **A.** Organisms share most of the same DNA. They differ, however, in the number of chromosomes found in most of their cells.

5. **A.** DNA codes for proteins, which direct a cell's activities.

6. Genes exist on the chromosomes. The set of genes residing on the chromosomes in an organism represents the organism's genotype. The observable expression of those genes is the organism's phenotype.

7. A homozygous pair of alleles is a pair of identical alleles, such as TT. A heterozygous pair of alleles represent different forms of the alleles, such as Tt.

8.

	R	R
r	Rr	Rr
r	Rr	Rr

9.

	WP	Wp	wP	wp
wp	WwPp	Wwpp	wwPp	wwpp

10. In humans, the female has two identical sex chromosomes, XX. The male has one X chrommomosome he inherits from his mother and one Y chromosome he inherits from his father. There are a number of genes residing on the X sex chromosome, including a recessive gene for red-green colorblindness. Because the trait is recessive, a female may carry the gene on one of her X chromosomes, but the gene will not be expressed. However, if that X chromosome is inherited by a male child, the child inherits only this form of the gene, and the gene is expressed. In other words, the male child will have red-green colorblindness.

Application of Science Practices

Interpretation

1. The superhero's phenotype will depend on the traits you chose. Remember that the phenotype can be considered the physical expression of the genotype.

2. Make sure the letters you choose clearly express the genotype and that you represent dominant and recessive alleles correctly.

Answer

Phenotypes can often indicate whether a person has dominant or recessive alleles for a trait. For example, if you see that a person has freckles, you know that he or she carries the dominant allele for that trait.

There are three possible genotypes the child might inherit if both parents have identical genotypes that include one dominant and one recessive allele. Using the Hh in item 2 above, the three possible genotypes (found by using a Punnett square) are HH, Hh, and hh.

Evidence

Use the example in your book to ensure you draw the Punnett square correctly. Remember to use capital letters for dominant alleles and lowercase letters for recessive alleles.

Answer Key

CHAPTER 6 Evolution

Lesson 6.1

Think About Science, page 201
1. Evolution is change over time.
2. The theory of evolution states that older species of living things give rise to newer species over time.

Think About Science, page 203
1. C.
2. B.
3. A.
4. D.

Think About Science, page 204
1. Mutation is a change in the genes that pass from parent to offspring.
2. Some mutations are harmful, but others have no effect or are beneficial. The beneficial mutations can lead to new, useful traits in the population.
3. Disease-causing bacteria gradually develop resistance mutations to every antibiotic to which they are exposed.

Vocabulary Review, page 207
1. fossil
2. adaptation
3. evidence
4. evolution
5. mutation

Skill Review, page 208
1. C.
2. Possible answers: All living things are related by similarities among their DNA and genes. Fossils are evidence of ancient organisms unlike those alive today. Mutations can change species over time.
3. (3) He noted that individuals competed with one another for limited resources, such as food.
4. (5) Mutations are a source of slightly different traits among some individuals of a particular species.

Skill Practice, page 209
1. C. Darwin proposed that living things evolved gradually, from simpler forms to more complex ones.
2. B. Darwin applied his observations on the Galápagos Islands to develop the theory of evolution. The idea that food supplies limit populations, however, came from other scientists.

3. C. In a pattern called punctuated equilibrium, rapid bursts interrupt periods of stability and cause change. This idea was developed after Darwin's time.
4. D. Darwin proposed a linear path of evolution, in which primitive species gave rise to more advanced ones.

Lesson 6.2

Think About Science, page 212
Seeds formed within a female reproductive organ.

Think About Science, page 215
The fewer divergences there are in evolutionary history, the more likely they are to have happened.

Think About Science, page 216
Five splitting events occurred. The first occurred at the root, the others occurred at branches B through F. Terminal taxon A represents the outgroup—the plesiomorph. Taxa C and D are sister taxa, which are more closely related to each other than to other taxa in the ingroup.

Vocabulary Review, page 217
1. H.
2. F.
3. A.
4. G.
5. D.
6. C.
7. I.
8. E.
9. B.

Skill Review, page 218
1. Phylogeny is the study of evolutionary history. Systematics is a method of classifying organisms, both past and present, based on evolutionary relationships. Cladistics is a systematic classification method that reveals evolutionary relationships, so it is a form of phylogenetic systematics.

(Skill Review, page 218, cont.)

2. The tiny hoof at the end of each toe in the horse's distant ancestor is an example of a plesiomorph. A series of genetic changes, or evolution, led to the divergent, or apomorphic, trait seen in modern horses—single-toed animals in which the toe is a hoof.

3. Scientists can use the outgroup taxon to determine which trait in a cladogram is the plesiomorph, or primitive trait, and which are the apomorphs, or divergent traits.

4. Taxa B through E

5. Taxa B and C

6. Three

7. Taxon A

Skill Practice, page 219

1. Cladistics is a method for classifying biodiversity. It focuses on the evolutionary relationships between traits. The method works on several assumptions. One is that all life appeared on Earth only once. Another is that splitting events lead to new species that have genetic differences from their ancestors. The third is that genetic changes in a population, or evolution, are always occurring.

2. A hypothesis is a prediction based on evidence. Cladistics, which examines the evolutionary history of organisms, can reveal new evidence that causes longstanding hypotheses to be rejected.

3. Steps in a flow chart could include: 1. Select a group of organisms. 2. Compare the organisms and list the traits that make them alike and different. 3. Use a stacked Venn diagram to group organisms that are most closely related. 4. Use the stacked Venn diagram to arrange the relationships in a cladogram. Use the plesiomorph, or primitive trait, as the root of the evolutionary tree, with the other traits forming the tree's evolutionary branches.

4. Answers will vary. Sample answer: produces eggs; exothermic or endothermic; feathered; has hair; gives birth to live young; and walks on four legs.

5. A homologous feature is one that is similar in shape and location. It is inherited from a common ancestor.

6. Answers will vary. Sample answer: Given Earth's immense biodiversity, both living and recorded in the fossil record, and given the distribution of life across such a vast planet, a common method of classification and related vocabulary allows scientists everywhere to share and analyze data that help explain the evolutionary connections among all life.

Lesson 6.3

Think About Science, page 221

Lamark and Darwin both believed that organisms adapt to changes in their environment by acquiring traits that help them to survive. Lamark believed organisms change during a lifetime, and only those who are able to adapt can survive. Darwin believed that organisms adapted to their environment over long periods of time and that genetic differences were the reason. Organisms with genetic variations that help them to survive changes in their environments have more offspring that carry the differences. Over time, as the population survives and thrives, these genetic traits become common among the population.

Think About Science, page 223

A geographical barrier is any physical obstacle to interbreeding. Sample examples: For animals as large as bears, mountains and lakes could be geographical barriers. For animals as small as worms, a road could be a geographical barrier.

Think About Science, page 225

If the adaptations were not successful, the animals would not live long enough to reproduce and pass on their genetic information.

Think About Science, page 227

Maggots show preferences for where they find mates and lay eggs. Males return to the fruits they grew up on to find mates. Females prefer to lay eggs in the same fruit they developed in. Consequently, apple maggots mate with other apple maggots and hawthorn maggots mate with other hawthorn maggots. Interbreeding stopped.

Vocabulary Review, page 228

1. speciation
2. gene flow
3. continental drift
4. fossil record
5. lineage
6. incipient species
7. hierarchy
8. natural selection

Answer Key

(Lesson 6.3, cont.)

Skill Review, page 228

1. A domain is the largest possible category of living thing. A species is the narrowest category of living things.

2. Species cannot continue unless they mate, or interbreed. If organisms are separated from a population and are unable to interbreed with the organisms left behind, they eventually become a new species.

3. If the seeds of the plants on the mountain are removed from plants in the original population, interbreeding becomes impossible. The plants become reproductively isolated and eventually become a new species.

4. Some of the transported organisms may not have the traits they need to survive in the new environment. They will die. Those with traits that allow them to adapt will live to adulthood, when they can reproduce and pass their traits to offspring.

Skill Practice, page 229

1. It may be possible to plant sweet vernal grasses that have a tolerance for heavy metals around abandoned mine sites. The plants will absorb the metals, which will leave the soil free of poisons. Afterward, other species of organisms may be able to live in the cleaner soils.

2. Answers will vary.

   A possible sequence of events is:

   1. Apple flies laid their eggs on apples, fruit from trees planted by immigrants.
   2. Apple maggots developed and hatched inside apples.
   3. Apple flies went in search of mates on apples, the same fruit they developed in.
   4. Apple fly females lay eggs on the same fruit they develop in, so they laid eggs in apples.
   5. Eventually, speciation occurred, meaning apple flies became a different species from their relatives, the hawthorn flies.

3. The fossil record shows physical evidence of organisms that lived over different periods of Earth's history. Scientists are able to gather clues about the environment and traits of organisms living in each period of time. By examining changes in fossil evidence over these time periods, they can build pictures of what changes occurred and when they occurred. They can also build lineages, showing where species split to become new species.

4. If a change, such as a natural disaster, a new organism that competes for resources, or an organism that causes disease, happens in an environment, genetic variation affects the consequences. If the level of genetic variation in a population is high, chances are good that some of the individuals in the population will survive and eventually reproduce. The population will continue. But if the level of genetic variation is low, there is a chance that too few or no individuals will have traits that allow them to survive and reproduce. The population will die.

Chapter 6 Review, pages 230–232

1. **D.** Fishes appeared in the fossil record before insects, amphibians, or dinosaurs.

2. **A.** The classification system developed by Carolus Linnaeus enabled scientists to group organisms into meaningful categories for study based on the characteristics shared by the species within each group.

3. **C.** Darwin observed structual similarities and differences among the species he studied. Genes, DNA, and radioactive isotopes were unknown to science at the time when Darwin carried out his research.

4. **B.** The ancestral finches of the Galapagos finches, flew from the South American mainland and landed on the remote islands of the Galapagos archipelago. Today's different finch species have evolved from different populations that were separated from each by a geographic barrier, a characteristic of allopatric speciation.

5. **B.** A cladogram begins with a point of origin at the base then shows how new groups arise from an ancestral group through time.

6. **B.** A *clade* is a group of one ancestor and all of the descendants of that ancestor.

7. **D.** A mutation is defined as a mistake in a gene that is passed from parent to offspring.

8. **C.** A cladogram shows how new species arise from existing species. In a cladogram, a parent gives rise to two new species at a splitting point.

9. *Natural selection* is the process in which organisms best adapted to their environment tend to survive and pass on their genetic characteristics. *Evolution* is a process of change over time. Natural selection is a mechanism by which evolution can occur, as those organisms that are best suited for their environments are most likely to live to reproductive age and thus pass on their characteristics to offspring.

(Chapter 6 Review, cont.)

10. Scientists think that about 200 million years ago Earth's continents were united in a single landmass called Pangaea. After the continents separated, the resulting geographic isolation led to the development of many species unique to one continent.

11. Answers will vary, but student's flow charts should show early variation in giraffe neck lengths, with those animals with longer necks being able to get more food and thus living long enough to reproduce and pass on the long-necked characteristic to offspring.

12. In a cladogram, an *outgroup* is a group that does not belong with the other groups that are being classified. An *ingroup* is one of the groups whose evolutionary origin from a common ancestor is being shown in the cladogram.

13. Both Lamarck and Darwin observed variation among the individuals of a species. However, Lamarck and Darwin drew different conclusions with regard to how this variation came about. Lamarck concluded that an inidividual organism could acquire a characteristic by certain behaviors during its life span (such as a giraffe stretching to reach leaves high on a tree), which could then be passed on to offspring. Darwin, however, concluded that variation among individuals would enable only those individuals who were best suited for their environments to reproduce. In this way, the characteristics that enabled that individual to survive to reproductive age would be passed on to its offspring.

14. Taxonomy makes it easier for scientists to study living things by providing a method for classifying closely related organisms into groups. It also gives each species a unique, two-part scientific name

15. Mosses would be placed closer to the base of the cladogram, below the ferns.

16. A difference in flowering times could lead to two different plant species because the two plant populations are reproductively isolated. Since one population blooms in late spring and the other population blooms in early summer, the difference in timing serves as a barrier to prevent gene exchange. With time, the two populations would become increasingly different and less likely to be able to interbreed, untimately leading to the two populations becoming two different species.

Application of Science Practices

Interpretation
Table 1 shows greater genetic variation.

Answer
Based on the data, a larger population allows for greater genetic drift.

Evidence
Construct similar tables to those created in the simulation, but choose only one color to be represented.

Answer Key

CHAPTER 7 Energy

Lesson 7.1

Think About Science, page 240
1. kinetic
2. potential
3. chemical
4. electrical
5. nucleus

Vocabulary Review, page 242
1. law of conservation of energy
2. expand
3. transformation
4. contract
5. efficient
6. energy

Skill Review, pages 242–243
1. **C.** The invention is a generator that uses the energy of human motion to replace batteries.
2. The caption adds specific information about how much power the generator can produce.
3. He must walk 10 minutes, because a 1-minute walk powers the cell phone for 30 minutes, a 10-minute walk would power it for 300 minutes, or 5 hours.
4. **D.** A leg's up-and-down motion is kinetic energy. The generator changes this energy into electrical energy, which is what powers cell phones and other devices.

Skill Practice, page 243
1. **B.** A rolling ball has kinetic energy, which is the energy of motion. This energy was slowly transferred as heat due to friction with the table.
2. **B.** The atoms and molecules in wood carry chemical energy that is released when the wood is burned. The energy is changed into heat and light.
3. **D.** Iron is a good conductor, so it can be used for the bulk of the pan—that is, the part where food is heated. Wood is a good insulator, so it can be used to make the handle.
4. **C.** Placing the heater at the bottom of the tank allows convection currents to form. The warm water rises from the bottom, and the cool water sinks from the top. In this way, heat is transferred throughout the tank.

Lesson 7.2

Think About Science, page 247
1. gamma rays, x-rays, ultraviolet rays
2. infrared waves, microwaves, radio waves
3. waves
4. shadow
5. refract
6. reflect

Vocabulary Review, page 249
1. electromagnetic spectrum
2. reflects
3. refracts
4. prism
5. ultraviolet
6. frequency

Skill Review, page 250
1. **B., C., and E.**
2. **B.** The main idea of the passage is that exposure to the Sun causes skin cancer, which can be prevented with sunscreen. Choice B discusses skin cancer and would support the main idea of the paragraph.
3. **A.** The frequency of Wave A is higher because its wavelength is lower. Therefore the sound from Wave A has the higher pitch.
4. **D.** In wave diagrams such as this one, the heights of the crests and troughs show the amplitude of the wave. Amplitude indicates loudness.

Skill Practice, page 251
1. **C.** Sound waves can travel through solids, liquids, and gases, but not empty space. Note that other answer choices are true statements about sound waves, but are true for light waves as well.
2. **A.** A prism is a thick, triangle-shaped piece of glass that refracts light. White light separates into colors when it passes through a prism.
3. **C.** A sound wave travels as regions of compressions and rarefactions through the air or another medium. The greater the amplitude of the waves, the louder is the volume.

(Lesson 7.2, page 251 cont.)

4. **A.** Both a lightning bolt and the thunder it causes are made at the same time. Because light travels faster than sound, you typically see the bolt before you hear the thunder. Only when the lightning is quite close will you observe both at the same time.

5. As sound waves move through the air, particles are compressed and then expand along the wave. Air particles are pushed tightly together in an area called a compression. Behind that area, is an area of low pressure, called a rarefaction, as the particles are spread out.

Lesson 7.3

Think About Science, page 254
1. B.   4. E.
2. D.   5. C.
3. A.

Vocabulary Review, page 256
1. electromagnet   4. generator
2. resistance   5. electricity
3. circuit   6. magnet

Skill Review, page 257
1. **A.** electrosurgery
   B. electrothermal
   C. electrostatic
   D. electroacoustics
   E. electromechanical
2. **B.** In a parallel circuit, current flows in two paths. If one path is broken, current will still flow through the other path.

Skill Practice, page 257
1. **C.** The conductor of an electrical cord is the metal wire that carries electricity. It is wrapped by an insulator, usually made of plastic. The insulator keeps the electricity inside the cord and prevents overheating.
2. **C.** Static electricity is a separation of charges. It often forms from rubbing two objects together, causing electrons to jump from one object to the other.
3. **A.** A generator changes mechanical energy, specifically the energy of its moving parts, into electrical energy. An electric motor changes energy in the opposite direction, from electric energy into mechanical energy.

4. **B.** All magnets have a north pole at one end and a south pole at the other end. When a magnet is cut, the two pieces keep their magnetic properties. This means that each piece forms a new pole, typically at the cut edge.

Lesson 7.4

Think About Science, page 259
Since the combination of ethanol and gas would be used as the energy source, less gasoline would be used than if the fuel consisted of only gas. Using less gasoline per gallon makes supplies of gasoline last longer.

Think About Science, page 261
Wind is produced by the sun's uneven heating of Earth's surface. So, wind power is available as long as the sun shines. The same is true for solar power.

Think About Science, page 263
Biomass and fossil fuels are made of organic matter. Biomass is renewable, but fossil fuels are not.

Vocabulary Review, page 266
1. crowdsourcing
2. nonrenewable
3. nuclear fission
4. magma
5. biomass
6. renewable
7. energy density
8. reservoir

Skill Review, pages 266–267
1. The Law of Conservation of Energy says that although energy can change form, the total amount of energy stays the same. It doesn't disappear and the quantity doesn't change.

2. | **Advantages** | **Disadvantages** |
   |---|---|
   | Cost effective | Quality will vary |
   | Find new talent | Ideas could be stolen |
   | Could create interest | Difficult to manage |

3. The human body is like a car in that it requires energy to do work. Food takes the place of gasoline. Potential energy is stored in the chemical bonds of food molecules. The digestive system breaks down those molecules into simpler molecules, yielding energy that the body can use to do work.

Answer Key

(Skill Review, pages 266–267, cont.)

4.

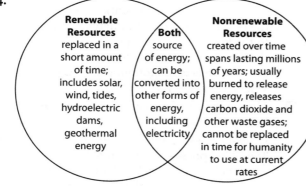

Renewable Resources
replaced in a short amount of time; includes solar, wind, tides, hydroelectric dams, geothermal energy

Both
source of energy; can be converted into other forms of energy, including electricity

Nonrenewable Resources
created over time spans lasting millions of years; usually burned to release energy, releases carbon dioxide and other waste gases; cannot be replaced in time for humanity to use at current rates

Skill Practice, page 267

1.

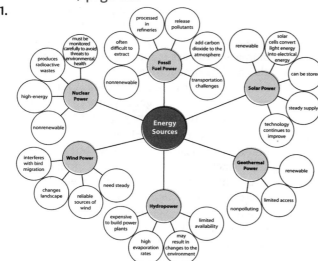

2. Answers will vary. You could suggest that they consider relying on a renewable resource instead. You could explain that biomass, wind power, solar power, and geothermal power are all renewable, making their supplies unlimited.

3. Relying on wind power is only practical where winds are constant throughout the year.

4. Answers will vary. You might support the plan if community leaders followed the strictest possible guidelines for managing the waste. Or you might object to the plan if you feared that radioactive waste would be released into the air, water, and soil, where it might harm living things.

Lesson 7.5

Think About Science, page 270
The chemical reaction is exothermic because it releases energy in the form of light and heat.

Think About Science, page 271
The phase change from ice to liquid water is like an endothermic reaction because heat must be absorbed for the change to occur, and that heat is stored in the liquid water.

Think About Science, page 272
More energy is required for an endothermic chemical reaction to occur.

Think About Science, page 273

Reactant 1 + Reactant 2 + Catalyst ⇒ Product 1 + Product 2

Vocabulary Review, page 274
1. chemical reactions
2. compounds
3. potential energy
4. catalyst
5. activation energy
6. exothermic
7. reactant
8. endothermic
9. product

(Lesson 7.5, cont.)

Skill Review, pages 274–275

1. Both processes represent a chemical reaction.
2. The first chemical reaction releases heat. It is an exothermic reaction. The second chemical reaction absorbs heat. It is an endothermic reaction.
3. An exothermic reaction occurs because the process releases heat.
4. The products of a chemical reaction have different properties than those of the reactants.
5. Manufacturers must provide a catalyst to encourage nitrogen to react with hydrogen.
6. Like a catalyst that causes a chemical reaction in the environment, an enzyme causes a chemical reaction in the body.

Skill Practice, page 275

1. The sentence summarizes the process of photosynthesis.
2. The chemical reaction is endothermic, because the reaction does not occur without the absorption of energy from sunlight. This energy causes the bonds of carbon dioxide and water to break and rearrange themselves to form sugar and oxygen.
3. The energy is stored in the chemical bonds of the product.
4.

Gravitational Potential Energy	Gravitational potential energy is stored in an object's position or structure. When the force of gravity acts upon the object, potential energy changes to kinetic energy.
Elastic Potential Energy	Elastic potential energy is stored in a substance's elastic structure. When pulled, the elastic potential energy is high. When released, the potential energy changes to kinetic energy.
Chemical Potential Energy	Chemical potential energy is stored in a molecule's structure, or the way its atoms are bonded. An attractive force holds the atoms together. During a chemical reaction, the attractive force is broken and rearranged, changing potential energy to kinetic energy.

5. Industries use catalysts to speed the chemical reactions required to make specific products. Faster reactions mean more product produced in less time.

Chapter 7 Review, pages 276–279

1. **A.** Potential energy is the stored form of energy. A bow and arrow has potential energy when the bow is stretched. The energy is converted to the kinetic energy of a moving arrow when the bow is released. The other two options describe objects in motion, which would have kinetic energy.
2. **D.** The amplitude of a sound wave indicates the loudness of the sound. Loudness is independent of pitch, meaning how high or low the sound is.
3. **A.** Two "like" poles—two north poles or two south poles—will repel each other when they are brought close together. To repel means to push away.
4. **A.** Biomass is carbon-based, organic matter derived from all living and once-living organisms. Grass clippings, for example, are biomass. Biomass is renewable. Fossil fuels are also carbon-based, organic matter derived from microscopic organisms that decayed hundreds of millions of years ago. Because fossil fuels take so long to form, they are nonrenewable.
5. **D.** A molecule's structure gives it chemical potential energy. The atoms within a molecule are attracted to each other. This force is a kind of potential energy that is converted to kinetic energy during a chemical reaction.
6. **A** Freon compresses easily, which causes a rise in temperature. The heated gas then releases its heat into the air, cooling and condensing into a liquid that travels to special tubing, where it expands, returning to a gas state and absorbing heat from inside the refrigerator in the process.
7. **C.** The inside of a refrigerator is cool because heat is absorbed by the tubes on its back. As described in the last paragraph, the heat exchange happens when the liquid refrigerant is allowed to evaporate.
8. **C.** To condense is to change from a gas into a liquid. This process happens in the condenser, and involves a release of heat.
9. Combing your hair strips electrons from the hair, leaving the hair with a positive charge. The electrons gather on the comb, giving the comb a negative charge. Opposite charges attract, so when the comb is brought near your hair, your hair moves toward the comb.

Answer Key

(Chapter 7 Review, cont.)

10. Sound waves are mechanical, back-and-forth waves that move through a medium, such as water. Sound waves bounce off hard surfaces, creating echoes. Scientists send sounds, or pings, from a loudspeaker dragged behind a boat. Some of the waves hit solid objects or the seafloor and bounce back as echoes. Microphones beneath the water pick up the echoes. Knowing the speed of sound in water, scientists can calculate the distance to the seafloor, eventually producing a map.

11. In an endothermic reaction, such as photosynthesis, more heat is absorbed than is released. Photosynthesis occurs with an input of energy, which comes from sunlight. This energy breaks the chemical bonds of carbon dioxide and water. The atoms rearrange themselves to form a molecule of sugar, and heat is stored in the molecule's chemical bonds. In an exothermic reaction, more heat is released than is absorbed as the bonds of reactants break and rearrange themselves to make new products.

12. Nuclear power plants use the heat released during nuclear fission to generate electricity. The heat is released when atoms of Uranium-235 split, a process called nuclear fission. There are different kinds of nuclear reactors, but in each case, heat released from a core of U-235 transfers to water, and the water changes to steam. Steam is used to spin turbines that drive generators, producing electricity.

Application of Science Practices

Interpretation
Your answers will depend on the information you collected and the choice you think is the best option. Be sure you can justify your recommendations based on the information you found.

Answer
Answers will vary based on location.

The source of energy you believe has more advantages will depend on the options available in your state.

Evidence
Be sure the evidence you gathered clearly supports your opinion of which source of energy has more advantages.

CHAPTER 8 Work, Motion, and Forces

Lesson 8.1

Think About Science, page 286

1. The sentence describes the airplane's position (above Minneapolis, Minnesota) and its speed (500 miles per hour), but not the direction of its velocity.

2. About 2 hours; 500 miles per hour × 2 hours = 1,000 miles

Vocabulary Review, page 288

1. distance
2. velocity
3. inertia
4. speed
5. motion
6. acceleration

Skill Review, page 288

1. Newton's first law, which states that an object in motion, such as a passenger in a speeding car, tends to continue moving.

2. Newton's third law, which states that an action, such as gases speeding out the bottom of a rocket, causes an equal and opposite reaction, such as the lifting of the rocket.

3. Newton's second law, which states that force, such as the force needed to lift objects, equals the mass times the acceleration.

4. **D.** Friction is the force that slows down motion between objects in contact with each other, such as a hockey puck and the ice beneath it.

Skill Practice, page 289

1. **B.** Ice is slippery because it provides little friction, which is the force that opposes motion. The ice will not resist your feet pushing against it, so you can slip or fall easily.

2. **A.** Acceleration is a change in velocity. It can be a decrease or an increase in speed. As it slows down and then speeds up again, the truck is experiencing a change in acceleration.

3. **D.** Force is the product of mass and acceleration. Note that the unit of force is kg × m/s^2, which is also called a newton (N).

4. **B.** Velocity is defined as both speed and direction. The truck maintains a constant speed but changes direction. This means that its overall velocity has changed.

Lesson 8.2

Think About Science, page 292

1. compound machine
2. force
3. friction
4. equilibrium
5. simple machine

Think About Science, page 294

1. inclined plane
2. fulcrum
3. wheel and axle
4. direction

Vocabulary Review, page 295

1. E.
2. C.
3. B.
4. F.
5. A.
6. D.

Skill Review, page 296

1. **B.** The ridge of a screw and the circular ramp of the garage have the same shape and work much the same way, but the ridge is much smaller.

2. **A.** All simple machines make work easier to do, often by reducing the amount of force needed to accomplish a task.

3. **D.** The two boxes have different masses, and this accounts for the different forces needed to move them. Note that force is not a property of the boxes.

4. **D.** Newton's second law of motion relates force and mass. It can be illustrated by comparing the forces needed to move objects that have different masses, such as the heavy book and a lightweight book.

Skill Practice, page 297

1. **C.** Speed is a measure of the change in distance over time, such as 5 kilometers per second or 25 kilometers per hour. Velocity, however, is a measure of both speed and direction.

2. **B.** The force applied to the rope is downward, while the force on the object at the other end of the rope is upward. The pulley changes the direction of the force but not its strength.

3. **C.** By combining gears of different sizes and orienting them in different ways, both the strength and direction of an applied force can be changed.

Answer Key

(Skill Practice, page 297, cont.)

4. **A.** For work to be done, a force must be applied across a distance. Only in choice A is this the case.

5. **D.** Friction is created whenever one object moves in contact with another. Even the air is an object, as it is made of matter. Only in deep space can friction be avoided.

6. **D.** The sloped sides of a butter knife act like a wedge, or two inclined planes placed back-to-back.

7. **A.** A lever is a long bar that moves about a fulcrum. A crowbar is an example of a very simple lever.

Chapter 8 Review, pages 298–299

1. **C.** Push against a wall, and the wall will push against you with a force that is equally strong but opposite in direction. This fact is explained by Newton's third law of motion.

2. **B.** When an object is not moving, the forces upon it are balanced. In the other answer choices, the rope's motion is changing, and thus the forces upon it are not balanced.

3. **C.** Using a simple machine makes work easier to do, but it does not reduce the amount of work. In this case, the ramp made lifting a barrel easier because it reduced the amount of force needed to lift it. However, simple machines do not reduce the amount of work needed to complete a task. Both Sonia and Dave did the same amount of work against gravity, just in different ways.

4. Speed is the change in the distance an object travels over time, and it is always measured in a unit of distance, such as a mile, over time, such as an hour. Velocity is an indication of both speed and direction. For example, a vehicle may travel 55 miles per hour in a northwesterly direction.

5. In science, the term *acceleration* indicates any change in velocity. It can mean both an increase in speed and a decrease in speed. So, when a driver steps on her car's brake, she is decreasing her speed, or changing acceleration in the opposite direction of its motion.

6. Newton's Second Law of Motion says that a change in the motion of an object depends on two things—the mass of that object and the amount of force applied to it. A compact car has less mass than an oversized pick-up truck. The driver of the compact car will exert less force to move his or her vehicle than will the driver of the pick-up truck.

7. Newton's Third Law of Motion says that for every action, there is an opposite and equal reaction. Burning gases exit a rocket and push down on a launch pad. There is an equal and opposite reaction that pushes the rocket upward.

8. Inserting wheels beneath each corner of a heavy piece of furniture reduces the amount of surface area pushing down on the floor. There is less friction between the two surfaces. With less friction, the furniture is easier to move.

9. In science, work is done only when a force moves an object over a distance. Pushing a brick wall doesn't move the wall. So, no work is done, despite exerting a force against it.

10. A handheld can opener is a compound machine that combines a wedge, a lever, and a wheel and axle. The handles are a lever. They open and close to grip and open a can. Turning the wheel at the top of the opener holds and turns the can. A wedge slices into the can's lid and moves as the wheel and axle move. Some can openers have a bottle opener at the end of one arm. The bottle opener is a lever that moves a load, or bottle top.

Application of Science Practices

Interpretation
When the balloon was pinched, the car moved.

Answer
The forces acting on the car include the force of gravity that keeps the car on the floor or the table, the force of friction between the car's "wheels" and the surface they are on, and the force of the air being expelled through the straw.

A smaller amount of air in the balloon would probably have led to the car traveling slower and/or covering less distance. A greater amount of air in the balloon would probably have led to the car traveling faster and/or covering more distance.

Evidence
When the balloon is pinched, air is released through the straw that pushes on the air around the car. The surrounding air pushes back with an equal and opposite force, which moves the car forward. This demonstrates Newton's Third Law of Motion.

CHAPTER 9 Chemical Properties

Lesson 9.1

Think About Science, page 306

1. B.
2. C.
3. A.

Think About Science, page 307

1. A physical property can be observed without changing the identity of matter and includes observations such as color, size, and shape.

   Chemical properties describe how a substance reacts with another.

2. An element is a type of matter that cannot be broken down into a simpler substance.

Vocabulary Review, page 310

1. matter
2. element
3. physical property
4. state of matter
5. chemical property

Skill Review, page 310

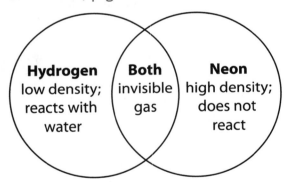

1. A. Aristotle, the Greek philosopher, described the properties of elements with words such as *warm, cool, wet,* and *dry*. The idea that elements have properties is one of the few Grecian ideas about elements that has proven correct.

2. B. X is not an element because it changed into two elements, hydrogen and oxygen, which are the elements that make it up. X most likely is water, although other substances are made of hydrogen and oxygen as well.

Skill Practice, page 311

1. C. The particles in a solid remain in fixed positions, but they do move back and forth—which is like standing in one spot but swaying slightly. Choice A models a liquid, while choice D models a gas.

2. B. How well a substance combusts, or burns, is an example of a chemical property in which the identity of the substance changes. The other answer choices describe physical properties, all of which are measured without changing the identity of the substance.

3. C. In a liquid, the particles stay in loose contact with one another because an attractive force holds them together. By raising the temperature to the boiling point, the particles gain enough energy to escape from these attractive forces and become a gas.

4. D. Boiling point depends on air pressure. With no air around it, liquid water will boil quickly at any temperature. Notice that the boiling point of water is the same for choices A and B, as it does not depend on the weather.

5. A. Changes of state are examples of physical changes. Although the temperature and other properties of water change when ice melts, the substance remains water. The other examples are all chemical changes.

6. B. The properties of salt are very different from those of the two elements—sodium, a metal; and chlorine, a nonmetal—that combine chemically to make salt. This shows that salt forms from a chemical change. The other answer choices are physical changes.

7. C. As the table on page 309 shows, the ten most common elements on Earth include metals and nonmetals, and they have a diverse set of properties. This table does not indicate facts about elements elsewhere in the universe.

8. D. Matter in stars is in the plasma state, and nuclear fusion reactions in stars release a huge amount of energy. Plasmas at high temperatures are not found on Earth, so options A, B, and C are incorrect.

Answer Key

Lesson 9.2

Think About Science, page 314
1. 8
2. ion
3. atomic number
4. electrons

Think About Science, page 315
1. Mendeleyev observed a relationship between the atomic mass of elements and their properties.
2. Elements are arranged in order of increasing atomic number and every element has a unique atomic number.
3. Elements in the same vertical column share many chemical properties.

Vocabulary Review, page 318
1. B.
2. E.
3. C.
4. D.
5. A.
6. F.
7. G.

Skill Review, page 318
Sentences 3, 5, and 7 present factual details in support of the main idea. Each of these sentences describes a fact about a model of the atom that a scientist developed.
1. C. Calcium has an atomic number of 20, so a calcium atom has 20 protons.
2. D. In this version of the periodic table, the atomic mass for an element is listed at the bottom of the box.
3. B. The atomic number for each element is circled. The elements are arranged in the periodic table according to atomic number.

Skill Practice, page 319
1. B. The nucleus is in the center of the atom. Its size is very small, less than 1% of the size of the whole atom.
2. C. All atoms of the same element have the same number of protons, but they may have different numbers of neutrons. These versions of an element are called isotopes. Different isotopes have different atomic mass.
3. B. Mendeleyev arranged the elements according to their atomic mass, not atomic number as done today. By including blanks in the table, he was able to keep elements with similar properties in the same group. The blanks represented elements yet to be discovered.

4. D. The periodic table is organized so that elements within the same column, or group, have similar properties. Iodine is in group VII, the halogens.
5. B. Atomic number determines the identity of an element. Carbon has atomic number 6, which means that every carbon atom has 6 protons in its nucleus. Different isotopes of carbon have different numbers of neutrons, and thus different atomic masses.

Lesson 9.3

Think About Science, page 323
1. covalent bond
2. compounds
3. formula

Vocabulary Review, page 324
1. C.
2. A.
3. D.
4. B.

Skill Review, pages 324–325
1. chemical bonds (As stated in the second sentence, chemical bonds form because atoms seek to become more stable.)
2. free radicals (Free radicals form when a bond breaks in a molecule.)
3. free radical (By "stealing" electrons, a free radical changes another atom or molecule into a free radical.)
4. cell and tissue damage (Free radicals make more free radicals, and this can damage the molecules that make up the cells and tissues of the human body.)
5. antioxidants (These compounds donate electrons and thus help neutralize free radicals in the human body.)
6. C. The word *volatile* is used in the first paragraph to describe sodium. A clue to its meaning is in the next sentence that describes sodium's explosive reaction with water.
7. D. The main idea of the passage is that a compound and the elements that form it have different properties. The elements oxygen and hydrogen have different properties from water, the compound they form.
8. D. In the passage, salt is used as an example of a compound that has very different properties from its component elements, which are sodium and chlorine. Describing the physical properties of salt serves to emphasize this point.

(Lesson 9.3, cont.)

Skill Practice, page 325

1. **D.** The structure shows one atom of carbon and four atoms of chlorine. Thus, the formula for the compound is CCl_4

2. **C.** Only covalent bonds join atoms into a molecule.

3. **B.** The molecular formula shows the names and numbers of each kind of atom in the molecule. However, the 24 atoms of this molecule could be arranged in many ways. A diagram of the molecular structure shows this arrangement.

4. **C.** The fact that argon does not form compounds means that its atoms are stable individually. This stability comes from a full outer shell of electrons.

Lesson 9.4

Think About Science, page 328

Sample Answers:

1. The same number of atoms of each element are present before and after a reaction.

2. In keeping with the law of conservation of matter, the same number of atoms of each element are shown on both sides of the equation.

Think About Science, page 329

1. **B.**
2. **D.**
3. **E.**
4. **A.**
5. **C.**

Vocabulary Review, page 330

1. base
2. law of conservation of matter
3. solution
4. balanced
5. acid

Skill Review, page 331

1., 2., and 3. possible answers
   lifetime; life, time; the whole span of a person's life
   everything; every, thing; all of something
   textbook; text, book; a book containing information
   teamwork; team, work; work done by a team
   timesaving; time, saving; something that saves time

4. the law of conservation of matter; the missing 48 grams must equal the mass of the gases that the log released as it burned

5. The more water that is added, the less concentrated the acid in the juice becomes. The pH of the juice will rise closer to the pH of water, which is 7.0.

Skill Practice, page 331

1. **D.** According to the law of conservation of matter, matter is neither destroyed nor created during a chemical reaction. This means that the same number of atoms will be present before and after the reaction. It does not mean that the matter will not change its state.

2. **C.** All solutions are thorough mixtures of more than one substance. You cannot point to individual pieces inside gasoline, as you could do in a jar filled with screws and nails.

3. **B.** Sodium chloride (NaCl) is an electrolyte, a substance that dissolves into ions when mixed with water. This mixture is a solution, and it conducts electricity very well. Although glucose and oxygen can dissolve in water, neither is an electrolyte.

4. **C.** According to the pH scale, milk is slightly acidic. Adding milk, a weak acid, to the solution would lower its pH.

Lesson 9.5

Think About Science, page 334

1. polymer
2. nitrogen, sulfur
3. lipids

Vocabulary Review, page 335

1. replicate
2. organic chemistry
3. biomolecule
4. hydrocarbons
5. structure

Skill Review, page 336

1. **D.** The passage is organized to present a logical argument. The information about landfills and burning garbage are facts to support the author's conclusion, which is that plastics should be reused and recycled.

2. **A.** All the facts in the passage are in favor of the low-carb diet, which reveals the author's bias. Although some of the other answer choices are correct statements, they do not answer the question.

3. **C.** Presenting both sides of an issue or question is the best way to avoid bias.

Answer Key

(Lesson 9.5, cont.)

Skill Practice, page 337

1. **A.** Carbohydrates are made of carbon in combination with water (H_2O), and thus include the elements carbon, hydrogen, and oxygen in a specific ratio.

2. **B.** Hydrocarbons, by definition, contain carbon and hydrogen only. Carbohydrates contain carbon, hydrogen, and oxygen, and proteins contain these and other elements.

3. **C.** Plastics are made of organic polymers, just as many biomolecules are. By definition, a polymer is made of repeated units. These units must contain carbon because only carbon can form long, stable chains of atoms.

4. **A.** Sulfur is found in proteins, but not carbohydrates and lipids. Thus, foods rich in proteins are the best source of sulfur.

Lesson 9.6

Think About Science, page 341

$C_6H_{12}O_6 + 6O_2 \rightarrow 6CO_2 + 6H_2O$

Think About Science, page 343

A single replacement reaction takes place when two compound elements exchange elements to form two new compounds. In the formula, element Y has replaced element A (in the compound AX) to form a new compound YX and the free element A. The elements A and Y are both positively charged ions.

Vocabulary Review, page 345

1. **A.** when the rate of reactants becoming products matches the rate of products becoming reactants

2. **C.** a reaction in which two or more simple reactants yield a more complex product

3. **G.** to combine fuel and oxygen in a reaction that releases heat

4. **B.** a reaction in which a complex reactant yields two or more simpler products

5. **I.** a reaction in which atoms are exchanged between two molecules or compounds

6. **D.** a reaction that cannot easily be reversed

7. **F.** a reaction in which the products can be transformed back into the reactants that made them

8. **E.** a reaction in which one type of atom replaces a similar type of atom

9. **H.** a number that describes the relative amount of each type of molecule

Skill Review, page 346

1. $2Fe + 3Cl \rightarrow 2FeCl_3$

2. The chlorine displaces, or takes the bromine atom in KBr.

3. For this displacement to occur, chlorine needs to be more active, or reactive, than bromine, the atom it displaces.

4. Acids release hydrogen ions (H+) into water, while bases release hydroxide ions (OH-) into water. A reaction between these ions will yield water: H+ + OH- $\rightarrow$ H_2O, which is neutral. So, when acids and bases react, their ions combine to yield a neutral product—water.

5. The reaction is a combination, or synthesis, reaction because the reactants combine to make a new product, zinc iodide.

6. The Law of Conservation of Mass states that matter is neither created nor destroyed, but it may change form. In the reaction, two solids produce a solid and release a gas of unused iodine. All of the matter is accounted for, although it has taken different forms.

Skill Practice, page 347

1. $2HCl + Mg \rightarrow MgCl_2 + H_2$

2. $2C_6H_5COOH + 15O_2 \rightarrow 14CO_2 + 6H_2O$

3. The stoichiometric coefficients are the numbers before each molecule. In this equation, those numbers are 2, 13, 8, and 10.

   The letters indicate states of matter. The letter *l* stands for liquid. The letter *g* stands for gas.

   The arrow means this is a net forward reaction that isn't reversible.

   It is a combustion reaction because a hydrocarbon (C_4H_{10}) reacts with oxygen to produce carbon dioxide and water.

4. Combination: A + B $\rightarrow$ AB
   Decomposition: AB $\rightarrow$ A + B
   Single replacement: A + CB $\rightarrow$ C + AB
   Double replacement: AE + FB $\rightarrow$ AF + XE

5. Nitrogen gas causes the bag to inflate.

6. The final product is glass.

7. All of the reactions are decomposition reactions because reactants decomposed, or broke apart, to produce new substances.

Chapter 9 Review, pages 348–352

1. B. All matter has two properties: mass and volume.

2. A. Only physical properties can be observed without changing the object. Chemical properties can only be observed during chemical changes.

3. C. The atomic number is the number of protons in the nucleus.

4. D. Chemical compounds are formed when two or more elements are combined chemically.

5. B. Atoms give up or gain electrons in an ionic bond. Atoms share electrons in a covalent bond.

6. C. Matter is neither created nor destroyed in a chemical reaction. The number of reactant atoms will always equal the number of product atoms.

7. A. Organic chemistry is the study of carbon and its compounds. Organic means "of or relating to carbon."

8. D. Carbohydrates are the organic polymers that all living things use for energy. Proteins and lipids are also organic polymers.

9. B. Chemical equations are used to describe what happens in chemical reactions.

10. B. The Bohr model is useful because it explains many events in chemistry. However, because of new discoveries about atomic particles, it is too simple to be accurate.

11. D. The role of muons and other newly discovered atomic particles remains unknown and, therefore, a question for scientists to research. The other answer choices present questions that scientists can answer already.

12. B. The path of motion of an electron is random and varied and in this way is like a moth around a lamp.

13. state P
    flammability C
    color P
    corrosiveness C
    density P
    boiling poin P
    chemical reactivity C
    shape P

14. atomic number of 55 __cesium__
    atomic mass of 10.8 __boron__
    chemical symbol Na __sodium__
    chemical symbol O __oxygen__
    atomic mass of 237.0 __neptunium__
    atomic number of 22 __titanium__

15. solute: drink crystals
    solvent: water
    solution: fruit punch

16. top: covalent
    bottom: ionic

17. The structure on the left is an organic compound, because it contains carbon.

18. $4Fe + 3O_2 \rightarrow 2Fe_2O_3$
    $2Al + 3CuSO_4 \rightarrow Al_2(SO_4)_3 + 3Cu$
    $Ba(CN)_2 + H_2SO_4 \rightarrow BaSO_4 + 2HCN$

19. synthesis reaction
    single replacement reaction
    double replacement reaction

Application of Science Practices

1. The chemical reaction yielded a gas.

2. The chemical reaction stopped when all the reactants that could combine had done so.

Answer

A chemistry experiment can help you solve a chemical reaction by revealing what the products of the reaction are.

____$CO2$____
carbon dioxide

__H_2O__
water

Evidence

The carbon dioxide is a gas, and the water is a liquid.

Answer Key

CHAPTER 10 Earth and Living Things

Lesson 10.1

Think About Science, page 359
Without decomposers to break down dead matter, dead matter would accumulate on Earth's surface. Nutrients would remain stored inside the dead matter, making them unavailable to living things.

Think About Science, page 363
Without nitrogen fixers, nitrogen would remain in forms that plants cannot use. Animals depend on plants or animals that eat plants for food. If plants couldn't use the nitrogen, then they would die. When they died, animals that depend on them would die.

Think About Science, page 365
As the wood weathers, it breaks into smaller and smaller parts. Eventually, decomposers will break down the small parts into the chemical substances stored in the parts, making those substances available to plants.

Vocabulary Review, page 365
1. detritivore
2. nutrient
3. weathering
4. producers
5. biogeochemical cycle
6. algae
7. nitrogen fixers

Skill Review, pages 365–366
1. Energy from the sun and gravity drive all biogeochemical cycles.
2. Detritivores eat waste and dead plants and animals. As they digest these materials, they break them into smaller forms. Waste from detritivores and the detritivores themselves become food for decomposers in the soil. Decomposers are mostly bacteria and fungi, which break complex molecules into simpler molecules that plants can absorb.
3. The carbon cycle depends on the presence of carbon dioxide in the atmosphere. Plants use carbon dioxide during photosynthesis. Plants use the carbon from this process to make compounds that they store in their bodies. When plants die, their bodies decompose, returning carbon compounds to the carbon cycle.
4. Algae live in soil, freshwater, and salt water. The world's population of algae produces billions of tons of oxygen equaling about 80 percent of the oxygen in Earth's atmosphere. Almost all living things depend on oxygen, making them dependent on algae.
5. Although 70 percent of Earth is covered in water, 97 percent of that water is salt water, which living things cannot use. Of the freshwater that is left, most is in the form of ice. That leaves about 1 percent of the world's water available for living things to use, and most of that water is stored underground.

Skill Practice, pages 366–367
1. The Law of Conservation of Matter tells us that matter cannot be created. Nor can it be destroyed, but it can change forms. The biogeochemical cycles show that existing matter changes form, but it is never lost. Nor is new matter ever created.
2. As an earthworm eats the plant and animal waste in the compost, it digests the small bits of matter that it eats. Digestion breaks down the matter even further. When undigested matter leaves an earthworm's body as waste, bacteria in the soil then digest the chemical substances in the waste, releasing these nutrients to the soil. Gardeners then have a natural fertilizer they can use to promote the growth of new plant life.
3. Carbon dioxide traps low-level energy that Earth reflects into space. If there is too little carbon dioxide in the atmosphere, the reflected heat gets away, and Earth cools. If too much carbon dioxide collects in the atmosphere, it stores the extra heat, making Earth warmer.
4. Nonliving nitrogen fixers include lightning and fertilizers. The energy in lightning splits N_2 and nitrogen atoms combine with oxygen in the atmosphere to make new nitrogen compounds. These compounds dissolve in rain and enter the soil.

   Living nitrogen fixers include bacteria. Some live in the soil. Others live in nodes on some plants' roots. These fixers convert N_2 into more complex nitrogen compounds that plants can absorb from the soil.

   Both: Living and nonliving nitrogen fixers create nitrogen compounds that plants can absorb to meet their needs.

(Skill Practice, pages 366–367, cont.)

5. Much of the phosphorus that living things need is stored in sediments, or fine particles of rock. Wind and rain carry sediments away. Sediments settle in the soil or dissolve in rainwater and move through the soil. Plants absorb phosphates from the soil.

6.

Carbon	Nitrogen	Phosphorus
Carbon is stored in plant and animal cells.	Plants use nitrogen compounds to build tissues like leaves and stems. Plants and animals use nitrogen to make DNA, proteins, and other materials. Plants use nitrogen to make chlorophyll.	ATP, or high-energy molecules that cells depend on to do work, contain phosphates. Phosphorus is also found in DNA, the genetic code of living things.

Lesson 10.2

Think About Science, page 369
The process that produced the fossil fuels people depend on today took hundreds of millions of years to occur. Fossil fuels cannot be replaced quickly enough to maintain steady supplies.

Think About Science, page 370
As pressure increases, temperature increases. There is a direct correlation between the two factors.

Think About Science, page 372
In the upper atmosphere, ozone reflects harmful ultraviolet radiation from the Sun. In the lowest layer of the atmosphere surrounding Earth, ozone is harmful. It reacts easily with other molecules, forming toxins that can lead to disease and make it difficult for people to breathe. The toxins also decrease plant growth and weaken plants, including crops and trees. When plants weaken, they become more vulnerable to disease.

Vocabulary Review, page 374
1. petroleum
2. toxins
3. plankton
4. particulates
5. biofuel
6. smog
7. acid rain
8. ozone
9. hydrocarbons

Skill Review, page 374
1. A molecule of methane contains one carbon atom and four hydrogen atoms. Molecules of carbon and hydrogen are called hydrocarbons.
2. Petroleum is transported to a refinery, where it is boiled. Different substances in the petroleum, such as diesel, boil at different temperatures. The vapor is carried away and condensed to form a liquid. The liquid is then sent to manufacturers.
3. Plankton are microscopic organisms that live in water. They have lived on Earth for hundreds of millions of years. As plankton died, their remains settled into the sediments. Bacteria in the soil decomposed the plankton, producing a mixture of hydrocarbon compounds in the soil. As layers of sediments put pressure on the older layers beneath them, temperatures rose. The heat altered the hydrocarbons, forming oil and later, natural gas.
4. If people detect a rotten-egg smell, they usually contact emergency services. Fire departments respond first, and they determine whether there is a natural gas leak and whether people are in danger.
5. Ozone reacts readily with other molecules. Too much ozone in the air we breathe can react with chemicals to produce toxins, or poisons. These poisons can lead to breathing problems, disease, and plant damage.

Skill Practice, page 375
1. As the length of the hydrocarbon increases, the state of the hydrocarbon changes from liquid to sticky tar.
2. Oil—As sediment layers press one upon another, pressure increases. As pressure increases, temperature increases. As temperature increases, the hydrocarbons in the sediments change state, from a liquid to a tar-like solid.

   Natural gas—The process remains the same as for oil, however, as more sediment layers settle on top, pressure increases. As pressure increases, temperatures rise. Additional heat changes substances in the tar-like, solid oil into gas.

Answer Key

(Skill Practice, page 375, cont.)

3. Trees, ferns, and other plants died and fell. Bacteria in the soil began the work of decomposition. Freshwater, salt water, or a mix of fresh and salty water flowed over the sediments. Layers of decaying plant matter buried older layers of decaying plant matter. The sediments squeezed together, increasing temperatures. The temperatures "cooked" the decayed matter, producing coal.

4. Coal provides about one-fourth of the supply of US energy, and its abundance makes it the world's leading fuel. Power plants exist specifically for using coal to generate electricity. As long as there are coal supplies that are easy to reach and facilities build to convert the fuel into electricity, it is economically practical to continue relying on coal as a source of energy.

5. Smog is a cloud of ozone, other compounds, and particulate matter, or bits of unburned material, fumes, dust, and pollen.

Chapter 10 Review, pages 376–377

1. **C.** Bacteria (along with fungi) act as decomposers, which break down the complex molecules in decaying matter into simpler molecules.

2. **D.** Evaporation is the process during which water changes to water vapor.

3. **A.** If all algae disappeared, the amount of oxygen on Earth will decrease. Algae produce most of Earth's oxygen, which they release as a byproduct during photosynthesis.

4. **C.** Decomposers release nitrogen from dead organisms.

5. **B.** Nitrogen in the air exists in a form (elemental nitrogen, N_2) that plants cannot use. Nitrogen-fixing bacteria and other bacteria in the soil change elemental nitrogen into a form that trees and other plants can take in through their roots.

6. **B.** Scientists identify rising levels of carbon dioxide as the main cause of the rising temperatures associated with global climate change. Burning coal and other fossil fuels releases this gas into the atmosphere.

7. **D.** Of these four biogeochemical cycles, only the phosphorus cycle does not include the movement of a chemical through the air.

8. **A.** Sulfur and nitrogen compounds can react with water vapor in the air to produce sulfuric and nitric acids. When these acids dissolve in rain or snow, they fall as a form of precipitation known as acid rain.

9. Answers will vary but should follow one or more of the pathways in the carbon cycle. For example, the carbon atom could become part a sugar molecule made by the corn plant during photosynthesis. Once the corn is eaten by an animal, the sugar would be broken down and used for energy by the animal, with the carbon being released as carbon dioxide gas exhaled by the animal.

10. Benefits: Coal is readily available and easy to transport. Drawbacks: The burning of coal releases carbon dioxide (which traps additional heat in Earth's atmosphere that can cause Earth's temperature to rise) and other chemicals associated with air pollution.

11. When oil is distilled, different hydrocarbons boil off and are collected at different temperatures.

12. Smog is a combination of smoke and fog that also contains solid particulates and ozone gas.

13. Both detritivores and decomposers help recycle waste products and dead organisms. Detritivores eat waste and dead organisms. However, decomposers actually break down the complex molecules in decaying matter into simpler molecules, which then become available in the soil for use by plants.

Application of Science Practices

Interpretation
Answers will depend on the natural disaster and your community's resources. Be sure to provide specific details about your community's plan for dealing with the natural disaster.

Answer
Answers will depend on your community's resources and the natural disaster(s) being considered.

Evidence
Answers will depend on your community's resources and the natural disaster(s) being considered.

CHAPTER 11 Earth

Lesson 11.1

Think About Science, page 383
Both Earth and an apple have a thin, hard, outer layer (the crust or apple skin), a softer inner layer (the mantle or apple flesh), and a central core (inner core or apple core).

Think About Science, page 387
1. C.
2. A.
3. B.

Vocabulary Review, page 388
1. rock cycle
2. sedimentary
3. outer core
4. igneous
5. inner core

Skill Review, pages 388–389
1. C. Metamorphic rocks form when an existing rock is changed by heat and pressure.
2. D. Igneous rocks form from the cooling of magma or lava. In order for a metamorphic rock to change into an igneous rock, the rock would need to melt at high temperatures and then cool.
3. D. Tectonic plates are composed of ocean and continental crust. The edges of the tectonic plates do not always follow the edges of the continents.

Skill Practice, page 389
1. A. Like an apple, the Earth has a relatively thin outer layer, a thick inner layer, and a central core.
2. D. The movement of tectonic plates causes many changes to Earth's crust, including earthquakes, volcanoes, and mountain building.
3. D. Igneous rocks are "fire-made." They form from the cooling and hardening of liquid rock material, such as the lava from a volcanic eruption.
4. C. Metamorphic rock forms when other types of rock are subject to extreme heat and pressure. These extremes are found deep underground, never on the surface.
5. B. First the sediments were squeezed and cemented into flat layers of sedimentary rock. Then forces from the movement of tectonic plates squeezed and folded the layers into a curved shape.
6. C. Cracks in Earth's crust are called faults.

Lesson 11.2

Think About Science, page 392
1. seafloor spreading
2. continental slope
3. mid-ocean ridge

Think About Science, page 393
1. wind
2. cold-water
3. cools
4. deep-water

Vocabulary Review, page 394
1. C.
2. A.
3. D.
4. B.
5. E.

Skill Review, page 395
1. trough: the lowest part of a wave
2. intertidal zone: the area between the high tide line and the low tide line on the shore
3. seafloor spreading: process in which older oceanic crust is pushed away by new ocean crust formed where two plates move apart
4. continental slope: a steep drop in the ocean floor that marks the boundary between continental crust and oceanic crust
5. wavelength: distance between two consecutive crests on a wave
6. A. 3
   B. 5
   C. 1
   D. 4
   E. 2

Skill Practice, page 395
1. B. Phytoplankton perform photosynthesis, the process of using sunlight to make food. No sunlight reaches the abyssal zone.
2. D. The ocean floor is smooth and flat in some regions and mountainous in others, and it has canyons, trenches, and other features similar to those on land.

Answer Key

(Skill Practice, page 395, cont.)

3. **C.** Ocean currents include surface currents, which form because of wind patterns, and deep-water currents, which form due to temperature and density differences in the water. Both types of currents flow throughout the world ocean.

4. **D.** Seafloor spreading has shown that the youngest rocks in the ocean floor are at the mid-ocean ridges. At these points, plates move apart, bringing new material up from the mantle. New crust hardens as lava spills onto the ocean floor.

5. **B.** The abyssal zone of the ocean extends below 2,000 meters. Sunlight does not penetrate into this layer of the ocean. Therefore, organisms of the abyssal zone are adapted to live without sunlight.

Lesson 11.3

Think About Science, page 399
1. thunder
2. tornado
3. hurricanes

Vocabulary Review, page 400
1. air pressure
2. forecast
3. weather
4. greenhouse effect
5. wind
6. meteorology
7. humidity

Skill Review, pages 400–401
1. Underline sentences 1, 2, 4, and 5.
2. **B.** Ozone protects people from skin cancer. Had scientists not discovered the damage to ozone and taken action to stop it, skin cancer would likely have become more common.
3. **C.** People who made and used CFCs and Freon likely did not know that their actions were harming ozone in the atmosphere.

Skill Practice, page 401
1. **A.** Carbon dioxide makes up less than 1 percent of the atmosphere, but as a greenhouse gas it helps keep the surface temperatures warm.
2. **A.** Thunder is the noise that lightning causes. Thunder forms because a lightning bolt heats the air that it travels through.

3. **D.** Hailstones form when strong winds blow upward. The stronger these winds blow, the larger the hailstones can become before they fall to the ground.

4. **C.** Scientists have learned how to predict the paths of hurricanes but not how to prevent or alter them.

Chapter 11 Review, pages 402–405
1. **A.** Earth's first volcanoes erupted frequently and at places across Earth's surface. Today, volcanic eruptions are rarer and occur at specific locations only, such as subduction zones formed by tectonic plates.

2. **C.** The fish fossil most likely formed after it was buried in sediment, and then both the sediment and fish remains turned to rock over time. The extreme temperatures needed to form igneous and metamorphic rock would almost definitely destroy all traces of a fish's remains.

3. **D.** Seafloor spreading begins at the mid-ocean ridge, which is a long chain of underwater volcanoes. The ocean floor spreads outward from the mid-ocean ridge and toward the edges of continents.

4. **D.** Most of the weather on Earth occurs in the troposphere.

5. **A.** Although Earth's rotation can affect the direction of winds, winds always blow from regions of high air pressure to low air pressure.

6. **A.** Energy moves from areas of greater concentration to lesser concentration. Warm water has more energy than cool water, so warm water flows toward cool water.

7. **B.** The world's tallest and longest mountain chain, called the mid-ocean ridge, formed where crustal plates are spreading apart. The ridge lines the floor of the Atlantic Ocean and extends around Africa, into the Indian Ocean, and across the Pacific Ocean.

8. **C.** A scratch test is used to test the hardness of minerals. Minerals are rated from 1 to 10, with 1 being the softest mineral and 10 being the hardest. The soft, powdery mineral called talc has a hardness of 1 on the scale. Diamond has the hardness of 10 on the scale.

9. **D.** Each increase of 1 on the Richter scale corresponds to the power increasing by 30 times. Therefore, an increase of 2 on the Richter scale corresponds to a power increase of 30×30, or 900 times.

(Chapter 11 Review, cont.)

10. **B.** A strong earthquake can cause little damage if it occurs in a remote area. Both strength and location affect the damage an earthquake causes.

11. **C.** A tsunami can strike suddenly, has tremendous power, and can cause great destruction.

12. **B.** The tsunami caused much destruction, and it can be inferred that lives could easily have been lost had people been at or near the water's edge at the time.

13. An early warning system allows people to prepare for tsunamis, but the tsunami strikes within a few hours after it forms. The system can help people escape, but property damage along the ocean shore would remain high.

14. **B.** As the table indicates, snow falls when the temperature in clouds is below 32°F and the temperature on the ground is below 37°F.

15. Due to the change in precipitation from snow to sleet, this tells us that the temperature at the ground has changed between 37°F and 39°F, but the temperature of the clouds has remained the same.

Application of Science Practices

Interpretation
Computations will depend on the data gathered. The fan speed and wind speed should vary proportionally with each other.

Answer
Answers will vary depending on the speed of the fan you are using.

Evidence
You will need to use all average speed measurements—low, medium, and high—in order to predict a speed for medium-low and a speed for medium-high.

Answer Key

CHAPTER 12 The Cosmos

Lesson 12.1

Think About Science, page 411
The Moon formed when a planetary body slammed into Earth about 4.5 billion years ago. During the collision, some of Earth's material flew off into space, where gravity later pulled it together to form the Moon.

Vocabulary Review, page 412
1. mantle
2. nebula
3. habitable

Skill Review, page 413
1. C. While the other listed events did occur, it was study of the collected Moon rocks that led scientists to the theory.
2. H.
3. F.
4. H.
5. F.

Skill Practice, page 413
1. D. Compared to today's Earth, the young Earth was much hotter, lacked an oxygen-rich atmosphere, and there was no life.
2. D. Although the other answer choices are correct statements, only choice D is evidence that the Moon was once part of Earth.
3. A. Both formations involved rocky material pulled together by the force of gravity.
4. B. The evolution of photosynthetic organisms introduced free oxygen into the atmosphere. Eventually the oxygen increased to its current levels.

Lesson 12.2

Think About Science, page 415
1. light-years
2. big bang
3. expanding

Vocabulary Review, page 416
1. light-year
2. nebula
3. assumption
4. supernova

Skill Review, page 417
1. B.
2. B.
3. D.

Skill Practice, page 417
1. A. The elements of a star react with one another throughout the star's life cycle. A young star is made mostly of hydrogen. Then hydrogen changes into helium, which changes into lithium and other elements. The amounts of these elements indicate the star's age.
2. D. Supernovas formed all the elements heavier than iron. Because heavy elements are part of Earth, a planet similar to Earth could only have formed after the first supernovas.

Lesson 12.3

Think About Science, page 419
1. galaxy
2. 200 to 400 billion; 120,000
3. elliptical galaxies; irregular galaxies

Think About Science, page 420
1. gravity
2. eight
3. star

Think About Science, page 421
1. Inner planets are smaller, warmer, and rocky and have few or no moons. Outer planets are gas giants with multiple moons; some have rings.
2. Asteroids are rocky objects that orbit the Sun. Scientists believe they are fragments left over from the formation of the solar system.

Vocabulary Review, page 422
1. C.   4. F.
2. E.   5. B.
3. A.   6. D.

Skill Review, pages 422–423
1. C.   3. C.
2. B.   4. D.

(Lesson 12.3, cont.)

Skill Practice, page 423
1. **A.** A spacecraft could enter the black hole in the middle of the Milky Way, but not leave it. The gravity at the core of a black hole is so strong that not even light or matter can escape from it.
2. **C.** All eight planets orbit the Sun in nearly circular paths. The other answer choices identify properties of one or more planets, but not all of them.
3. **D.** Pluto was once thought of as unique. Then scientists discovered objects much like Pluto, which led them to revise their definition of a planet. They developed new criteria in response to the new discoveries.

Lesson 12.4

Think About Science, page 425
1. rotation
2. revolution
3. northern

Think About Science, page 426
Earth's temperature, atmosphere, and energy from the Sun allow it to sustain life.

Vocabulary Review, page 427
1. interactions
2. revolution
3. rotation
4. tide
5. phases

Skill Review, page 428
1. **B.**
2. **B.**
3. **D.**

Skill Practice, page 429
1. **A.** Because of Earth's tilted axis, seasons in the Southern Hemisphere occur in the opposite pattern from the Northern Hemisphere. When the Northern Hemisphere has summer, the Southern Hemisphere has winter.
2. **B.** A new moon occurs when the Moon is directly between the Earth and the Sun in its orbit. The Moon is not visible from Earth at that point. Tides are caused by the gravitational pull of the Moon.

3. **D.** The Moon lacks water, air, living things, and other agents that change objects on Earth's surface. A flag on the Moon can be inferred to last with little or no change.

Chapter 12 Review, page 430–431
1. **C.** A light-year is a vast distance. It equals the distance that light travels in one year, which is over 9 trillion kilometers. Scientists use light-years to measure distances between objects in the universe.
2. **A.** The big bang was the event that created the Universe. The Universe has been expanding ever since and, according to modern theory, will continue to expand throughout time.
3. **D.** The eight planets are classified into two groups—the inner planets and outer planets—according to their distance from the Sun.
4. **A.** Moon phases are caused by the relative positions of Earth, the Moon, and the Sun. On a given day or night, the same Moon phase can be observed from all regions of Earth.
5. **A.** Earth is made of about 90 chemical elements. These elements formed as nuclei fused to form heavier and heavier elements. Only hot stars, like the one that exploded in a supernova that eventually formed our Sun and solar system, are hot enough to create the elements found on Earth.
6. **C.** The details of the passage all support the main idea, which is that Venus cannot be a home for living things as we know them.
7. **B.** As the passage describes, solar flares release particles and radiation. Particles are a form of matter, and radiation is a form of energy.
8. **B.** The passage identifies sunspots and solar flares as the cause of disruptions to radio transmissions.
9. 1. Billions of years ago, a huge star exploded in a supernova. 2. The debris crashed into a nebula. 3. Extreme temperatures cooked the dust and made droplets. 4. Gravity pulled the droplets together, making lumps. 5. Temperatures in the central lump grew high enough for the lump to self-ignite, or burn. This became the Sun. 6. The remaining lumps grew to form the planets.
10. In the fairy tale, Goldilocks entered a strange home and discovered three bowls of porridge on a table. She tasted each and found that one was too hot, one was too cold, and one was just right for eating. Earth is in the Goldilocks Zone because it has just the right conditions for life to exist.

Answer Key

(Chapter 12 Review, cont.)

11. Unlike the other planets in the solar system, Pluto has a less than spherical shape, and its orbit occasionally puts it in the path of another planet. These factors require a reevaluation of Pluto's status.

12. Edward Hubble found evidence that the universe continues to expand outward in all directions. Even today, evidence supports his conclusion of an expanding universe. Scientists continue to look for evidence, which may also support the conclusion, or may cause it to be revised.

13. A drawing would show Earth tilting toward the Sun during summer in the Northern Hemisphere and tilting away from the Sun during winter in the Northern Hemisphere.

14. **B.** If an unidentified object passed by the planet, the gravitational attraction between the two would affect the planet's motion, perhaps dramatically enough to shift its orbit. An asteroid strike could also affect a planet's orbit.

15. Although the Sun has far greater mass than the Moon, it is also much farther away. The Sun is about 93 million miles from Earth, while the Moon is less than 240,000 miles away. Moon's proximity, or closeness, to Earth, gives it greater influence in forming Earth's tide cycles.

Application of Science Practices

Interpretation
Answers will depend on the data gathered. Be sure to measure the crater diameter for each trial before you determine the average diameter.

Answer
The diameter of the impact crater increases proportionally as the diameter of the meteorite increases.

Evidence
Dropping each meteor a fourth time should provide measurements consistent with the first three trials.

Glossary

A

AIDS (ADZ) acquired immunodeficiency syndrome, a disease where the body is severely unable to have immunity

abiotic (AY bye ah tic) features that are unrelated to living or dead organisms such as air, water, or minerals

abundance (ah BUN dents) a large number or amount of something

abyssal plain (uh BI suhl playn) the wide, flat region of the center ocean floor

abyssal zone (uh BI suhl zone) open ocean below 2,000 meters

acceleration (ak SEL uh ray shuhn) any change in velocity

accountability (uh COUNT uh bill it ee) being held responsible for the outcome of an event

acid (ASS id) a substance that has a sour taste, such as lemon juice or vinegar, and that produces salt and water when added to a base

acid rain (ASS ihd rayn) rain containing dissolved sulfur and nitrogen that has lowered the pH of the water

acquire (uh KWY er) to gain or achieve something

activation energy (ak tih VAY shuhn EN ur jee) the minimum amount of energy needed to start a reaction

active transport (AK tiv TRANSS port) process in which cells move molecules from less crowded areas outside the cell to more crowded areas inside the cell

adapt (uh DAPT) change

adaptation (ad apt TAY shuhn) a change in a species in response to its environment that improves its chances of survival

aerobic (air ROH bic) involving or requiring oxygen

affixes (A ficks es) a word element, such as a prefix or a suffix, added to a base or root word, to modify the word's meaning

air pressure (AIR presh ur) a measure of how strongly the atmosphere pushes against Earth

algae (AL jee) plantlike organisms ranging from single-celled organisms to large seaweeds that carry out photosynthesis

alkaline (AL kuh line) having a pH of more than 7

allele (uh LEE uhl) any of the alternative forms of a gene

amino acids (uh MEE noh ASS idz) the building blocks of proteins

amphibians (am FIB ee uhnz) a group of vertebrates that undergo metamorphosis

amplitude (AM pli tood) height of a wave

anaerobic (an air ROH bic) an organism or a process that requires the absence of oxygen

analyze (AN uh leyez) to pick apart and closely look at the various parts to better understand the whole

anatomy (uh NAT uh mee) the study of the structure of living things

anthracite (AN thruh seyet) the hardest of the four types of coal that is dark black or gray and shiny; once widely used for home heating and cooking

antibiotic (an ti beye OT ik) a chemical that slows the growth of or kills bacteria in the body

antibody (AN tih bah dee) proteins produced by the immune system to kill harmful microbes or prevent them from reproducing

appendage (uh PEN dehj) an external structure such as an arm, leg, or tail

apply (uh PLYE) use for a purpose

arachnid (uh RAK nid) invertebrates, such as spiders, that have eight legs, segmented bodies and an exoskeleton

Archaebacteria (ar kee bak TIHR ee uh) single-celled organisms thought to be the first organisms on Earth

arthropods (AR thruh podz) invertebrates with an exoskeleton and legs that bend

asteroids (ASS tuh roidz) small rocky objects that orbit the Sun mostly between the orbits of Mars and Jupiter, too small to be called planets

assumption (uh SUMP shuhn) beliefs or ideas thought to be true without evidence

atmosphere (AT muhss fihr) the layer of mixed gases that surrounds a planet

atom (AT uhm) the smallest particle of an element that retains the characteristics of that element

atomic mass (uh TOM ik mass) the average mass of atoms of an element; calculated using the relative abundances of the isotopes of the element

atomic number (uh TOM ik NUHM bur) the number of protons in the nucleus of an atom

author's purpose (AW thurz PUR puhss) the reason a work is written

B

balanced (BAL uhnst) an equation where the number of atoms for each element of the reaction is the same on both sides of the equation

base (BAYSS) a substance that has a soapy or slippery feel, such as a detergent, and that produces salt and water when added to an acid

bathyal zone (BA thee uhl zone) the area of the open ocean from the surface to about 2,000 meters down

bias (BEYE uhss) favoring one side over another about a topic based on personal ideas or experiences

biochemistry (beye oh KEM is tree) the study of the atoms and molecules that form cells

biodiversity (beye oh duh VURS it ee) the variety of living things on Earth

biofuel (BEYE oh few uhl) fuel composed of biological matter

biogeochemical cycle (beye oh JEE oh KEM uh kuhl SEYE kuhl) the movement of nutrients through the living and nonliving parts of the ecosytem

biomass (BY oh mass) organic material used as fuel to power generators

biome (BEYE ohm) a large group of ecosystems that have similar climates

biomolecules (beye oh MOL eh kyoolz) organic molecules that organisms make and use

bioremediation (beye oh ree mee dee AY shuhn) a branch of biotechnology that applies biological processes to solve problems in industry, medicine, and the environment

biotechnology (beye oh tek NOL uh jee) the use of microbes to make useful materials for human consumption

biosphere (BEYE uhs fihr) the part of Earth in which life exists

biotic (bye AH tic) living organisms or materials that come from living organisms

bituminous (bit TYOO mihn us) soft, dark brown or black coal widely used to generate electricity

black hole (blak hohl) area in space where the force of gravity is so strong that light cannot escape

bloom (bloom) 1. A single flower or cluster of individual flowers that appears to be one flower 2. The production and opening of a plants flowers

bonding (BON ding) the joining of atoms or molecules to form chemical compounds

botany (BOT uh nee) the study of plants

C

calorie (KAL uh ree) a unit used to measure the energy value of different foods

caption (KAP shuhn) an explanation of a visual

carbohydrates (kar boh HEYE draytss) starches and sugars that are the main energy source for the body

carbon cycle (KAR buhn seye kuhl) the movement of carbon from inorganic to organic compounds and back again

cardiac muscles (KAR dee ak muhss uhlz) involuntary muscles that control the heartbeat

carpel (KAR puhl) the female reproductive structure of a flower consisting of the ovary, stigma, and style

carrying capacity (KAIR ee ing kuh PASS eh tee) the largest number of an organism that a habitat or ecosystem can support

cartilage (KAR tuh lij) a tough, flexible material that covers bones and joints

catalyst (KAT uh list) a substance that starts or increases a reaction without being affected itself

cause (kawz) the reason something happens

cell (sel) the smallest unit of structure in a living organism

cellular respiration (SEL you lar res peh RAY shun) the production of energy by the addition of oxygen and the removal of carbon dioxide from organic compounds in cells

cell membrane (sel MEM brayn) the flexible outer surface of a cell

cell wall (sel wawl) the outermost structure of a plant cell, just outside the cell membrane

cellulose (SEL yoo lohs) a sturdy, insoluble carbohydrate fiber that forms cell walls

cerebellum (sair uh BEL uhm) the part of the brain that controls muscle coordination and balance

cerebrum (suh REE bruhm) the outer portion of the brain that is responsible for thought

chemical bonds (KEM uh kuhl bondz) the force that joins atoms or molecules to form chemical compounds

chemical energy (KEM uh kuhl EN ur jee) energy released when two substances are combined in a reaction

chemical equilibrium (KEM uh kuhl ee kwuh LIB ree uhm) when a reversible reaction reaches a state of balance and every molecule of a product that reverts back to the reactants is replaced with a new molecule of the product

chemical potential energy (KEM uh kuhl puh TEN shuhl EN ur jee) the energy stored within the molecular bonds of a substance

chemical properties (KEM uh kuhl PROP ur teez) those things that describe how a substance reacts with another substance

chemical reaction (KEM uh kuhl ree AK shun) an interaction between two or more chemicals that produce one or more new chemicals

chloroplast (KLOR uh plast) a cell structure that contains chlorophyll

chromosome (KROHM uh sohm) genetic material that determines the traits of an organism

circuit (SUR kit) the path that electrons follow; it must be complete if electricity is to flow clade (klayd) a monophyletic group or a set of organisms related by a common evolutionary ancestor

cladistics (klah DIST iks) a systematic method for making and testing predictions about evolutionary relationships among living things

claim (klaym) a statement or conclusion that has been reached based on an investigation or research

classify (KLASS uh feye) to organize organisms into groups based on their structure, growth, and function

climate (KLEYE mit) the average weather patterns from year to year in one area

collaboration (kul LAB OR a shun) working with others as a team

combust (kom BUHST) to burn or be consumed by flames

comets (KOM itss) small objects made of dust and frozen gas that orbit the Sun

commensalism (kahm MENTS uh liz uhm) a form of symbiosis in which the host is not affected by the foreign organism

communicable disease (kuh MYOON ik uh bull dih ZEEZ) an infectious disease that can be transferred from one person or organism to another

community (kuh MYOO nuh tee) all the organisms living in a certain area

compare (kuhm PAIR) to examine two or more things to see how they are alike

compound (KOM pound) two or more elements combined chemically to form a single substance

compound machine (KOM pound muh SHEEN) a machine that is made of two or more simple machines

compound word (KOM pound wurd) a word that is made up of two smaller words

compression (kom PRESH uhn) squeezing, pushing, or forcing something together in a smaller space

compression wave (kuhm PRESH uhn wayv) a sound wave of high pressure that vibrates in the direction of a line

concept (KON sept) an idea or abstract thought

condensation (kon den SAY shuhn) the process by which water vapor forms droplets of liquid water

conduction (kuhn DUHK shuhn) the method in which heat is transferred by objects that are in contact with each other

conservation (kon sur VAY shuhn) the practice of saving nonrenewable resources

consumers (kuhn SOO murz) animals that eat either producers or other animals

context (KON tekst) the words and sentences surrounding an unknown word

context clues (KON tekst klooz) words, phrases, and sentences that surround an unfamiliar word

continental drift (kon tih NEN tul drihft) the gradual movement of the continents due to convection in Earth's mantle

continental shelf (kon tuh NEN tuhl shelf) a relatively flat region of the ocean floor that extends from the edges of continents

continental slope (kon tuh NEN tuhl slohp) steep drop at the edge of the continental shelf

continuity (kon tuh NOO uh tee) the process by which life continues as it is

contract (kuhn TRAKT) get smaller

contrast (kuhn TRAST) to examine two or more things to see how they are different

convection (kuhn VEK shuhn) the process by which heat moves through air or water

covalent bonding (koh VAY luhnt BOND ing) the bonding of elements by sharing pairs of electrons

convection current (kuhn VEK shuhn KUR uhnt) the circular movement of a fluid such as water or air, which rises as it heats, then sinks back down as it cools

crest (krest) the high point of a wave

cristae (KRISS tee) the folds in the inner membrane of the mitochondria containing the matrix

criteria (kreye TIHR ee uh) standards on which a judgment is based

crust (kruhst) the thin outer layer of Earth

crustal plate (KRUHST uhl playt) part of Earth's outer surface that shifts during an earthquake; also called tectonic plate

crowdsourcing (KROWD sor sing) use of large numbers of people through social media to contribute or collaborate

current (KUR uhnt) a stream of water that moves through the ocean

cycle (SEYE kuhl) a never-ending series of changes that occur in a specific order or pattern

cytosol (SY toh sahl) the liquid part of the cytoplasm

cytoplasm (SEYE tuh plaz uhm) living material in a cell

D

deceleration (dee sel uhr AY shuhn) a negative change in velocity

decomposers (dee kuhm POHZ urz) bacteria, fungi, or protozoa that feed on dead organisms

decomposition reaction (dee komp uh ZISH uhn ree AK shuhn) a reaction that causes a reactant to be broken down into multiple products; $AB \rightarrow A + B$

deduce (dee DUS) reason

deformation (dee for MAY shuhn) a change in the shape or size of a rock

degradation (deh gruh DAY shun) the breaking down or lowering of quality of a habitat or ecosystem due to pollution or invasive species

deoxyribonucleic acid—DNA (dee OK see reye bo noo KLEE ik ASS id) the chemical basis for heredity in humans

destruction (de STRUK shun) to ruin or break apart; to remove necessary parts of a habitat through artificial means, such as cutting trees, draining wetlands, or constructing buildings

detritivore (deh TREYE tih vohr) an organism that eats waste and dead organisms

diagram (DEYE uh gram) a drawing or a graphic illustration that presents information

diffusion (di FYOO zhuhn) movement of molecules in and out of cells without the use of the cell's energy

digest (deye JEST) to break down food

dihybrid (DEYE heye brid) a cross using a Punnett square to predict the possible outcomes for two genetic traits

disruption (dis RUP shuhn) breaks or interruptions of normal events

distance (DIS tuhnss) a measurement of change in position

distill (dis TIHL) the process of boiling and condensing crude oil to separate out the usable parts

distillation (dis tuh LAY shuhn) process by which hydrocarbons are separated

distinguish (diss TING gwish) to identify or recognize as different

diversify (dih VER sih feye) to increase the variation of something

DNA replication (dee en ay re pluh KAY shuhn) the process of copying DNA molecules

dominant (DOM uh nuhnt) tending to overpower the effects of, as in a dominant gene for brown eyes over blue eyes

double replacement reaction (DUH buhl ree PLAYS muhnt ree AK shuhn) atoms of two reactants displace atoms, creating two new product compounds; $AY + XB \rightarrow AB + XY$

drug (druhg) a kind of chemical that affects the body, mind, or behavior

E

earthquake (URTH kwayk) the sudden shaking of the ground caused by the release of pressure and movement of plates within Earth's crust

echo (EK oh) a reflected sound wave

ecological succession (ee kuh LO ji kuhl suhk SESH uhn) a series of changes in the species of a community, typically in response to a disturbance in the environment

ecologist (ee KOL uh jist) a scientist who studies the interactions between biotic and abiotic factors in an ecosystem

ecology (ee KOL uh jee) the study of how organisms interact with one another and the environment around them

ecosystem (EE koh siss tuhm) the system of interacting organisms and their environment

ectoparasite (EK toh PAIR uh seyet) a harmful foreign organism that lives on another organism and feeds off of it

effect (uh FEKT) the result of something that happens

efficient (uh FISH uhnt) having little heat loss during transformation

elastic potential energy (ee LAST ik puh TEN shuhl EN ur jee) the stored energy in an object due to being stretched or compressed

electric current (i LEK trik KUR uhnt) the continuous flow of electric charges in a circular path

electrical energy (i LEK tri kuhl EN ur jee) the movement of electrons within matter

electricity (i lek TRISS uh tee) the energy caused by the flow or separation of charged particles (electrons)

electrolyte (i LEK truh leyet) a liquid that conducts electric currents

electromagnet (i lek troh MAG nit) a device that acts like a magnet when an electric current passes through it

electromagnetic spectrum (i lek troh mag NET ik SPEK truhm) the range of frequencies of waves of electromagnetic radiation including visible light, ultraviolet radiation, and x-rays.

electron (i LEK tron) a negatively charged particle orbiting the nucleus of an atom

element (EL uh muhnt) a substance that cannot be broken down into a simpler form

embryo (EM bree oh) a vertebrate in its early stage of development

endangered (en DANE jurd) a species with so few members that it is at risk of becoming extinct

endangered species (en DAYN jurd SPEE sheez) species that has very few individuals left alive and could die out completely

endoparasite (EN doh PAIR uh seyet) a harmful foreign organism that lives inside another organism and feeds off of it

endoplasmic reticulum (en duh PLAZ mik ri TI kyuh luhm) a cellular structure that leads from the cell membrane to the nucleus

endothermic (EN doh THUR mik ree AK shuhn) the absorption of heat, usually through a chemical reaction

energy (EN ur jee) the ability to do work

energy density (EN ur jee DEN sih tee) the amount of energy per unit of a fuel

engine (EN juhn) a compound machine that has a source of energy other than human muscle

environment (en VEYE ruhn muhnt) the living and nonliving surroundings of an organism

enzymes (EN zeyemz) proteins that bring about specialized chemical reactions

equation (i KWAY zhuhn) an expression that shows the elements that combine in a chemical reaction and the products that are formed

equilibrium (ee kwuh LIB ree uhm) a physical state in which forces and changes are equal and opposite

erosion (i ROH zhuhn) the removal of soil or sediment from one location to another by forces such as water, wind, and gravity

ethanol (ETH uh nahl) ethyl alcohol produced from the fermentation of glucose by yeast

etymology (et iM awl O GE) the study of the origin of words

eubacteria (YOO bak tihr ee uh) a group of complex single-celled organisms; most bacteria are in this group

eukaryote (yoo KAR ee ote) a cell that contains a nucleus; an organism with cells each containing a nucleus and membrane-bound organelles

eukarya (yoo KAR ee ah) all organisms with cells each containing a nucleus and membrane-bound organelles

evaporation (i vap uh RAY shuhn) the process in which liquid water changes into water vapor

evidence (EV uh duhnss) the observations and data from experiments

evolution (ev uh LOO shuhn) the theory that complex forms of life develop from simpler forms

exceed (ek SEED) to surpass or go over a defined amount

excretory system (EK skruh tor ee SIS tum) the group of organs that remove waste from the body

exoskeleton (eks oh SKEL uht uhn) an outer skeleton that is made of nonliving material

exosphere (EK soh sfihr) outermost layer of Earth's atmosphere

exothermic (ek soh THUR mik) the production of heat, usually from a chemical reaction

expand (ek SPAND) increase in size

extinct (ek STINGKT) no longer present on Earth

eye (eye) a low-pressure center of a hurricane characterized by few clouds and calm winds

F

F generation (EFF jehn uhr ay shuhn) filial generation; the offspring whose possible genotypes are produced in a Punnett square

fact (fakt) a conclusion, based on evidence, that scientists agree on

factual details (FAK chuh wuhl DEE tayls) words and phrases that support the main idea and provide accurate information

fat (fat) oily or greasy matter that provides a concentrated dose of energy

fault (fawlt) a fracture in Earth's crust along which movement has occurred.

fermentation (fur men TAY shun) the process of breaking down organic compounds with enzymes in the absence of oxygen

ferns (furnz) a group of simple, nonflowering plants that have true roots, stems, and leaves but no true seeds

fetus (FEE tuhss) a developing human from three months after conception to birth

flammable (FLAM uh buhl) having the ability to ignite and burn

flower (FLOU ur) the reproductive organ of a plant consisting of petals, sepals, carpals, and stamen

folding (FOHLD ing) deformation in rock layers caused by tensional (squeezing) forces

food chain (food chayn) a group of organisms arranged in an order showing how each organism feeds on and obtains energy for the one before it

food web (food web) the many food chains within a community

force (forss) anything that affects the motion of an object

forecast (FOR kast) predict

formula (FOR myuh luh) a way of expressing the number and types of atoms in a chemical compound

fossil (FOSS uhl) the preserved remains of an organism

fossil record (FOSS uhl REK urd) the collected evidence regarding the history of life provided by the study of fossils

fragmentation (frag men TAY shun) breaking apart into smaller, disconnected pieces

frame of reference (fraym uhv REF uh renss) the background against which motion occurs

frequency (FREE kwuhn see) the number of waves that pass a point in a given amount of time

friction (FRIK shuhn) the resistance to motion between two surfaces moving over each other

fulcrum (FOOL kruhm) the point around which a lever turns

function (FUHNGK shuhn) the purpose served by a form; to work

fungi (FUHN jeye) a group of multicelled organisms that are rooted in one place like plants but lack the chlorophyll necessary to produce food

fusion (FYOO zhuhn) a process in which two or more atomic nuclei combine to form a heavier nucleus, which releases energy

G

galaxy (GAL uhk see) a large group of stars and other interstellar matter

gas (gass) a substance that has no definite volume and no definite shape

gears (gihrz) a compound machine that can change the direction of a force and can either increase the force or change its speed

gene (jeen) each small section of DNA; genes determine such characteristics as eye color, the shape of blood cells, and whether the right or left side of the brain is dominant

gene flow (jeen floh) the transfer of genes within a population

generator (JEN uh ray tur) a machine that uses mechanical energy and magnets to produce electricity

genetics (juh NET iks) the study of the inherited characteristics of plants and animals

genetic variation (jen EH tik vayr ee AY shuhn) having different alleles of a gene within a population

genotype (JEE no type) the set of genes that are responsible for a given trait

geologic time (jee uh LO jik teym) divisions of time in relation to events in Earth's geologic history

global climate change (GLOH buhl KLEYE mit chaynj) significant environmental issue involving rising temperatures across Earth; also called *global warming*

glossa (GLOS uh) front-most section of a bee's tongue

glossary (GLOSS uh ree) a list of words and their definitions, usually located at the back of a textbook

glycolysis (gly CALL uh sis) the breaking down or disintegration of carbohydrates or sugars with enzymes

gravitational potential energy (grah vih TAY shuhn uhl puh TEN shuhl EN ur jee) the stored energy in an object resulting from its position or distance from the Earth or other large body

greenhouse effect (GREEN house uh fekt) a condition in which the atmosphere traps the Sun's heat and prevents it from escaping into space

greenhouse gas (GREEN house gass) gas in Earth's atmosphere that absorbs and releases infrared radiation, warming the planet; includes water vapor, CO_2, and ozone

groups (groopss) columns on the periodic table of elements

H

habitable (HAB uh tuh buhl) capable of being lived in

habitat (HAB uh tat) a place where a person or group lives

heading (HED ing) a title

heart attack (HART uh tak) sudden failure of the heart

heredity (hair RED eh tee) the transmission of traits from parents to offspring

heterozygous (het uh roh ZEYE gus) having one dominant and one recessive allele for a trait

hierarchy (HI er ark ee) a system of ranking groups one above another

homologous (hum ALL uh gus) something that is similar in form or function

homozygous (ho mo ZEYE gus) have identical alleles, both dominant or both recessive, for a trait

hormones (HOR mohnss) chemicals produced by endocrine glands; hormones control growth, how the body uses energy, and the ability to reproduce

host (hohst) an organism that another organism lives or feeds on

hot spots (hot spotss) weak spots in Earth's crust through which magma may erupt

Human Genome Project (HYOO muhn JEE nohm proj ekt) research project that identified all of the genes in human DNA; completed in 2003

humidity (hyoo MI duh tee) a measure of the amount of water vapor in the air

humus (HYOO muhss) decayed plant and animal matter; helps the soil hold water and increases fertility

hybrid (HEYE brid) a plant or animal that results from breeding two different purebred strains

hydrocarbons (HEYE droh kar buhnss) compounds, such as fuels, that are made up of hydrogen and carbon

hydrologic cycle (heye droh LAHJ ik SEYE kuhl) the water cycle; the movement of water through living and nonliving parts of the ecosystem

hypothesis (heye POTH uh siss) a reasonable explanation of evidence or a prediction based on evidence

I

igneous (IG nee uhss) a type of rock that forms from melted rock material

immunity (i MYOO nuh tee) protection from disease

implied (ihm PLEYED) an idea that is suggested without being directly stated

inclined plane (IN kleyend playn) a simple machine that consists of a flat surface that has one end higher than the other to magnify force

inert (in URT) condition of atoms that do not form compounds with other elements; they are chemically inactive

inertia (in UR shuh) the tendency of an object to resist a change in its motion

infectious disease (in FEK shus dih ZEEZ) bacteria or viruses that cause illness in the body

ingroup (INN groop) a group of organisms that are classified as being relatively closely related

initiative (in ISH ya tiv) taking steps in achieving a goal

inner core (IN ur kor) the part of Earth that extends about 800 miles to Earth's center

inner planets (IN ur PLAN itss) small, solid, rocky planets closest to the Sun; Mercury, Venus, Earth, and Mars

insipient species (in SIP ee uhnt SPEE sees) a group separated from the larger population that becomes a separate species over time

instinct (IN stingkt) an automatic behavior

insulator (IN suh lay tur) a material that does not conduct electricity very well

insulin (IN suh lin) a hormone created in the pancreas that helps the body regulate glucose in the blood

interact (in ter AKT) to affect one another through action

interaction (in tur AK shuhn) the effect or relationship two or more objects have

interdependence (in tur di PEN duhnss) the dependence of one living thing on another living thing

interpret (in TUR prit) to explain or make sense of

intertidal zone (in tur TEYE duhl zohn) the area between the high tide line and the low tide line on the shore

intravenous (in truh VEE nuhss) delivered directly into the veins

invasive species (in VAY siv SPEE sheez) a harmful organism that has been introduced to a new region where it does not have natural predators to control its population

invertebrates (in VUR tuh britss) animals without backbones

ion (EYE uhn) an atom with either a positive or a negative charge

ionic bond (eye ON ik bond) a bond formed when atoms give up or gain electrons

isotope (EYE suh tohp) an atom of the same element that has different numbers of neutrons

J

jargon (JAR guhn) special or technical vocabulary

K

kinetic energy (ki NET ik EN ur jee) the energy of a moving object

kinesiology (ki NEZ E ol O gee) the study of human movement

L

labor (LAY bur) the contraction and relaxation of the smooth muscles of the uterus resulting in birth

lava (LAH vuh) magma after it reaches Earth's surface

law of conservation of energy (law uhv kon sur VAY shuhn uhv EN ur jee) the idea that the total amount of kinetic and potential energy remains the same

law of conservation of mass (law uhv kon sur VAY shuhn uhv mass) the scientific principal that states that a chemical reaction can rearrange atoms, but cannot change the overall number of each atom

law of conservation of matter (law uhv kon sur VAY shuhn uhv MAT ur) a fundamental scientific principle stating that matter cannot be created or destroyed in a chemical reaction

leaves (leevz) the part of a plant in which glucose is made

lever (LEV ur) a simple machine consisting of a rod or bar that turns around a point to move a load

life science (LEYEF SEYE uhnss) the study of living things

ligaments (LIG uh muhnts) tough strands of tissue that connect bones

light-year (leyet yihr) the distance that light travels in one year

lignite (LIG neyet) soft, brownish coal that is relatively young and made of incompletely decayed plant matter

limiting factor (LIM eh ting FAK tor) a factor or element that restricts the growth, population, or distribution of an organism in an ecosystem

lineage (LIN ee uhj) the line of descent from an ancestor

lipid (LI puhd) fats, oils, waxes, and steroids; made from atoms of carbon, hydrogen, and oxygen

liquid (LIK wid) a substance that has a definite volume but no definite shape

locate (LOH kayt) to find or pinpoint

M

magma (MAG muh) melted rock located within Earth's crust

magnet (MAG nit) an object that attracts iron; creates an external magnetic field

main idea (mayn eye DEE uh) a short statement of what a passage is about

mammals (MAM uhlz) a group of vertebrates that have hair and give birth to live young

mantle (MAN tuhl) a thick layer of very dense rock that lies under Earth's crust

marrow (MAR roh) the jellylike substance consisting of nerves and blood vessels inside human bones

marsupial (mar SOO pee uhl) a type of mammal that nourishes its live-born young in a pouch

mass number (mass NUHM bur) total number of protons and neutrons in the nucleus of an atom

matrix (MAY triks) the fluid substance in the mitochondria that contain enzymes

matter (MAT ur) anything that has mass and occupies space

mechanical energy (muh KAN uh kuhl EN ur jee) energy that comes from motion; it can be either kinetic or potential

medulla (muh DUH luh) the part of the brain that controls involuntary actions such as the heartbeat, breathing, and digestion

melanin (MEL uh nihn) the pigment that determines skin, hair, and eye color

menstrual cycle (MEN stroo uhl seye kuhl) a period resulting in the growth and release of a mature egg

mesosphere (MEZ uh sfihr) layer of the atmosphere that extends from the top of the stratosphere to about 50 miles above Earth

metamorphic (met uh MOR fik) sedimentary or igneous rock that changes due to high pressure and high temperature

metamorphosis (met uh MOR fuh siss) a process of change in an organism from one life stage to another

meteorology (mee tee uhr OL uh jee) the study of Earth's atmosphere

microbes (MYK rohbz) very small organisms such as bacteria, algae, amoebas, or fungi that cannot be seen without a microscope

microbiology (meye kroh beye OL uh jee) the study of microbes

mid-ocean ridge (mid OH shuhn rij) mountain ranges on the ocean floor formed at divergent plate boundaries

mineral (MIN ur uhl) a naturally occurring, inorganic solid with a regular crystalline structure and chemical composition

mitochondria (meye tuh KON dree uh) sausage-shaped structures that trap the energy from food and release it to the cell

model (MOD uhl) a drawing, physical object, or plan that represents the actual thing

molecule (MOL uh kyool) the smallest particle of any material that has the chemical properties of that material

mollusks (MOL uhskss) a group of invertebrates that have an outer shell

molting (MOHLT ing) process in which arthropods shed their exoskeletons to grow

monohybrid (MAHN oh HEYE bred) a cross using a Punnett square to predict the possible outcomes for one genetic trait

monophyletic group (MAHN oh feye LEH tik groop) a cluster of organisms connected by a common evolutionary ancestor

mosses (MAW siss) a group of simple, nonflowering plants that have a two-stage life cycle

motion (MOH shuhn) movement

multimedia (MULT ih MEE dee uh) multiple forms or communication including textbooks, the internet, software, CDs, DVDs, podcasts, television, etc.

mutate (MYOO tayt) to change, as in a gene or a chromosome

mutation (myoo TAY shuhn) a change in the genetic information within a cell

mutual (MYOO choo uhl) directed toward and received by multiple organisms in a relationship equally

mutualism (MYOO choo uh liz uhm) beneficial association between different kinds of organisms

mutual symbiosis (MYOO choo uhl sim bye OH sis) organisms that rely equally on each other to carry out life functions and each benefit from the relationship

N

natural resources (NACH ur uhl REE sorss uhz) those things in the environment, such as air, water, and food, that people need and use to survive

natural selection (NACH ur uhl suh LEK shuhn) the theory that nature acts as a selective force in which the least fit die and the most fit survive

nebula (NEB yuh luh) cloud of gas or dust in space

neritic zone (nuh RI tik zohn) the area from the shore to the point where the ocean floor takes a steep drop

net forward reaction (net FOR werd ree AK shuhn) a chemical reaction that progresses from reactants to products; symbolized by a single arrow →

neutral (NOO truhl) having no electrical charge

neutron (NOO tron) a particle found in the nucleus and that has no electrical charge

niche (nich) a population's role or job in a community

nitrogen cycle (NEYE truh juhn seye kuhl) the movement of nitrogen from the air to the soil, into living things, and back into the air

nitrogen fixers (NEYE troh jen FIKS uhrs) bacteria that convert N_2 into more complex nitrogen compounds that can be used by plants and other organisms

nonrenewable resources (non ri NOO uh buhl REE sors uhz) resources that cannot be easily replaced, such as topsoil, oil, and coal

nuclear energy (NOO klee ur EN ur jee) energy derived from splitting or changing the nucleus of an atom

nuclear fission (NOO kleer FIZZ uhn) the splitting of the nucleus of an atom, releasing large amounts of energy

nucleic acids (noo KLEE ik ASS idss) the substances that control a cell's activities

nucleolus (noo KLEE uh luhss) a dense, round body that makes specialized cell structures called ribosomes

nucleotides (NOO klee uh teyeds) the building blocks of nucleic acids, including DNA; includes adenine (A), guanine (G), cytosine (C), and thymine (T)

nucleus (NOO klee uhss) the control center in a cell; the central core of an atom

nutrient (NOO tree uhnt) a substance necessary for an organism to live, grow, reproduce, and maintain healthy life functions

nutritionist (noo TRI shuh nist) one who studies what foods the body needs to stay healthy

O

offspring (OFF spring) the young created through reproduction

opaque (oh PAYK) light does not pass through; description of a material that absorbs light

opinion (uh PIN yuhn) a personal belief that is often based on a person's own value system

orally (OR uh lee) taking a drug by ingesting by mouth.

ores (orz) naturally occurring minerals that are of value and are mined and processed for use

organs (OR guhnz) a group of tissues working together to perform a complex function

organ system (OR guhn SISS tuhm) a group of organs that work together to perform a life process

organelles (or guhn ELLZ) the "little organs" or specialized structures in a cell that each perform a specific function

organic chemistry (or GAN ik KEM is tree) the study of carbon compounds present in life forms

organism (OR guh niz uhm) any living thing

organization (or guh nuh ZAY shuhn) classifying systems to understand their similarities and differences

oscilloscope (os SIL uh skohp) instrument that produces a picture of a sound wave

outcome (OUT kuhm) result

outer core (OUT ur kor) the layer of Earth that is about 1,400 miles thick and is made up of melted iron and nickel

outgroup (OUT groop) organisms that are not as closely related to the organisms defined as the ingroup

over-the-counter (OH vur thuh KOUN tur) drugs that can be purchased in pharmacies without a prescription

oxygen (OKS uh juhn) composes 21% of Earth's atmosphere; a necessity for most living organisms

ozone (OH zohn) O_3; a naturally occurring gas composed of three atoms of oxygen; in the upper atmosphere, it protects the Earth from harmful radiation from the Sun, whereas lower, it can be harmful

P

***P* generation** (pee jehn uhr AY shuhn) parent generation; the set of parents whose genotypes are crossed in a Punnett square in order to predict the possible genotypes for the offspring

Pangaea (pan JEE uh) the single landmass of all the continents that existed 200 million years ago

parallel circuit (PAIR uh lel SUR kit) electric current that flows in at least two different paths

parasites (PAIR uh seyetss) harmful organisms, such as ticks, fleas, and lice, that live inside the body of a larger animal and feed off it

parasitism (PAIR uh suh ti zuhm) the harmful attachment of organisms, such as ticks, lice, and fleas, to other creatures

particulate (par TIK yoo luht) tiny pieces of ash, unburned fossil fuels, fumes, dust, and pollen

pathogens (PATH oh jenz) disease-causing organisms such as some bacteria

pattern (PAT urn) regular predictable movement

period (PIHR ee uhd) a row on the periodic table of elements

periodic table of the elements (pihr ee OD ik TAY buhl uhv thuh EL uh muhntss) the organization of the elements according to mass and the number of electrons in their outermost shell

peripheral nervous system (puh RIF ur uhl NUR vuhss SISS tuhm) all nerves in the central nervous system located outside the brain and spinal cord

petals (PET uhlss) colorful, leaf-like parts of a flower; attractive to specific insects, birds, and mammals that are essential for the reproduction of the plant

petroleum (peh TROLL ee uhm) crude oil; unrefined oil made of hydrocarbons

pH scale (pee aych skayl) measure of the relative amounts of hydrogen ions and hydroxide ions in a solution; a way to measure the strength of an acid or a base

phase (fayz) a particular appearance of the Moon or planet at any given time

phases of the Moon (FAYZ uhz uhv thuh moon) the repeating position of the moon in relation to Earth and Sun from month to month

phenomenon (feh NOM meh non) a fact or event that is observed or experienced

phenotype (FEE no type) the observable traits of an organism

photon (FOH ton) a bundle of energy released by an atom

photosynthesis (foh toh SIN thuh siss) the process by which plants use light energy to make food

phylogenetic systematics (FEYE low jen EH tik sis tum MAT iks) the way scientists reconstruct events to show how evolution has led to the species they study

phylogeny (feye LAHJ uh nee) the study of how organisms are related to each other through evolution

physical change (FIZ uh kuhl chaynj) change in a physical property such as state, size, or shape

physical properties (FIZ uh kuhl PROP ur teez) those things that can be examined without changing the identity of the matter

physiology (fi zee AH luh jee) the study of the functions and vital processes of organisms

phytoplankton (FEYE toh plank tuhn) a kind of algae that live on the ocean surface

pigment (PIG muhnt) colored substance

pistil (PISS tuhl) the female reproductive structure of the flower

pitch (pich) the measure of the frequency at which a note vibrates

placenta (pluh SEN tuh) the organ inside the uterus of a pregnant mammal that delivers nutrients to the fetus via the umbilical cord.

planet (PLAN it) a large body orbiting a star in an elliptical path

plankton (PLANK tuhn) small organisms that live in water including animals, protists, archaea, algae, and bacteria

plasma (PLAZ muh) a light-colored watery liquid that contains some proteins, minerals, vitamins, sugars, and chemicals; it provides transport throughout the body; the fourth state of matter

platelets (PLAYT litss) tiny particles in the blood that form blood clots

pollen (POL uhn) the male reproductive material of flowering plants

pollination (pol uh NAY shuhn) the process by which plants transfer reproductive materials from one plant to another

pollution (puh LOO shuhn) the contamination of the environment with substances in a quantity great enough to affect an ecosystem

polymer (POL uh mur) a chemical compound made from repeated units

population (pahp yoo LAY shuhn) all the organisms of one type in a defined area

potential energy (puh TEN shuhl EN ur jee) the stored energy in an object; a result of its position

precipitation (pri sip i TAY shuhn) water returning to Earth's surface, usually in the form of rain or snow

precise (pri SISE) a strictly exact description or measurement

predators (pred uh turss) life forms that live by eating other life forms

predict (pre DIKT) to foretell or know what will happen in advance

prescription (pri SKRIP shuhn) a doctor's order for a drug that is carefully regulated

prey (pray) organisms that predators hunt and kill for food

principle of parsimony (PRIN sih pul uhv PAR sih moh nee) the preference of scientists to seek the simplest explanation

prism (PRIZ uhm) clear triangular piece of glass or plastic that refracts light into its separate wavelengths

probability (prah buh BIL uh tee) the number of desired outcomes out of all possible outcomes

procedure (pro SEE jur) a series of steps or actions needed to perform a task

process (PRAH sess) a series of actions that occur for a purpose

produce (pruh DOOSS) to make

producers (pruh DOOSS urs) green plants that manufacture food by photosynthesis, directly or indirectly

product (PRAH dukt) a substance that is created by a process or reaction

productivity (pro duc TIV uh tee) the rate or speed at which work is completed

proteins (PROH teenz) nutrients that are broken apart in the body into amino acids

protists (PROH tistss) a group of single-celled organisms with a nucleus

proton (PROH ton) a positively charged particle found in the nucleus of an atom

pulley (PUHL ee) a simple machine that changes the direction of a force

punctuated equilibrium (PUHNGK choo ay tid ee kwuh LIB ree uhm) times of stability interrupted by periods of rapid change

Punnett square (PUN net skwayr) a tool that helps to determine the possible combinations of alleles among offspring

purebred strains (PYOOR bred straynz) plants or animals bred only from members of the group with a recognized trait; offspring of many generations that have the same traits

pyruvate (peye ROOH vayt) an acid that is the product of glycolysis

R

radiant energy (RAY dee uhnt EN ur jee) nuclear energy that travels through space as waves

radioactive dating (ray dee oh AK tiv DAYT ing) method for calculating a rock's absolute age by identifying and precisely measuring the proportions of radioactive material in the rock's minerals

range (raynj) the difference between a minimum and a maximum in a set of data

rarefaction (rer uh FAK shuhn) the space between sound waves; the opposite of the compression of the sound waves

reactant (ree AK tahnt) a substance involved in a chemical reaction

recessive (ri SE siv) tending to be suppressed, as a recessive gene for blue eyes versus brown eyes

reconcile (REH kon sile) to ensure that an explanation is supported by data or evidence

reflect (ri FLEKT) to change direction of a wave in such a way that the wave bounces off an object back in the direction from which it came

refinery (ree FEYEN uh ree) a factory where petroleum is processed and distilled

reflex (REE fleks) a type of reaction in which signals are immediately transmitted to the spinal cord and the brain

refract (ri FRAKT) to change direction of a wave as it passes from one medium to another

reinforce (ree in FORSS) strengthen

relate (re LAYT) to make a connection between two ideas or to share an idea; to connect ideas and meanings

renewable resources (ri NOO uh buhl REE sorss uhz) resources that can be replaced within an average lifetime, such as animals, crops, and trees

repel (ri PEL) to push away from

reproduction (ree pruh DUHK shuhn) the process by which plants and animals have offspring

reptiles (REP teye uhlz) a group of vertebrates related to the dinosaurs that once roamed Earth

research (REE surch) careful study or investigation of a subject

reservoir (REZ urv wahr) a large container or basin that is used to hold or store material such as water

resistance (ri ZISS tuhnss) slowing down of the movement of electrons

resources (REE sorss uhz) materials that people use

respiration (ress puh RAY shuhn) the process of cells breaking down simple food molecules to release energy, usually involving the exchange of oxygen and carbon dioxide

respiratory system (RES per uh tor ee SIS tum) the group of organs that exchange oxygen and carbon dioxide in the body

respond (re SPOND) to react to a prompt or stimulus

reversible reaction (ree VERS uh buhl ree AK shuhn) a reaction that can occur in either direction; products can be combined to re-create reactants

revolution (rev uh LOO shuhn) one complete pass of a planet around the star it orbits; equal to one year

ribosomes (REYE buh sohmz) tiny dots on the endoplasmic reticulum that combine amino acids into proteins

rock cycle (RAHK seye kuhl) the transition of rocks between sedimentary, metamorphic, and igneous rock

roots (rootss) the part of a plant that anchors it into the ground

root hairs (root hairz) threadlike structures that grow off the surface of a root

rotation (roh TAY shuhn) the turning of a planet or other body on its axis

S

STD (eS T DEE) sexually transmitted diseases; a variety of diseases contracted through direct sexual interaction

salt (sawlt) a compound that results when an acid is mixed with a base

satellites (SAT uh leyetss) objects, such as moons, that orbit planets

scan (skan) to read quickly to find specific information

scavengers (SKAV uhn jurz) animals that feed on dead organisms

scratch test (skrach test) a test for hardness of minerals in which one mineral is used to scratch another

seafloor spreading (SEE flor spred ing) process by which new rocks are formed on the ocean floor as two tectonic plates move apart in a divergent plate boundary

sedimentary (sed uh MEN tuh ree) rock formed from loose material that becomes tightly packed over time

seizure (SEE zhuhr) temporary electrochemical changes in the brain, often resulting in either physical shaking or loss of consciousness

sequence (SEE kwuhnss) the order in which events occur or in which they should be done

series circuit (SIHR eez SUR kit) electric current that flows in one path only

simple machine (SIM puhl muh SHEEN) a machine that performs a single function

single replacement reaction (SING uhl ree PLAYS muhnt ree AK shuhn) one reactant atom replaces another to create a new product compound; $A + CB \rightarrow C + AB$

skeletal muscles (SKEL uh tuhl MUHSS uhlz) voluntary muscles that allow humans to change the position of their body

smog (smahg) a combination of smoke and fog; a haze of harmful air pollution

smooth muscles (smooth MUHSS uhlz) involuntary muscles that are controlled by the nervous system

solar system (SOH lur SISS tuhm) the system consisting of the Sun, eight planets, their moons, and smaller bodies such as dwarf planets, asteroids, and comets

solid (SOL id) a substance that has a definite shape and a definite volume

solstice (SOHL stuhss) the moment that marks the beginning of summer and winter on Earth; the time when Earth is most inclined toward or away from the Sun

solute (SOHL yoot) the substance that dissolves in a solution

solution (suh LOO shuhn) a mixture formed by one substance dissolving in another

solvent (SOL vuhnt) the substance in which a solute is dissolved

speciation (spee see AY shuhn) to become separated into distinct, separate species or organisms capable of reproducing

species (SPEE sheez) the smallest group in the classification of organisms that can interbreed

speculation (spek yoo LAY shuhn) an opinion or guess based on incomplete evidence

speed (speed) the rate of change in distance over time

sponges (SPUHNJ ez) simple invertebrates that live in water, have no head, arms, or legs, and do not move from place to place

squall lines (SKWAWL leyenz) irregular rows of thunderclouds

stamen (STAY muhn) the male reproductive structure of a flower

star (star) a massive body of a gas-like substance held together by gravity that radiates heat and light

state of matter (stayt uhv MAT ur) one of four states of anything with mass and volume: solid, liquid, gas, or plasma

static electricity (STAT ik i lek TRISS uh tee) buildup of electric charges on the surface of objects

stationary (STAY shuh nair ee) nonmoving

stems (stemz) the part of a plant that provides support and transports food and water

stoichiometric coefficient (stoy kee oh MET rik koh uh FISH uhnt) a measurement of the quantities of reactants and products

stomates (STOH maytss) tiny openings in leaves that allow the movement of oxygen and carbon dioxide

storm surge (storm surj) offshore rise in ocean water caused by hurricanes and other low pressure storms

stratosphere (STRAT uh sfihr) layer of the atmosphere that extends from the top of the troposphere to about 20 miles above Earth; contains the ozone layer

stroke (strohk) sudden failure of blood supply to the brain

structures (STRUHK churz) forms

subbituminous (suhb bit TOO mihn uhs) soft, black coal that is relatively young but made of plant matter more fully decayed than that of lignite

subduction (suhb DUHK shuhn) process in which one tectonic plate slips beneath another; can create volcanoes and earthquakes

summarize (SUHM ur eyez) to retell the facts or major events contained in a passage

summary (SUHM ur ee) brief statement of the most important ideas of a paragraph or passage

supernova (soo pur NOH vuh) the explosion and death of a large star

support (suh POHRT) to strengthen, corroborate, or give evidence for an idea

supporting details (suh PORT ing DEE taylz) phrases and sentences that support the main idea

symbiosis (sim bye OH sis) organisms that rely equally on each other to carry out life functions

symbol (SIM buhl) a letter, letters, or shape that represents an object or idea

symptoms (SIMP tuhmz) the effects that someone feels from an illness

synthesize (SIN thuh seyez) to blend ideas

synthesis reaction (SIN thuh sis ree AK shuhn) two reactants combine to form a single product; $A + B \rightarrow AB$

systematics (sis tum MAT iks) the study and classification of organisms according to their biological diversity and evolution

T

table (TAY buhl) a graphic organizer that presents information in rows and columns

taproot (TAP ruht) the largest part of the root system of some plants that grows deep into the ground vertically

taxon (TAKS ahn) one hierarchical grouping in the study of taxonomy, such as species, genus, family, order, class, phylum, kingdom, or domain

taxonomy (tak SAHN uh mee) the science of biological classification of organisms

tectonic plate (tek TAHN ik playt) also crustal plate; any of the several large pieces of the Earth's crust

theme (theem) common underlying topics or connections

theory (THIHR ee) a logical explanation of events and evidence from the natural world

threatened (THRET end) a species that is presently abundant, but that could become extinct if steps are not taken to protect its habitat

thermosphere (THUR muh sfihr) layer of Earth's atmosphere above the mesosphere

thrive (threyev) grow

tides (teyedz) the bulge of water on Earth as a result of the pull of the Moon's gravity

time line (teyem leyen) graphic organizer that shows events in time order

tissue (TISH oo) a group of specialized cells working together

tolerance (TAHL uhr unts) the upper and lower limits of a factor that an organism can accept and still survive

toxin (TAHK sin) poisons, usually produced by living things

trachea (TRAY kee uh) the windpipe; carries air from the rear of the mouth to the lungs

trait (trayt) characteristic displayed by an organism

transcription (tran SKRIP shuhn) the first step of decoding DNA, in which DNA serves as a model to make a molecule of RNA

transformation (transs for MAY shuhn) change in form of energy

translation (transs LAY shuhn) the process by which RNA acts as a model to make proteins

translucent (transs LOO suhnt) scattering light as it passes through

transparent (transs PAIR uhnt) allowing light to pass through

troposphere (TROH puh sfihr) the layer of the atmosphere that extends from the ground to about 7.5 miles above Earth; where humans live and where most weather occurs

trough (trawf) the lowest part of a wave

tundra (TUHN drah) a flat, treeless Arctic biome characterized by permafrost and limited plant and animal populations and little precipitation

turbine (tur BEYEN) spinning paddles or fan blades moved by wind or water that provide energy

U

ultraviolet (UV) rays (uhl truh VEYE uh lit RAYZ) high-energy rays in sunlight that are powerful enough to burn or tan skin

unstated assumption (uhn STAY tuhd uh SUHMP shuhn) an assumption that is implied or suggested but is not expressed directly

V

vacuoles (VA kyuh wohlz) cellular organs that have a variety of functions, such as digesting food and disposing of wastes

velocity (vuh LOSS uh tee) rate of change of direction; defined by speed and direction

Venn diagram (VEN DEYE uh gram) type of diagram used to compare two things

vertebrates (VUR tuh britss) animals that have backbones

visual (VIZH oo uhl) an image that helps illustrate written information, such as a graph, chart, drawing, diagram, or photograph

visual spectrum (VIZH oo uhl SPEK truhm) portion of the electromagnetic spectrum that is visible to humans

volcano (vol KAY noh) an opening in Earth's crust that allows magma, ash, and gas to escape; also the mountain that builds up around this opening

volts (volts) measurement of the flow of electricity

voluntary (VOL uhn ter ee) able to be controlled, as in voluntary muscles

W

water cycle (WAW tur SEYE kuhl) the continuous movement of water between the surface and atmosphere and back

wavelength (WAYV length) the distance between two waves

weather (WETH ur) the condition of the atmosphere at any given time and place

weathering (WETH ur ing) process that breaks down or changes rock on or near Earth's surface

well-balanced diet (wel BAL uhnst DYE eht) eating a variety of foods to get calories from protein, carbohydrates, and fat in proper proportions

wetland (WET land) area where the soil is saturated all or part of the year, such as a marsh or a swamp

wheel and axle (weel and AK suhl) a simple machine that combines a wheel with the idea of a lever

wind (wind) the movement of air from places of high pressure to places of low pressure

work (work) the result of energy that can be defined as force multiplied by distance

worm (wurm) a complex invertebrate with a simple brainlike structure and digestive tract

Y

yield (yeeld) to produce or create

Z

zoology (zoo OL uh jee) the study of animals

Index

Introduced species, as ecosystem disruption, 112–113
Invasive species, 112
Invertebrates, 106, 148–153
    arthropods, 151
    cnidarians, 149
    definition of, 148
    insects, 151–153
    mollusks, 151
    sponges, 149
    worms, 150
Involuntary muscles, 20
Involuntary nervous system, 20
Ion(s), 312, 322
Ionic bonds, 320, 322
Isotopes, 314
IT (information technology) jobs, 39

J

Jargon, 98, 391
Jellyfish, 149
Joints, 18, 156
    arthritis in, 39
    muscles controlling, 20

K

Kelp forest food web, 99
Kinetic energy, 238, 258, 269
Kingdoms, 140
Kudzu, 113

L

Labor (birth), 34
Lactic acid fermentation, 70, 71
Lakes, 394
Lamarck, Jean-Baptiste, 221
Land pollution:
    from burning fossil fuels, 373
    and weather changes, 397
Land use, as environmental issue, 122
Larvae, 153
Larynx, 28
Lava, 385, 392
Law of Conservation of Mass, 341
Law of Conservation of Matter, 326, 362
Leadership, 190, 334
Leaves, 55, 56
Lever(s), 292–293
Ligaments, 18
Light waves, 247
Light-year, 414
Lightning, 363
Lignite, 371
Limiting factors, 96
Lineage, 224
Linnaeus, Carolus, 211, 220

Lipids, 334
Liquids, 304, 306
List (text structure), 264
Living things (*See also* Animals; Organisms; Plants; Species)
    as biofuel, 373
    carbon stored by, 360
    conditions for life, 373
    on Earth, 426
    nutrients for, 358
Lobsters, 141
Low-energy waves, 246
Lunar cycle, 426–427
Lungs, 28

M

Machines, 292–295
    compound, 292, 294–295
    simple, 292–294
Maggots, 227
Maglev trains, 256
Magma, 262, 385, 392
Magnetism, 255–256
Magnets, 255
Make Predictions (core skill), 178, 270, 371
Male reproductive system, 33
Mammals, 156, 163
Mantle, 382, 410
Marrow, 16
Marsupials, 163
Mass:
    atomic, 314
    Law of Conservation of, 341
Mass extinctions, 207, 287
Mass number, 314
Matrix, cell, 64
Matter, 304–309
    biogeochemical cycles of, 358–364
    chemical reactions, 326, 328
    compounds, 320–323
    definition of, 304
    elements, 307–309, 312–317
    Law of Conservation of Matter, 326, 362
    properties of, 307
    solutions, 328–330
    states of, 304–306
Meaning:
    determining, 20, 111, 211, 323, 341, 391, 394
    of terms, 63, 286
Measurements, 359
Mechanical energy, 238
Media literacy, 113, 223, 253, 254, 363
Medulla, 32
Melanin, 187
Mendel, Gregor, 176, 178, 186
Mendeleyev, Dmitry, 315
Menstrual cycle, 33

Nutrients, 40
    availability of, 362
    definition of, 358
    recycling of, 359
Nutrition, 40–41
    balanced diet, 41
    for building bones, 16
    and cirrhosis, 39
    during pregnancy, 34
Nutritionists, 40

O

Occam's razor, 415
Ocean floor, 392
Oceanic crust, 385, 392
Oceanography, 390–394
Oceans, 92
    and carbon cycle, 360
    currents in, 393
    floor of, 392
    and oxygen cycle, 361
    and phosphorus cycle, 364
    tides, 427
    waves of, 394
    zone of, 390, 391
Octopuses, 151
Offspring, 184
Oil, 263
    crude, 368–369
    land and water pollution from, 373
Omnivores, 90
Orally, drugs taken, 44
Opaque (term), 247
Open ocean, 391
Orchids, 106
Order (text structure), 264
Ores, 387
Organ, 143
Organ system, 143
Organelles, 64
Organic chemistry, 332–335
Organic polymers, 334–335
Organisms (See also Animals; Plants)
    as biofuel, 373
    in biomes, 91–92
    carrying capacity for, 96–99
    crude oil from, 369–370
    DNA of, 181
    ecology as study of, 93
    in ecosystems, 88 (See also Ecosystems)
    in energy cycles, 90
    habitats of, 96–99
    hosts, 103
    in lakes, 394
    multicellular, 143–145
    in oceans, 390
    phenotypes and genotypes of, 184

scavengers, 111
simple, 140–143
symbiosis among, 102–106
Organization, text, 18, 19
Origin of Species, The (Charles Darwin), 200
Oscilloscope, 248
Outer core, 382
Outer planets, 421
Outgroups, 214
Over-the-counter drugs, 42
Oxpeckers, 104
Oxygen:
    in atmosphere, 398, 426
    needed for life, 412
Oxygen cycle, 361
Oysters, 151
Ozone, 121, 372

P

P generation, 186
Pacific Ocean, 390
Pangaea, 222, 384
Parallel circuit, 254
Parapatric speciation, 224, 226
Parasites, 104–105
    foodborne and waterborne, 105
    roundworms as, 150
Parasitism (parasitic symbiosis), 104–105
Parsimony, principle of, 215
Parsing, 98
Particulates, 372
Pathogens, 105
Penicillin, 43, 204
Penis, 33
Percolation, 362
Period (in periodic table), 315
Periodic table of the elements, 315–317
Periods (in fossil record), 260
Peripatric parapatric, 224
Peripatric speciation, 226
Peripheral nervous system, 30
Petals, 56
Petroleum, 368–369
pH scale, 330
Phase(s):
    definition of, 426
    of the Moon, 426–427
Phenomenon, 265
Phenotype:
    in codominance or incomplete dominance, 189–190
    definition of, 184
    in Punnett square, 186–187
Phosphate compounds, 364
Phosphorus cycle, 364
Photon, 246